The workshop was

sponsored by
the German Marshall Fund of the United States, Washington D.C.
and
the Gottlieb-Daimler- and Karl Benz-Stiftung, Ladenburg, FRG
and
organized by
Pace University School of Law,
Center for Environmental Legal Studies,
Professor Richard L. Ottinger
White Plains, N.Y., USA
and
Fraunhofer-Institute for
Systems and Innovation Research (ISI),
Dr. Olav Hohmeyer
Karlsruhe, FR Germany

Olav Hohmeyer · Richard L. Ottinger (Eds.)

External Environmental Costs of Electric Power

Analysis and Internalization

Proceedings of a German-American Workshop,
Held at Ladenburg, FRG, October 23-25, 1990

With 44 Figures

Springer-Verlag

Berlin Heidelberg New York
London Paris Tokyo
Hong Kong Barcelona
Budapest

Dr. Olav Hohmeyer
Fraunhofer-Institut für System-
technik und Innovationsforschung
Breslauer Str. 48
D-7500 Karlsruhe 1

Prof. Dr. Richard L. Ottinger
Pace University School of Law
Center for Environmental Legal Studies
78 Broadway
White Plains
N. Y., USA

ISBN-13: 978-3-642-76714-2 e-ISBN-13: 978-3-642-76712-8
DOI: 10.1007/978-3-642-76712-8

Softcover reprint of the hardcover 1st edition 1991

2142/7130-543210

Preface

The Persian Gulf crisis demonstrated the imperative of reducing the world's dependence on imported oil. At the same time, environmental imperatives, particularly the phenomena of global warming, acid rain and urban smog, require decreased dependence on fossil fuels of all kinds, while the reactor accident at Chernobyl has raised the awareness of the hazards of nuclear power.

In the face of these imperatives, it is apparent that business-as-usual in the energy field is no longer tolerable, ignoring substantial costs to society from traditional energy production and use. Since the late 1980's, a growing part of the scientific community and public officials dealing with energy problems are recognizing a need for measures to take into account all costs of energy systems, including environmental externalities.

Research on the quantification of social or external costs started only a few years ago and still faces a number of unsolved basic problems. At the same time public officials see the need for action and administrators are starting to look for means to account for the elements of social or external costs known so far. These attempts are aiming at a direct inclusion of such cost elements in the market prices of energy or in the planning process for new energy supply systems.

To internalize the social costs of energy, various OECD countries are starting to adopt a variety of pollution fees and taxes to address global warming. In the electric utility industry in the United States, environmental externality costs are starting to be considered in selection of new electricity sources. More stringent pollution standards are being adopted and the U.S. Clean Air Act revisions enacted in November, 1990, provides for "emissions trading rights" to put the costs of air emissions into the marketplace.

In order to accurately set pollution fees and taxes, to incorporate environmental costs into utility selection and to set pollution standards, information about the social costs of energy production would be invaluable.

In this situation of growing international concern, the workshop brought together the leading scientific, economic, utility and governmental experts in the fields of quantifying the social costs of energy and accounting for these costs in energy policy and planning. Each prepared a paper on the subject which was discussed at the workshop and is included in these Proceedings.

As there have been few opportunities for an intensive international exchange of ideas in these areas, the workshop served as a first forum for a thorough international discussion of them. The workshop led to recommendations to hold a conference in a year, broadening the base to include environmental ministers, to provide an update on the considerable activitites in the field being pursued nationally and internationally. In particular, the EC and the U.S. Department of Energy have embarked on a major joint study of the environmental costs of the entire fuel cycles for energy production. A number of other significant studies have been initiated both in the EC and the United States.

Because of the importance of the third world countries, particularly to the global warming problem, it was also decided to try to initiate an international conference that would include their representatives in the discussion, and to explore the interest of the World Bank in co-hosting such a conference.

The workshop conferees also agreed to establish mechanisms for continued exchange of information and research results among the participants.

It is thus hoped that the workshop will have initiated a major thrust for the advancement of knowledge in the field of quantifying the social costs of energy and seeking means to apply this knowledge to internalize the social costs of energy.

On behalf of all participants the editors want to aknowledge their great appreciation for the generous financial support of the German Marshall Fund of the United States and the Gottlieb Daimler- and Karl Benz-Stiftung, which made this workshop possible.

As the manuscripts of the papers presented at the workshop varied significantly in layout and typing, we went through the process of retyping practically each manuscript to arrive at a more uniform and legible text. In this context the editors have to thank Monika Silbereis, Dorothea Schlick, Renate Klein, Roselinde Welke and Ilse Gottschalg for their efforts in retyping all texts. During the workshop Monika Silbereis assissted by Sonja Mohr and Irene Nootbaar ran the secretariat of the workshop very efficiently. We want to thank them and Mrs. Petra Jung from the Gottlieb Daimler-Stiftung for their invaluable support in organizing the workshop making it a very effective and enjoyable event for all participants. Last but not least we want to thank the city of Ladenburg for letting us use their fine conference facilities in the Ladenburg city hall allowing to have lively discussions despite of the substantial number of participants.

Pace University
Prof. Richard L. Ottinger

FhG/ISI
Dr. Olav Hohmeyer

Contents

Page

Page

1. INTRODUCTION

The international workshop which is summarized in this volume concentrated on the social or external environmental costs of electricity production and use. This rather limited scope should not be interpreted as a value judgement on the relative importance of this specific instance of external effects in market economies. It merely mirrors the fact that during the rather short period of empirical research conducted until today this has been the first main focus of our research and discussions. There are other instances of external effects than health and environmental damages like non sustainable resource use, certain macroeconomic effects, or governmental subsidies, which need to be addressed just as well. Furthermore, external effects play an important role with other types of energy or in other sectors of the economy like chemical industry or the transport sector. Some of these effects are at least equally important as the instances or external costs addressed in the workshop. Thus, most participants of the workshop think that it is mandatory to substantially broaden our research on and the discussion of the problems of external effects of economic activities in the future. Nevertheless, the ideas and results presented at the workshop on the narrower subject may serve as a starting point for the research on the more general problem.

Although the papers have been edited to a limited extent and retyped for printing, the texts still vary in a number of specific elements of their layout, citation procedures and language proficiency. To arrive at a more uniform text would have taken considerably more time and financial resources than we had available.

The order in which the papers are presented in this book resembles to a large extend the structure of the workshop.

First a short overview of the present knowledge in the field of damage research was given. These papers constitute **Chapter 3** of this book. Starting out with a presentation on the health damages due to air pollution by *Wiegand*, which unfortunately can not be included in the book, because we never received a manuscript, the workshop touched on the damages to plants presented by *Guderian*, the damages to materials caused by air pollutants summarized by *Weltschev*, and the possible impacts of energy consumption on global warming presented by *Bach*. *Hodas* gave the results of some US attempts to monetize the costs of global warming based on the compensation costs of afforestation programs. These presentations all emphasized the importance of the actual damages which have to be ascribed to airpollutants. Decisive action to tackle the problems seems to be asked for today, although we do not precisely know all of the cause effect chains and the complex causalities of damages due to a multitude of different emissions and emission sources.

Chapter 4 presents eight papers on the monetization of environmental damages, starting with some fundamental considerations on quantification, valuation and monetisation by *Bernow et al.*, addressing the problems of complexity of the analysis of physical damages versus the somewhat smaller problems of monetisation in the paper by *Ewers*. *Chernik* and *Caverhill* discuss four possible ways to monetize environmental damages, strongly advocating a control cost approach as the preferable alternative. *Dudek* addresses the possibility of monetisation through the trading of emission certificates, a procedure heavily relying on economic mechanisms to find prices, while *Kümmel and Schüsseler* present a valuation methodology based on physical considerations centred around the heat emissions of pollution control technologies. To complete the picture *Buchanan* presents the general advantages of contingent valuation approaches to the problem of monetisation and gives an overview of the results of a contingent valuation study on electric power production conducted on behalf of Bonneville Power. Finally *Wooley* gives a comparison of the estimated external costs of waste-to-energy plants and coal fired power plants and *Fritsche* discusses the incorporation of externalities in least-cost planning based on a multi-attribute approach including mainly control costs and standards.

The papers of **Chapter 5** focus more on the impacts of internalisation of external costs, once these are identified. The presentations by *Hohmeyer* and *Kallenbach* gave a summary of two very different perspectives on this matter in West Germany. While Hohmeyer derives rather large differences in the external costs of competing energy systems, putting renewables at a substantial disadvantage, Kallenbach presents findings of a different study arriving at very small differences and negligible impacts on the competitive position of the energy technologies considered. It is up to the reader and future additional research on the matter to decide these questions. *Weisheimer* presents some first results on the external costs of electricity production in the former GDR, pointing out the substantial difficulties in deriving such results for planned economies and at the same time showing the great importance of such analyses specifically for economies which have paid little attention to the problems of environmentally damaging production. Finally *Weil* gives a utility commissioners perspective on the impacts of environmental costs and argues strongly for their consideration in long-range utility planning as well as in short term dispatch decisions.

Chapter 6 presents six papers which are focussing on the question of 'How to internalize external costs?' from a more theoretical perspective. *Putta* discusses the possibility of including environmental costs in electric resource planning based on a mixed ranking system. *Sanghi* discusses a necessary CO2 tax to achieve a target reduction of 5 to 20% in

New York State by the year 2008. He shows the substantial impact which different uses of these tax revenues (spending on reduction measures versus general spending) have on the necessary tax level. *Beldock et al.* argue strongly that external costs should be included in integrated resource planning approaches, which they sketch drawing on the theory of technical innovation processes. *Peterson* criticises the cost oriented approaches based mainly on control cost estimates. In his view quantitative reduction targets for the overall emissions of certain pollutants lead to better results than taxes or adders based on control costs. *Brick* makes the point that it is not sufficient to include external costs in planning decisions but that these costs have to be included in the electricity prices in order to give the right price signals to the final consumers. Only in this way the necessary adaptation process of the energy demand to the real costs of energy will be achieved. Brick argues strongly to include more than just external environmental costs. *Hennike* cautions against the belief that an exclusive reliance on internalization strategies of monetized external costs can solve all problems involved. He shows that the electricity markets are heavily monopolized and that many other obstacles hinder the proper operation of these markets. Thus, he concludes that for countries like Germany a major restructuring of the electricity sector and the use of demand side management measures are required just as well.

Chapter 7 gives an overview of presently employed strategies for the internalisation of environmental costs of electricity in the US and West Germany and a short outlook on the new German study on the external effects of the energy sector. In the first paper of this chapter *Ottinger* gives a very comprehensive overview of all state activities aiming at the inclusion of external environmental costs of electricity in the US today. *Burrington* shows the historic development of the treatment of environmental costs by different utility commissions in the USA. *Cavannagh* reports on the special situation in the state of California. Here a state law was enacted connecting the allowed profit levels of utility companies to the success in reducing the electricity consumption in their service areas. *Munts* reports on the experiences of the Public Services Commission of Wisconsin, which has played a leading role in the inclusion of environmental considerations in utility planning processes. She concludes from the Wisconsin experience that politicians can and need to act on the basis of incomplete information because 'it is better to be roughly right than precisely wrong'. *Wilson* reports on the practical inclusion of external cost considerations in the electrical utility planning for the South Coast Air Quality Management District, one of the most polluted districts in the USA. *Walmet and Wolcott* report on the latest activities of the New York State Energy Research and Development Administration after the Pace Study has been completed. The new York Public Services Commission has ordered the electric utility companies to finance an extensive follow up study at the cost of 1.6 million US$. This study is expected to arrive at results which can be directly integrated into the established least-cost planning process and an existing demand-side management model.

This study will be completed in 1992. *Spreer* and *Rave* each report on the basic philosophies of the energy policies in two federal states of Germany, each trying to move towards a far more efficient use of energy and the large scale use of renewable energy sources before we know much about the actual external costs of different energy systems. In both instances the state governments are convinced of the necessity for such action on the basis of qualitative arguments. Thus, the painstaking efforts of quantifying and monetizing environmental and other costs are not necessary preconditions for decicive political action in these states. *Keppler* gives a short introduction to the major German follow up study on the problem of external costs, which was commissioned by the German Minister of Economic Affairs in late 1990. Results from this study are expected to be available by the middle of 1992. The final paper of this volume is an essay by *Moths* on the basic principles of our economy and the function of scientific reports addressing politically relevant questions. He concludes that market economies pay high bonuses on handing on costs to third parties (what he calls the 'privy-type philosophy') and that scientific reports analysing such patterns may turn out to be major 'accidents' for policy makers and administrators. Thus, the focus of such reports need to have the entire economy in view and the scientists involved should be prepared to meet massive resistance from politicians and administrations.

2. SOME CONCLUSIONS FROM THE WORKSHOP

Although the workshop reflects a rather early stage in the discussion of the external costs of energy and although it merely addressed the questions relating to the external environmental effects of electricity, the editors think that some first conclusions may be drawn from the papers presented and the discussions among the workshop participants following these presentations. We would like to summarize the results of the workshop in the following points.

1) The analysis of environmental damages due to energy use is rather difficult and the problems of cause effect relationships and damage attribution will remain to be substantial.

2) Nevertheless, external environmental costs of electricity seem to be of a considerable order of magnitude. Although this view is shared by the vast majority of the participants there is an oposing minority view of this problem.

3) The debate on the best way of measuring damages, valuing and monetizing external effects is quite intense and will certainly continue for a considerable period of time.

4) The impact of neglecting external environmental costs on the choice of energy technologies may be substantial.

5) Once external costs are quantified a number of ways to internalize these into the market prices or into planning decisions are available.

6) In the USA a number of states have already taken action to include external cost considerations in planning decisions.

7) In Germany two federal states have demonstrated that a more qualitative orientation of energy policy with respect to environmental damages and resource considerations may be superior to a policy purely oriented towards monetized cost figures.

8) Political action can and should be taken today instead of waiting until the remaining uncertainties with external cost data are resolved. As the Public Service Commissioner of Wisconsin, Mary Lou Munts, has put it **'It is better to be roughly wright than precisely wrong'**.

9) However, parallel to such first political and regulatory action there is a substantial need for more research on many aspects of the external costs of energy use. This should address more than the problems relating to electricity or energy and more than just the external effects based on environmental damages.

3. SUBJECT AREA 1:

RECENT RESULTS OF THE RESEARCH ON ENVIRONMENTAL DAMAGES

AN INTRODUCTORY OVERVIEW

3.1 Tasks and Problems of Analysis and Evaluation of Emission Impacts on Terrestrial Ecosystems

R. Guderian
Institut für Angewandte Botanik
Universität Essen
4300 Essen, FRG

The evaluation and internalisation of external costs of energy production has to be based on an understanding of the consequences on the environment that result from the generation, conversion and consumption of electricity. This will be discussed in this article in relation to air pollutants particularly, by pointing out those questions and problems which may arise when the effects of energy production on terrestrial ecosystems are used as a basis for the estimation of external costs of energy production (compare also Guderian 1986).

This requires knowledge of two aspects:
- characterization of exposure to air pollutants
- assessment and evaluation of risk and effects

1. Current levels of atmospheric pollution and their effects on terrestrial ecosystems

As a consequence of a large number of emission sources and, until recently a widespread policy of solving pollution problems by dilution of pollutants rather than reducing the quantity emitted into the environment, this has resulted in

- the presence of mixtures of potentially toxic air contaminants
- long-lasting exposures
- relatively low concentrations of both primary and secondary air pollutants
- distribution over large areas

The current effect situation is mainly characterized by
- effects of combinations of potentially toxic compounds
- direct and indirect effects at different levels within ecosystems
- accumulation of pollutants
- chronic effects that may result in a reduction of function and performance of ecosystems
- environmental stress by atmospheric pollution in combination with other abiotic factors or pathogens
- the absence of locally determined gradients of pollution due to distribution of pollutants over large areas.

2. Classification and evaluation of effects on terrestrial ecosystems

When the effects of exposure to air pollutants are determined and evaluated this has to be in relation to the different functions of whole ecosystems and in relation to their individual compartments that act as receptors, such as air, soil, plants and animals. Ellenberg (1973) has described an ecosystem as "the interrelationship of all organisms and their inorganic environment which although open to external influences may be able to a certain degree of self-regulation". A consequence of this property is the ability to tolerate a certain degree of stress without resulting in the loss of function or performance of the whole system. It is the aim of environmental protection to recognize the limits of stress beyond which effects occur and to avoid them. Thus, air pollutants produced during the generation of electricity in conventional power plants are known to be harmful to whole ecosystems and this will be described in more detail concerning effects on plants.

2.1 Classification of effects

Apart from particulate air pollution, which may be damaging to plants after desposition to plant surfaces by preventing the exchange of heat, access of light or blocking of stomata, the toxicity of gaseous air pollutants is normally related to uptake of these pollutants.

Depending on the specific effect of individual pollutants, as well as the dose absorbed with time, a variety of consequences may occur (TABLE 1).

Table 1: Classification of effects of air pollution on plants in terrestrial ecosystems

Organizational level

Cell	Tissue	Organism	Community
Increased membrane permeability	Change in photosynthesis, respiration, and transpiration Alterations in the partitioning of metabolites	Changes in plant growth	Reduced plant growth and productivity Fluctuations in composition and reduction in species abundance
Altered enzyme activities	Alterations in growth and development of individual organs	Increased susceptibility to biotic and abiotic stresses	Changes in stand structure
Increased stress ethylene production	Pigmentation, bleaching, and chlorosis	Disturbances in fruit production	Disruption of food chain
			Changes in plant succession Possible changes in nutrient cycling Risks to consumers and decomposers
Ultrastructural changes in organelles	Necrosis	Reduced yield and quality	Impairment of ecosystem productivity, including its stability and capacity for self-regulation
Changes in cellular metabolism			
Altered cellular structure	Reduction in rhizobium induced nodulation	Altered plant competitive ability	
Disrupted cellular functions	Disturbance of mycorrhizal development	Death of plants	
Cell death	Death or loss of plant organs		

Depending on the degree of alterations of structure and function of the cell, this may result in effects at the next hierarchical level within ecosystems, i. e. organs, and a reduction in photosynthesis, increased respiration or interference with water balance or metabolic processes may be the consequence. Thus, changes in gas exchange often lead to effects in leaves but also in other parts of the plant as well. Changes in root/shoot ratios have often been reported in this context where differential partitioning of assimilates between different plant organs results in a stronger reduced growth of the roots in comparison to the shoot. Eventually, this will not only result in reduced growth of the whole plant but may be enhanced through reduced nutrient uptake because of profound effects on mycorrhizal colonization of roots and a larger degree of infection by pathogens or saprothytic organisms. Thus, primary effects may be accompanied by secondary effects, which occur due to decreased resistance of plants against biotic or abiotic influences as a consequence of exposure to air pollutants. Eventually, such a stressful exposure to atmospheric pollution will result in harmful effects at all compartments within an ecosystem: microlimate, soil, primary producers, consumers as well as decomposers. Due to long-lasting exposures of large areas to complex mixtures of pollution, individual pollutants and their effects may eventually accumulate to levels where thex reduce the ability of individual parts or the whole system to function properly.

In close to natural environments, such as forest ecosystem, acid deposition requires special attention; not such because of direct effects but because of indirect ones. Anthropogenic inputs of acidity, which amount to 7 kmol ha^{-1} a^{-1} (Ulrich 1983), are causing soil acidification, leaching of alkali metals and alkaline earth metals (Bosch et a. 1983) as well as the release of heavy metals in soils. In humid ecosystems, this leads to the enhancement of naturally widespread trends towards "retrogressive development" (Iverson 1964), i. e. the development of ecosystems with permanently reduced productivity as a consequence of soil acidification, slower decomposition processes of humus and a diminishing quality of humus. It is now generally accepted that such influences of atmosperic pollution on soils may also result in increased sensitivity of trees towards gaseous air pollutants causing an increased formation of damage symptoms.

Generally, such effects of air pollutants on plant communities can induce two different processes: changes in structure and function due to direct and indirect effects which may result in the total destruction of the original community. In this context it has to be born in mind that because of differential effects on individual plant species with different sensitivities, changes in the interspecific competition may cause the plant community to be more sensitive than its individual parts (Küppers 1986). In parallel to such degradations and their gernerosions (Scholz 1984), spontaneous or man-supported processes occur in which either original, i. e., adaptively resistent members of the remaining community, and also immigrants undergo secondary succession. While at relatively low concentrations plant species with short generations cycles may become more resistant populations (Bradshaw combined with less growth and differently adapted plant communities (Roose et al. 1982). Accordingly, inputs into ecosystems and the rate at which they take place must be restricted to such levels that still allow plant communities to fulfil their functions within natural or close to natural ecosystems, including normal evolutionary mechanisms to occur.

The consequence of such generosions, i. e. reduced numbers of species as well as varieties, are obvious when this is compared to an exponentionally grwong humen populations. This is particularly well exemplified by a statement made by Jean Mayer during the International Conference on Crop Productivity in Michigan 1975 (Michigan State University 1975): "This means that we have to find in the next 25 years, food for as many people again as we have been able to develop in the whole history of man til now." It is paramount for the maintenance of world nutrition to preserve the naturally occuring gene-pool, not even accounting for ethical, aesthetic or ecological-scientific interests.

Negative impacts on consumers or decomposers are not so much due to direct but more to indirect effects, caused by changes in the composition and availability of individual parts within the foodchain. While accumulating substances like heavy metals or fluorides are mainly damaging because of increasing concentrations in food, individual gaseous compounds are deleterious because of reductions in biomass or litter production.

These effects on individual organisational parts of ecosystems may individually or in combination with others result in losses or reductions of performance of the whole ecosystem, including its ability for selfregulation and stability.

2.2 Evaluation of effects

Those effects on individual levels within ecosystems as have been outlined in part one have different consequences on the functions and the untilization of the affected plants. According to a previously made segregation (Guderian et al. 1960) all air pollutant-induced reactions had been divided into "injury" and "damage" (Table 2). While injury comprises all reactions, damage is only concerned with the reduction of the intended use of a plant. These useful values, characterized by economic production, ecological functions, ideal values, and vegetation as a gene pool, can be reduced through the direct or indirect impact of air pollutants. In agriculture, forestry as well as horticulture, economic losses may be caused not only primary and secondary effects on plants but also by indirect effects, i. e. costs that are caused by changes in management practise (Brandl and Matthies 1984). However, these effects on individual farmers do not necessarily have to result in negative consequences on the whole economy. As "human activities and even life as a whole may result in environmental changes (Moller et al. 1981). Economists interpret damage to the environment as a non-optimal allocation (mis-allocation) of environmental use (Brabänder 1983). However, if such environmental utilization results in widespread destruction of forests or negative effects on the productivity of forests - as is currently the case for forest decline - the aim of continuity in forestry practise is no longer met. This has not only detrimental economic consequences on forestry and enterprises associated with it, but also detrimental effects on the ecological ability of forests including their importance for environmental stability as well as their recreational function.

3. Assessment and evaluation of risk and effects

In order to provide useful guidelines for practical air quality control, the objectives of what is to be protected have to be clear. Furthermore, it must also be possible to record damage according to its nature, intensity and cause.

3.1 Items and aims of protection

There are many economic, ecological, scientific or ethical reasons that can be used as arguments for the protection of all organisms within the environment. There is little discussion over the relevance of this demecological postulate to protect and maintain all members of an ecosystem. However, there are differential species-specific sensitivities to pollution and they may be greater within one species than between species. It is generally recognized that practical air quality control can never guarantee the protection of all individuals (VDI 1988). However, it is unclear which part of the community may be excluded from protection. If the protection of all varieties of one species as one individual may be defined as a unit with properties particular only for itself.

In a sy-ecological context, it is even more difficult to find answers and a concensus than at the aut- or dem-ecological level. However, they are particularly important since communities may respond more sensitively to air pollution than the individuals that define a species (Guderian and Küppers 1980). Like individuals, varieties or ecotypes of one species objectives for protection have to be defined and also account for those parts that are particularly sensitive. If this is to be the case, then the difficulty remains how these may be found. Considering the large number of biocoenoses, it is questionalbe if the small number of specimen that can be part of any study reflect their differential sensitivity.

Table 2: Consequences of air pollution impact on plants within ecosystems

Injury

All reactions of air pollutants

Damage

Effects on useful value

Primary damage			Secondary damage	
Economic performance		Ecological functions	Ideal value	Genetic resources
Direct	Indirect			
Growth and yield through	Reduced selection	Reduction in the	Reduction of	Gene erosion
reductions	of plant species,	areal functions of	scientific,	reduced abundance
Reduced quality loss	Changes in	of vegetation	ethical and	distribution or
of foodstuffs,	cultivar selections	Stabilization of	aesthetic value	of plant species
fodder, raw	and cultural	water relations and		and ecotypes
materials, and genetic	practices	climatic regulation		Reduction of
ornamental plants	Increased	Nutrient cycling		resources for
Impaired seed	expenditures for	Reduction of soil		breeding purposes
production and	cultural and	erosion requirements		
plant quality	maintenance needs	Reduction of noise		
		and air pollution		
		Welfare effects		

On the level of the ecosystem it is often postulated that the equilibrium and stability of an ecosystem have to be guaranteed. However, this is in contrast to nature and evolutionary processes which are characterized by constant change and instability as well (Gigon and Bolzern 1988). Therefore, if the maintenance of existing structures are defined too strictly it becomes difficult to maintain credibility. This does not have to mean that stability of biological and ecological systems have to be excluded as aims for practical air quality control. However, this has to be based on guidelines that are defined in relation to time and space. Ecological stability on the other hand has to be understood as a dynamic system that consists of constancy, cyclic stability, resistance stability and elastic stability (Gigon and Bolzern 1988). Consequently, air quality control has to be mainly concerned with stability in relation to resistance, i. e. the degree of stress, that may be tolerated by this system without involving change.

3.2 Risk assessment and estimation of damage

If it is difficult to define aims of protection for the current air pollution situation, this is even more true for risk assessments which are concerned with the extent and nature of pollution problems and their causes. Two methods may be suitable:

- the correlation of estimated damage with pollutant concentrations and time of exposure
- comparison of measured concentrations of pollutants with dose-response relationships from controlled environment experiments

The first of these two procedures contains two requirements, the quantification of damage and the determination of its cause.

3.2.1 Problems of diagnosis

Normally, the single reactions caused by air pollutants are not specific enough for the diagnosis of potential causes. Consequently, it is necessary to use a different approach which involves the negation of other potential causes of biotic or abiotic origin in combination with other evidence proving the involvement of pollutants. This may be done using a variety of biological and chemical methods, possibly in relation to soil science.

In areas with high and severely damaging level of air pollutants direct effects predominate and other factors are of small significance in comparison to air pollutants. By contrast, at relatively low levels of pollution significant effects are often only noticed after long-term exposure, possibly only in combination with other (a)biotic stresses. In this context, air pollutants may be directly responsible, contribute to the effects of other stressors or induce higher sensitivity to other stresses (Manion 1981). In this case it is almost impossible to factor causes of plant damage make it difficult to discern effects from other potential causes such as climate and reduce the ability of other investigations to prove the involvement of atmospheric pollutants. This is reflected in the difficulties in establishing the causes of current forest decline problems although it is now widely accepted that combinations of low concentrations of different air pollutants in interaction with other stress factors are likely to be responsible or contribute to this problem.

3.2.2 Evaluation of the nature and extent of damage caused by atmospheric pollution

It has already been described that long-lasting exposures to different toxicologically active compounds result in effects on various parts within ecosystems. Consequently, it is difficult and costly to characterize and quantify the various ways in which individual pollutant classes may affect all of them and to recognize the resulting damage. In this context, important aspects interfering with function of an ecosystem have to be defined. This may be done by comparison with a control. However, in the field it is almost impossible to find areas which can be used as controls because of differences in microlimate or soils. Therefore, regional comparisons have to be ruled out and we have to increasingly rely on changes that have occured on individual sites with time. From this it follows that the establishment of sites where longterm studies are able to investigate successional aspects of plant communities and their vitality are paremount. Simultaneously, they would also provide sites where changes in soils or populations of animals could be monitored.

In comparison to natural ecosystems, it is much easier to describe and quantify potential influences of air pollutants or agricultural ecosystems and this will be discussed in more detail in section 4. In forestry, pollution related damage may be monitored by ecological inventories, aerial observations and monitoring of annual increments of growth. While this only provides information of general growth conditions, dendrochronology allows to establish the degree of damage and at what time the problem became obvious.

3.2.3 Risk assessment

The potential hazards are estimated by comparing levels of wet and dry deposition with experimentally or epidemiologically identified doseresponse relationships of atmospheric pollution. The results of such investigations are summarized to provide guidelines by groups of experts within organizations such as VDI, WHO or the EC. In this way, specialists working for the UNECE have defined critical levels for trace gases, such as O_3, SO_2, NO_2 and NH_3, which are meant to protect vegetation. Critical levels are defined as those concentrations of air pollutants, above which direct adverse effects on plants may occur and are meant to offer protection to the most sensitive plant species and communities in any ecosystem (Table 3).

For various reasons, such values can only serve as guidelines but are a suitable basis on which to evaluate potential hazards. However, they are problematic because they have to try and map risks on an international level for different ecosystems which are not necessarily comparable because of difference in climate, soil and management. Consequently, much work is still required in integrating knowledge of levels of pollution with sensitivity of plants and other receptors in relation to regional peculiarities to allow better predictions of potential risks.

With increasing times of exposure the effects on soil and via soil become more and more important and are defined by the pollutant deposited with time on an area basis (e. g. mol m^{-2} a^{-2}). Like critical levels of gaseous components, wet depositions will also vary for different areas. For forest soils critical loads are quantitative estimates of potentially acidifying depositions below which no harmful effects on structure and function of ecosystems will occur (UNECE 1988b). In different soils this may be influenced by climate, hydrology, texture of soil and its chemistry as well as biological processes.

Table 3: Critical levels for sulphur dioxide, ozone and nitrogen dioxide for protection of vegetation (UNECE 1988a)

Air Pollutant	Dose µg/m^3	Duration	Note
SO_2	20	Annual mean	Symptoms may occur in sensitive plants, but it is unclear wheather these symptoms lead to growth declines or long term changes in plant vigour
	>30	Annual mean	Growth decline
	30	Annual mean	Productivity decline in sensitive forest stands
O_3	300	0.5 h	
	150	1.0 h	
	110	2.0 h	
	80	4.0 h	
	60	8.0 h	
	50	Vegetation/ Growing period mean	Average of 7 h mean/day, 9.00 - 16.00 H
NO_2	<30	Annual mean	Physiological/biochemical change
	30	Annual mean	Growth decline

The importance of vegetation and its usage for the critical levels/critical loads concept will be exemplified for N-containing air pollutants. For N-containing air pollutants three possible modes of action have to be considered (Guderian and Tingey 1987):

- direct effects on above-ground vegetation
- indirect effects after N-inputs to soils
- role of nitrogen oxides in the formation of photochemical oxidants

With regard to direct effects of nitrogen (NO_2) or ammonia (NH_3) neither of these two trace gases are important for vegetation but may become important when critical levels for ozone or sulphur dioxide are exceeded (UNECE 1988b). At the present time, it is unknown by how much emissions of nitrogen oxides have to be reduced in order to prevent the formation of concentrations of ozone that are toxic to vegetation.

If N-containing air pollutants are of greater risks via direct or indirect pathways is dependent on the ecosystem concerned. Indirect effects may be related to acidifying properties in terrestrial and aquatic ecosystems - 30-40 % of total deposition in Germany - and extra inputs of nitrogen as a nutrient in soils. While nitrogen inputs are generally considered to be beneficial to agricultural ecosystems, in other ecosystems this will depend on the amount of nitrogen deposited (Table 4). In many areas of Europe critical loads for nitrogen are exceeded and N-deposition of 20 kg ha-1 a-1 are common. These few examples already document the important questions alone in mapping which have to be addressed by those working in emission effects research, i. e.

- evaluation and characterization of objects requiring protection
- indication of exposure situations with regard to direct and indirect modes of action and
- characterization of site-specific situations

Progress in the next two decades in these areas will probably determine the rate of success of practical air quality control in Europe.

4. Procedures for the quantification and establishment of cost for agriculture and horticulture

As has already been said, it is impossible for many aspects of effects of air pollutants to be evaluated in relation to costs to natural ecosystems. However, this is somewhat easier for agriculturally managed ecosystems and will be detailed below.

Table 4: Critical N loads for ecosystems varying in productivity (kg N $ha_{-1}yr^{-1}$; UNECE 1988b)

Ecosystem	critical load
deciduous forests^{++}	5 - 20^{+}
coniferous forests^{++}	3 - 15^{+}
dwarf shrub vegetation	3 - 5^{+++}
grassland (e. g. mesobrometum)	3 - 10^{+++}
raised bog	3 - 5^{+++}

+ in mature forests the critical load may approach O

++ declining systems should approach O

+++ without major removal of N by management

Considering a critical nitrate concentration for groundwater of 50 mg l^{-1} (EEC standard) and a precipitation surplus varying between 100 and 400 mm yr^{-1} leads to critical loads of up to 10 to 40 kg N $ha^{-1}yr^{-1}$.

The main interest in agriculture[1)] is focussed on economically important aspects of plant growth and reductions in quality or quantity of perfomance in agriculture and horticulture can be distinguished between

- assessment in affected areas on a local or regional basis and
- estimates based on dose-response-relationships on a regional basis or related to larger areas
- assessments in affected areas using open-top chambers

When damage is assessed locally in the field there is always the risk of errors in establishing its causes, while exact assessments on a regional basis are difficult because of the lack of controls. Unfortunately, it is impossible to relate visible leaf injury to reductions in growth. This problem is enhanced in the absence of visible symptoms (compare 3.2.1 and 3.2.2).

Dose-response realtionships suffer from similar disadvantages as they are only estimates. Furthermore, the information on pollutant concentrations on an area basis are only approximations, while climatic and edaphic conditions and the cultivars used may differ from the model.

---

1 for assessments of damage in forestry compare H.J. Ewert in this issue

In order to provide more reliable sources of information on a national level a special program, the National Crop Loss Assessment Network (NCLAN), was founded in the U.S. (Heck et al. 1982). NCLAN consists of different governmental and other institutions cooperating in assessing exposure situations in the field and to develop models to estimate long-term economic consequences of pollution in agriculture as related to plant production. Thereby, it has been possible to establish dose-response relationships for O_3, SO_2 and NO_2, alone and in combination, and the yield of the most important crop species. Now they can be used to estimate primary economic impacts of atmospheric pollution on crops. These studies were carried out in important agricultural areas of the U.S. using open top chambers. Based on these relationships and data on yield, Adams et al. (1984b) has calculated that for the six most important crop plants (corn, soybean, wheat, cotton, millet and barley) which cover 75 % of farmed land in the U.S. and hay a 10 %, 25 % or 40 % reduction in concentrations of ozone will result in economic surplus (producer surplus plus consumer surplus = economic surplus) of the magnitude of 0.756 (+ 0.7 %), 1.937 (1.4 %) and 2.859 US-Dollar (2.0 %), respectively. By comparison, a 25 % increase in concentration of ozone would result in losses amounting to 2.363 billion US-Dollar equivalent of 1.7 % of the current value. A similar program to NCLAN is presently conducted for the European Community.

Summary

Current levels of complex atmospheric pollution mixtures which are present throughout large areas of Europe endanger both economic as well as ecological performance of primary producers, consumers as well as decomposers. Above certain levels they may impair the function of whole ecosystems. The choice of criteria to assess effects on plants has to depend on the different functions of vegetation which may serve economic, recreational or ecological functions and also serve as a gene-pool for future generations. From this it is obvious that only certain aspects may be valued monetarily. For the situation in Germany, it is at present not sufficiently clear to what extent vegetation may be affected by levels of atmospheric pollution. Therefore, the foundation of a special program to investigate economic impacts of atmospheric pollution on agriculture and horticulture is presented. Such assessments, using open-top chambers similar to the American National Crop Loss Assessment Network (NCLAN), can be expected to provide reliable predictions of reductions of yield and crop quality as well as quantitative relationships between levels of air pollutants and their effects on plants.

References

Adams R M, Hamilton S A, McCarl B A, 1984:
The economic effects of ozone on agriculture.
U.S. Environmental Protection Agency, Corvallis, Oregon 97 333, EPA-600/3-84-090

Bosch C, Pfannkuch F, Baum U, Rehfuess K E, 1983:
Über die Erkrankung der Fichte (Picea abies Karst.) in den Hochlagen des Bayerischen Waldes, Forstw. Cbl. 102: 167-185

Brabänder H D, 1983:
Ökonomische Evaluierung von Immissionsschäden am Wald. AFZ 30: 776-778

Bradshaw A D, 1976:
Pollution and evolution, 135-139. In: Mansfield T A (ed): Effects of Air Pollution on Plants. Society of Experimental Biology, Seminar Series I, Cambridge, University Press, Cambridge

Brandl H, Matthies F, 1984:
Bewertungssystematik für Waldschäden durch Luftverunreinigungen. Forst-Holzwirt 39 (1): 78-12

Ellenberg, 1973:
Ziele und Stand der Ökosystemforschung. In: Ellenberg, H (Hrsg.): Ökosystemforschung. Heidelberg, Berlin, New York: 1-31

Gigon A, Bolzern H, 1988: Was ist das biologische Gleichgewicht? Überlegungen zu einem Phänomen, das es strenggenommen gar nicht gibt. Aus Forschung und Medizin (Schering) 3:18-28

Guderian R, 1986; Möglichkeiten zur Erfassung und Bewertung von Immissionswirkungen auf Pflanzen. In: UBA-Berichte 7/86: 101-121, Erich Schnmidt Verlag Berlin

Guderian R, van Haut H, Stratmann H, 1963:
Probleme der Erfassung und Beurteilung von Wirkungen gasförmiger Luftverunreinigungen auf die Vegetation. Zeitschrift f. Pflanzenkrankheiten u. Pflanzenschutz 67: 257-264

Guderian R, Küppers K: Responses of plant communities to air pollution. Proceedings of Symposium on Effects of Air Pollutants on Mediterranean and Temperate Ecosystems, 187-197, Riverside, California, 1980

Guderian R, Tingey D T, 1987:
Notwendigkeit und Ableitung von Grenzwerten für Stickoxide. UBA-Berichte 1/87, Erich Schmidt Verlag Berlin, 178 S.

Heck W W, Taylor O C, Adams R, Bingham G, Miller J, Preston E, Weinstein L, 1982:
Assessment of corp loss from ozone. J. Air. Pollut. Contr. Assoc. 32: 353-361

Iversen J, 1964:
Retrogressive vegetational succession in the Post Glacial. Brit. Ecol.Soc. Jubilee Sympos. J. Ecol. 52: 29-70

Küppers K. 1986:
Wirkungen von Ethen auf Pflanzengemeinschaften. Verhandlungen der Deutschen Gesellschaft für Ökologie (Hohenheim 1984), Band XIV 1986, 361-371

Manion P D, 1981:
Tree Disease concepts. Prentice-Hall Inc. New Jersey

Möller H, Osterkamp R, Schneider W, 1981:
Umweltökonomik, Königgsstein

Roose M L, Bradshaw A D, Roberts T M, 1982:
Evolution of resistance to gaseous air pollutants. 379-409. In: Unsworth M H, Ormrod D P (ed): Effects of Gaseous Air Pollutions in Agricultrue and Horticulture. Butterworth Scientific, London.

Scholz F, 1984:
Wirkungen von Luftverunreinigungen auf die genetische Struktur von Waldpopulationen. Forstarchiv 55: 43-45

Ulrich B, 1983:
Stabilität von Waldökosystemen unter dem Einfluß des sauren Regens, AFZ 26/27: 670-677

UNECE, 1988a: Critical Levels. Workshop Bad Harzburg: 146 S.

UNECE, 1988b: Critical Loads. Workshop Skokloster: 28 S.

VDI, 1988: Zielsetzung und Bedeutung der Richtlinien: Maximale Immissions-Werte. VDI 2310, Bl. 1: Berlin (Beuth) 1-8

3.2 Economic Losses to Society Due to Material Damages of Environmental Pollution in the Federal Republic of Germany

Margit Weltschev
Federal Institute for Materials Testing, Unter den Eichen 87,
D - 1000 Berlin 45

1. Introduction

In order to be able to assess the damage caused to materials by acid deposition the first question to be considered is:
Which materials are affected?

A wide range of materials are exposed to the external environment. These materials have different susceptibility to attack by acid rain and different stock at risk in the Federal Republic of Germany. In an initial selection process we have evaluated materials according to the criteria shown in Table 1.

On the basis of this evaluation we have focussed on five principal materials of concern:

- zinc and galvanized steel
- stone
- organic coatings
- concrete
- brickwork.

The damage-prone objects under consideration to estimate the pollution-induced damage are:

- transmission towers
- highway bridges
- railway bridges and railway transmission towers
- telecommunication system
- industrial plants (fuel tanks, sludge towers)
- buildings
- cultural asset.

Table 1: Material Selection Criteria

Material	Susceptibility to Acid Rain	Value to the FRG
Brickwork	? (Uncertainty)	Very high, (Economic value high as large stock at risk).
Concrete	? (Uncertainty)	Very high, (Economic value high as large stock risk).
Stonework (limestone, calcareous sandstone, marble)	High. (Severely affected by SO_2).	High. (Small to medium stock at risk, but high cultural value).
Carbon steel	High. (Severely affected by SO_2).	Very low. (Little carbon steel exists uncoated in the environment).
Stainless steel	Very low. (Excellent resistance to pollutant attack).	Medium.
Nickel and nickel-plated steel	High. (Corrode at high rates in SO_2-polluted environment).	Very low. (Very restricted use).
Zinc and galvanized steel	High. (Corrosion markedly increased in polluted environment).	Medium. (Stock at risk small to medium but some uses of economic importance, e.g. transmission towers).
Aluminium	Low. (In general excellent resistance; pitting in very polluted environments but only initially).	Medium. (Various commercial uses).
Copper	Medium. (Increased corrosion in polluted environments, but rates decrease in long term (patina layer formed)).	Low. (Stock at risk low).
Lead	Very low.(one of the most resistant commercial materials in the environment)	Medium /low. (Stock at risk low but important uses, e.g. telephone cables).
Paint	? (Uncertainty)	Very high. (Very large stock at risk).
Wood	? (Uncertainty)	Low.(Little uncoated wood used in the environment).

2. Atmospheric Corrosion

Atmospheric corrosion is generally an electrochemical process. It takes place in corrosion cells with anodes and cathodes. The cells can operate only in the presence of an electrolyte, which means that atmospheric corrosion occurs only when the surface is wet. The time of wetness varies with the climatic conditions at the site. Each metal has a particular critical humidity in the range of 60 to 80 per cent relative humidity, above which corrosion tends to accelerate. Hygroscopic aerosols, such as sulphates and sea salts, cause metal surfaces to remain wet at a lower relative humidity than would otherwise be the case. For example, the critical relative humidity of steel in clean air is about 70 per cent. In industrial environments containing sulphur and nitrogen oxides, the critical relative humidity for steel falls to about 60 per cent. *The relative length of time a metal surface is wet* is the most important variable affecting metal corrosion. However, corrosion rates correlate well with SO_2 concentrations. The contribution of nitrogen oxides to damage of metals is still unresolved. The effects of chlorides, in general, assume importance only in coastal regions and in the vicinity of roads which have been highly salted, where they can contribute more to weathering and corrosion than SO_2.

3. Effects of Air Pollutants on Zinc and Galvanized Steel

Steel is usually used only with a *protective coating* such as *organic coatings* (paints) or *zinc galvanizing*. In moist but unpolluted environments, zinc hydroxide is formed with the electrochemical reactions taking place on the surface. The zinc hydroxide formed reacts with the dissolved CO_2 to produce sparingly insoluble zinc carbonate $ZnCO_3$. Build-up of these insoluble corrosion products, $Zn(OH)_2$ and $ZnCO_3$, creates a protective layer that inhibits further corrosion. In the presence of sulphur compounds, however, the condensed moisture on the surface is acidic, and the $ZnCO_3$ film is depassivated. The zinc sulphate ($ZnSO_4$) formed is soluble and easily washed away. The amount of metal lost from the surface is *strongly associated* with the *SO_2 air concentration*.

Experimental work has focussed on the corrosion of zinc in the presence of SO_2. The majority of field and laboratory investigations have shown zinc corrosion to be a linear function of both SO_2 and time of wetness. The research behind each dose-response function has been carried out under a range of different conditions. Some are field experiments, others controlled chamber studies, some have only monitored zinc corrosion at one particular site, others are nationwide surveys.

The bulk of the evidence on damage to zinc is express in terms of ***zinc corrosion rates***. If an assumption is made about the typical thickness of the zinc coating on galvanized steel, it is possible to convert corrosion rates to "time of loss of protective coatings". Once the protective coating is lost it will need replacing or painting, imposing a cost on the owner. An increase of lifetime is therefore a benefit (Figure 1 and 2).

The majority of *galvanized sheet* and *general galvanized* steel is *painted*, rather than replaced, when the galvanized coating has been lost. The shorter the lifetime of the galvanized structure or component the sooner it has to be painted and the greater the number of times it has to be repainted in its "expected lifetime" (Figure 1 and 2). The depth of coating which needs to be destroyed before painting becomes necessary depends on the particular use of the galvanized steel. The time intervals for first painting and repainting of galvanized transmission towers and railway transmission towers are shown in Table 2. New galvanized transmission towers are still painted to achieve a long repainting intervall.

Figure 1: Relative life expectancies of different zinc coatings /1/

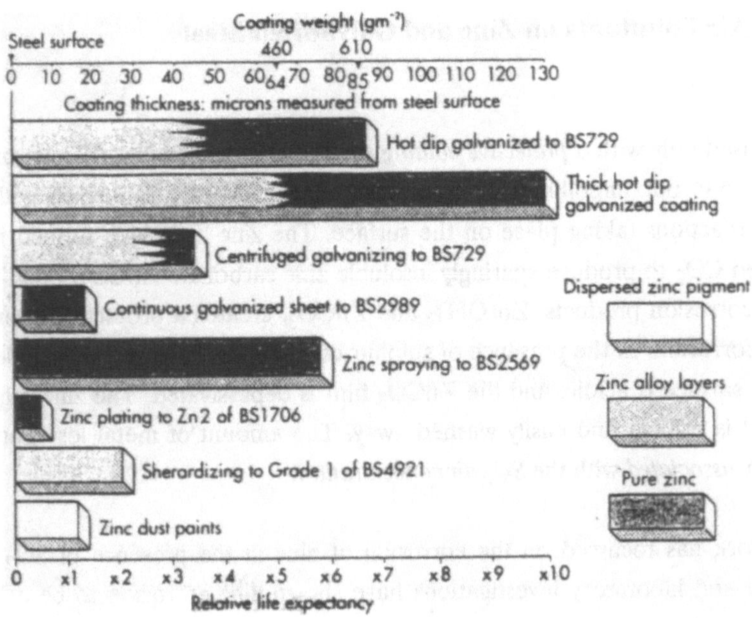

Figure 2: Life of galvanized coatings in different environments /1/

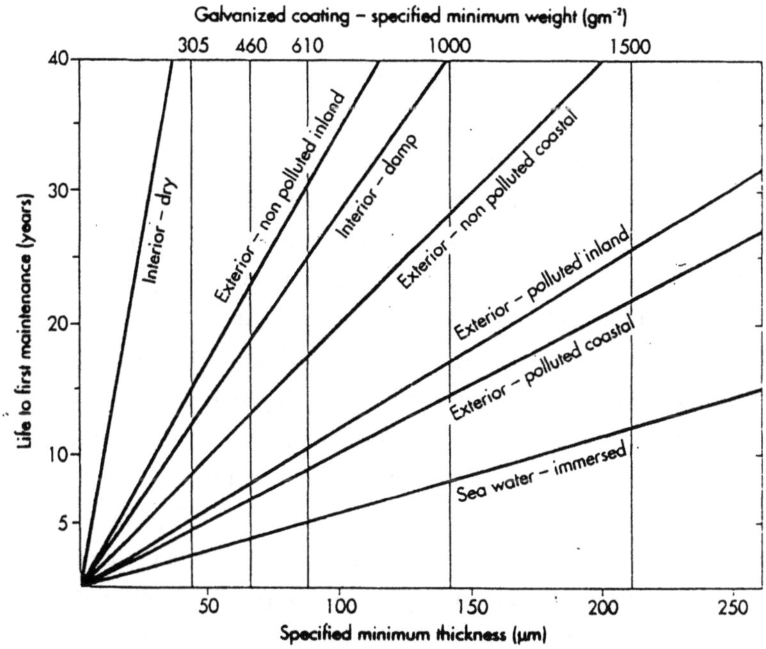

Table 2: Repainting intervals for galvanized transmission towers (years)

	non-polluted areas	polluted areas
galvanized transmission towers		
- first painting	7 - 10	1 - 3
- repainting	15 - 25	10 - 18
railway transmission towers		
- first painting	20 - 25	10 - 15
- repainting	22,5	12,5

These are empirical values of the main energy companies and the Federal Railway Central Agency in Germany.

4. Effects of air pollutants on paints

Paint formulations consist of pigment and vehicle. Pigments are finely ground solids (such as titanium oxide and zinc oxide) which provide colour, hiding power and durability. The vehicle provides the film-forming properties of the paint and contains resin binders, solvents and additives. Vehicles most often used in metal paints are:
oil vehicles, oil-alkyd vehicle, latex and acrylic emulsions, epoxy vehicles and polyurethan vehicles. These different paint types have different susceptibilities to air pollutants, but the effects of SO_2 are minor compared to the effects of wetness and sunlight. Research up to now has focussed on unflawed paint coatings, and the new line of thinking, which recent research supports, is that SO_2 may be important in the degeneration of paint systems (paint and paint substrate) exposed in the real environment, where the presence of flaws and scratches in almost all cases allow pollutants to attack the substrate. It is now thought that *damage to the substrate* or to the interface between substrate and paint is the critical factor, not damage to the paint itself. However, there is still large uncertainty over the effects of SO_2 on paint systems, little is known at present how paint protects metals and the mechanisms that take place at the substrate/paint interface. Dose-response functions derived for paints and for paints in inert substrates therefore taking no account of corrosion of the substrate/paint interactions that together cause paint decay. These functions are all in terms of erosion rate (μm/year). Paint failure tends to occur well before linear erosion of the film as a result of blistering, blooming, cracking and chalking. Current work in Sweden is using blistering rates to assess paint-system decay. The expert's views, based on a survey of painting frequencies of a number of companies in different SO_2 environments are shown in Figure 3.

Figure 3: Expert opinion on paint damage by SO_2 (for paint/metal substrate systems) /2/

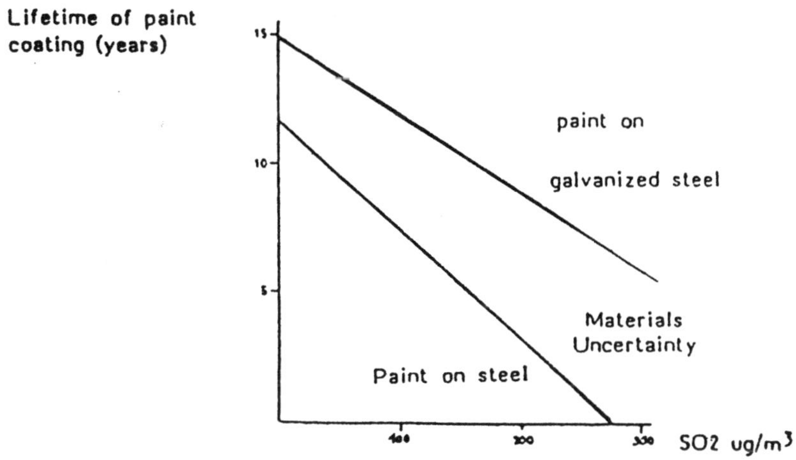

Repainting intervals are generally longer in rural than in urban atmospheres. An inquiry made to petroleum companies, energy companies and Federal Railways directions confirmed these differences of repainting intervals.

Table 3: Repainting intervals of fuel tanks, railway bridges and railway transmission towers consisting of steel (years)

	non-polluted areas	polluted areas
fuel tanks	25	8 - 15
railway bridges	25	20
railway transmission towers	20	15

The effect of atmospheric pollutants may vary with the formulation and the formulation can vary over the lifetime of any brand of paint. The datas are often estimate rates in default of experience. A very important aspect is the development of improved painting systems due to increased air pollution levels.

5. Effects of air pollutants on concrete and brickwork

The effects of SO_2 on concrete are insignificant compared to carbonation by carbondioxid and chloride attack on reinforcement bars inside concrete.

Apart from the deterioration processes inherent in the quality of reinforced concrete (for example high chlorine content, high porosity, alkali-silica reactions), the processes involving air pollutants are shown in Figure 4.

Figure 4: **Distribution of salts in concrete after reaction with acidic gases /3/**

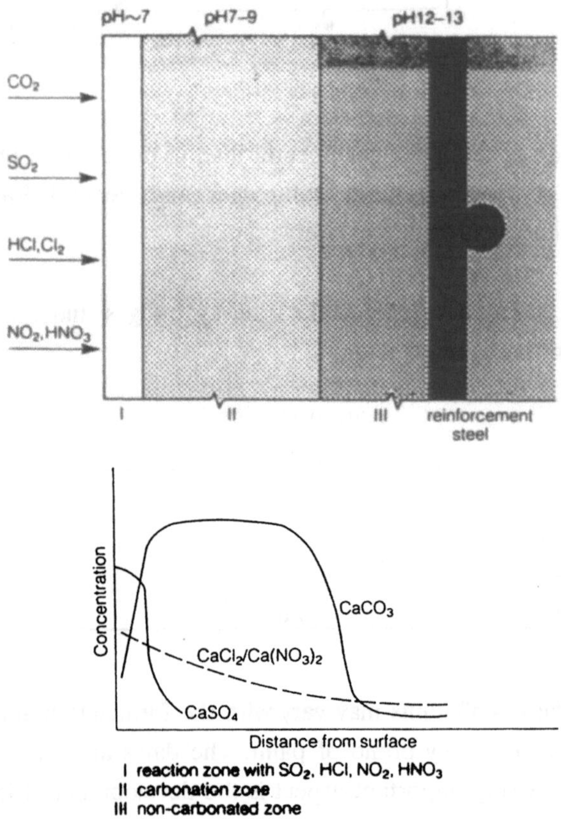

In well prepared concrete, the high alkalinity of the cement matrix, shown as zone III, protects the reinforcement steel from corrosion. Loss of alkalinity and hence corrosion of the steel occurs when the alkalinity is consumed by acidic species. The main process of atmospheric CO_2, depicted in zone II. Reaction products of SO_2 (gypsum) remain at the surface (zone I) and can be washed away by rain on the wetted surface areas. Chlorides are able to penetrate the carbonation zone and even the unreacted zone III and will initiate the corrosion of steel. Since the corrosion products of steel occupy a greater volume than the original material, cracks will develop rendering the system more open to attack by SO_2. If the concrete layer covering the steel is thick enough and the concrete is of good quality the acidic pollutants will only influence the surface.

Brickwork is made up of bricks and mortar. Brick, which is a calcium-aluminium silicate ceramic, is unaffected by SO_2 attack. But SO_2 may damage the mortar, which is made up of sand, calcium hydroxide and other carbonate phases which are susceptible to sulphates. This has yet to be researched, however, and at present no quantitative information exists on the effects of SO_2 on brickwork.

6. Anthropogenic air pollution as a factor of stone decay

It is evident that sulphur compounds attack natural stone which contains carbonate. Sulphur compounds react with the carbonate ($CaCO_3$) to form gypsum ($CaSO_4 \cdot H_2O$). There are two different mechanisms of pollutant transportation in the air, including adsorption on the surface of the natural stone:

1. dry deposition and
2. wet deposition

In areas with high pollution levels dry deposition is predominant. In clean air areas, however, especially in those with high rainfall rates, wet deposition is the predominant mechanism. These two different kinds of deposition lead to different kinds of natural stone deterioration.

Figure 5: Scheme of possible reactions to remove SO_2 and SO_4^{2-} from atmosphere /4/

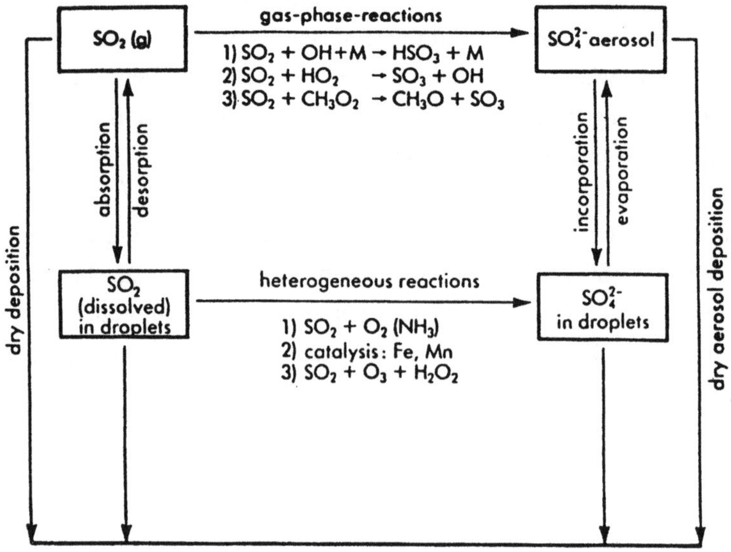

Dry SO_2-deposition leads to acid formation on the surface and within the pores under the stone surface. Within the surface zone the stone structure will become more and more dense. Simultaneously, the complex crystallization and recrystallization processes of gypsum exert pressure within the stone leading to deterioration of the stone. During wet deposition SO_2 is partly oxidized to sulphuric acid within the aerosol particles which will be accumulated in rain clouds. In rain-exposed areas of stone the reaction product gypsum is washed away, resulting in a weight loss. Stone deterioration derived from wet deposition generally is lower than that caused by dry deposition.

Influences of anthropogenic air pollution together with natural erosion lead to a damage situation in which the different factors cannot be properly distinguished, because neither sufficient material data are available nor the correlation of the different factors is known as yet. A further problem with the factor correlation is that the different factors may vary from stone type to stone type and their influence may vary in a non-linear manner during the damaging process. The uncertainty surrounding natural stone is substantial because of the lack of quantitative data available at present. Large uncertainty bands are obtained for the effect of SO_2 on unsheltered stone. Also included in these bands is the uncertainty regarding the role of carbonaceous particles in stone decay. The uncertainty range for marble is less than for limestone and sandstone because its material properties vary less. A large proportion of the value of cultural resources lies in aesthetic criteria and in the express in monetary terms, especially since the value is reduced by attempts at restoration, even when the damage is reversible.

The issue of concern to historic buildings and monuments is their cultural value. It is therefore necessary to find some indicator of "cultural value" which is useful for assessing the damage caused by air pollution.

Mr. Heinz will explain the economic methods of the monetary valuation of damage costs of transmission towers, railway bridges, highway bridges, fuel tanks and buildings.

References

/1/ Wall, A.J.: Applications for Zinc: Developments and Prospects. Metals and Materials 3, S. 134-137 (1989)

/2/ Henriksen et al., 1985

/3/ Lanting, R.W., Delft, Netherlands, TNO, Division of Technology for Society. Private communication (July 1986)

/4/ Perseke, C.: Die trockene und feuchte Deposition säurebildender atmosphärischer Spurenstoffe, Diss. Universität Frankfurt a.M. (1982)

3.3 Global Warming
Climate Protection Measures of the Enquête-Commission of the German Parliament

Wilfrid Bach
Center for Applied Climatology and Environmental Studies
Climate and Energy Research Unit
Department of Geography
University of Münster
Robert-Koch-Str. 26
D - 4400 Münster

1. Climate change as part of the complex environmental problems

The climate problem must be viewed within the context of the accelerating destruction of our natural environment. Industrial and agrarian society has, by wastefully exploiting natural resources and thoughtlessly applying technology, attained a potential for inflicting damage on a scale that is seriously endangering the sustained survival of man. The chemical catastrophes of Seveso, Bhopal and Basle, the nuclear reactor catastrophes of Three Mile Island and Chernobyl, and the recurrent famines in the developing countries are early warning signs that have briefly shocked society. But no less alarming are the preserving threats, such as air, water and soil pollution, the contamination of food, the destruction of forests and the ozone layer, the depletion of genetic resources and gene manipulation, the long-term storage of radioactive and chemical wastes, the stockpiling and proliferation of atomic, chemical and biological weapons, the degradation of landscapes, rapid population growth, the growing scarcity of food, and last but not least the epidemic spread of allergies, viral infections and cancer (Fig. 1).

For a long time, nature was able to absorb this ruthless pillaging of resources and the destruction of the environment. But once a certain "threshold" has been exceeded - and that could very soon be the case - it will respond in a way that must have catastrophic consequences for man. For the world-wide threat of ozone depletion and climate change, there is still a faint glimmer of hope, because the influencing factors are known, and it appears that the willingness exists (in the case of ozone) or seems to grow

Figure 1: Mankind as instigator and victim of environmental hazards

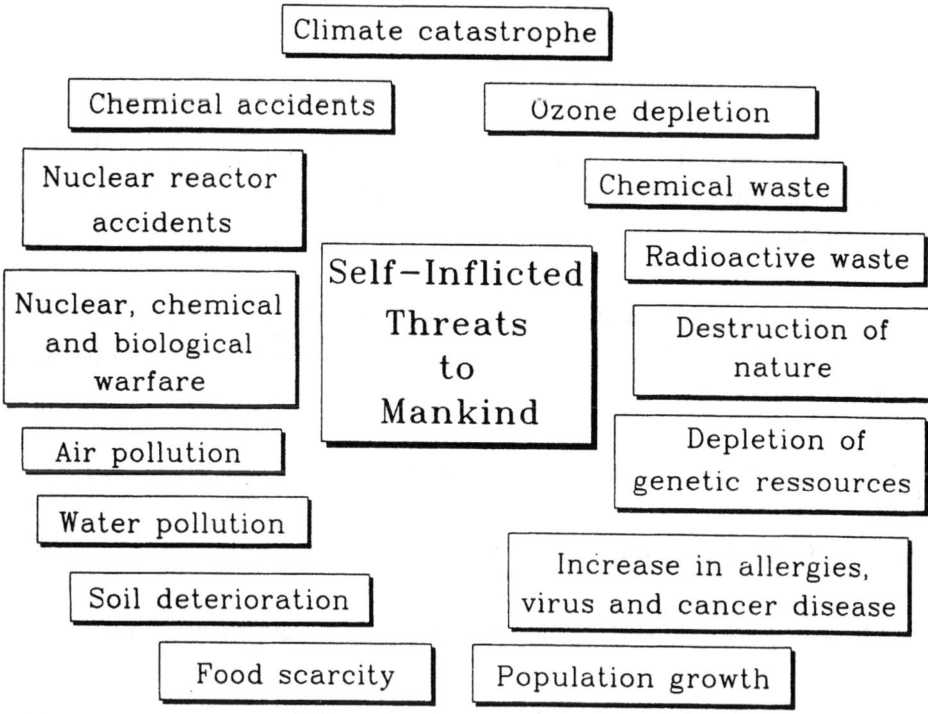

Source: Bach (1990)

(in the case of climate) to implement countermeasures. If joint action is soon taken, then we can perhaps utilize what little time is left to just barely ward off the global climate threat that is being posed to society by the additional man-made greenhouse effect with all of its far-reaching regional and local repercussions. The final decade of this century is our last chance to abandon the self-destructive logic of the present economic system. A system that is environmentally, climatically and socially benign and that takes into account the aspirations of future generations, requires a complete re-structuring of the economic and political institutions.

In the following, the climate-influencing factors and their possible impacts are described. The climate strategy developed by the Enquête-Commission of the German Parliament is summarized. The recommended measures are evaluated in terms of their effectiveness for mitigating the additional man-made greenhouse effect.

2. Climate-influencing factors

The most important climate-influencing factors are:

- The extremely wasteful use of fossil fuels in the industrialized nations, leading to greenhouse gas emissions from coal, oil and gas which are too large and too fast to be absorbed by the natural system.

- The manufacture and release of artificial products such as CFCs and halons, which are not only effective greenhouse gases, but whose waste products are also depleting the stratospheric ozone layer.

- The destruction of forests, other vegetation, and soils.

- The highly mechanized agro-business.

- The release of gases from waste dumps and the emissions from cement production, etc.

Table 1 shows that currently ca. 46 % of the greenhouse gases (37 % CO_2, 9 % CH_4 and O_3) are emitted by the energy and transportation sectors. Roughly 24 % is accounted for by CFCs and halons produced by the chemical industry. Another 18 % results from the destruction of tropical forests (12 % CO_2, 6 % CH_4, N_2O, CO etc.). Agriculture contributes around 9 %, for the most part in the form of CH_4 (from rice paddy fields and cattle feedlots) and N_2O (from artificial fertilizers). The remaining 3 % is accounted for by emissions from waste dumps and cement production. As Table 2 shows, in the past, CO_2 was the most prominent greenhouse gas with a share of ca. 60 %. During the last decade, emissions of CFCs have rapidly grown to take second place. The Montreal Protocol and the agreement of London should reduce the influence of the CFC-emissions on climate and the ozone layer during the decades to come.

Table 1: World-wide contribution of the various sources to the additional greenhouse effect in the 80's

Sector	Share of trace gases	Sources
Energy and Transport	46% total 37% CO_2 9% CH_4 and O_3 (O_3 formed by precursors NO_x, CO and NMVOC)	Emissions from coal, oil and gas; in the conversion sector from electricity and district heat as well as refineries; in the end-use energy sectors residential, commercial, industrial, and transport
Chemical products	24% CFCs, halons etc.	Emissions from CFCs, halons, FCs
Tropical deforestation	18% total 12% CO_2 6% other trace gases such as CH_4, N_2O, CO	Emissions from burning and decomposing tropical forests, and from soils
Agriculture	9% above all CH_4 and N_2O	Emissions from rice paddy fields and ruminants (CH_4) and artificial fertilizer (N_2O)
Others	3%	Emissions from cement production and waste dumps

After: UNEP/WMO (1990) and EK (1990)

Table 2: Relative contribution of the global concentration of greenhouse gases to the additional greenhouse effect

Greenhouse gases	Cumulated from 1860 - 1980[1] %	Actual values for 1980 - 1990 %
Carbon dioxide (CO_2)	60	46
Methane (CH_4)	14	9
Ozone in the troposphere (O_3)[2]	10	14
Chlorofluorocarbons (CFCs)	9	24
Nitrous oxide (N_2O)	3	4
Water vapor in the stratosphere (H_2O)	4	3

1) Calculated with the Muenster Climate Model.

2) Refers to the lower atmosphere of the Northern Hemisphere; data are not derived from a complete climate/chemistry model and are therefore uncertain.

After: EK (1990)

3. Climatic Impacts

The observed global warming is due to natural processes and to the increasing concentrations of greenhouse gases in the atmosphere as a result of man's activities. Because of the large heat capacity of the oceans, however, the warming effect is delayed by several decades, so that statistical proof of the additional greenhouse effect induced by human activities can only be obtained after a certain time lag. Since 1860 the mean global surface temperature has risen by about 0.6 °C. If the current trend continues, the global warming could exceed the critical rate of 0.1 °C per decade by a factor of between 3 and 5. There is reason for concern that the earth's ecosystems, having already been placed under considerable stress by pollution, especially tropospheric ozone, acid rain, and heavy metals, etc., will be unable to cope with this additional climatic stress. The vegetation zones are not likely to shift polewards because of the rapidity of the changes, and consequently there is a high probability that vegetation which is adapted to present site conditions will die out. As the indigenous vegetation disappears, new plant forms that are better-adapted to the changed conditions may eventually take their place.

With increased warming there will be more drought, floods, and subtropical storms. The warming will cause not only the snow and the sea ice to shrink and to become thinner, but also the Arctic permafrost soils to thaw. This will reduce the earth's reflection, enhance solar absorption, release great quantities of CH_4 and CO_2, thereby fuelling the warming process as a consequence of these feedback processes.

The sea level has already risen by about 15 cm over the course of the last century. If the present warming trend continues, an additional sea level rise on the order of 30 to 110 cm can be expected. It is almost certain that during the next few decades some 200,000 Maledivians and a few hundred thousand people living on atolls in the Pacific will have to be resettled. At a sea level rise of only 0.5 m millions of ecological refugees from Bangladesh, Egypt and other threatened, densely populated coastal regions are to be expected.

At the present rate of world population growth, the anticipated impact on the agricultural sector needs to be taken especially seriously. Agricultural yield will be reduced in vulnerable regions. World market prices for agricultural products will rise. This will result in more famines and conflicts over the distribution of food. The climatic changes triggered by human activities will thus have a destabilizing effect on world peace.

In order to ward off this threat, it is imperative to act immediately to reduce emissions of greenhouse gases. For this purpose, the Enquête-Commission of the German Parliament proposes the following climate strategy.

4. Recommendations for International and EC-Wide Action to Reduce the Additional Greenhouse Effect and to Protect the Earth's Atmosphere

In order to protect the earth's atmosphere, binding international agreements are called for. At the same time, the different types of problems regarding ozone-depleting substances, energy and climate, and protection of the tropical forests require very specific strategies for action in each individual case. Consequently, the Commission believes that the quickest way of arriving at international agreements is to take a sectoral approach. It is also of the opinion, however, that the framework for an overall protection strategy of the earth's atmosphere must be set up right at the outset to ensure a sustained ecological and economic development.

4.1 Measures and reduction targets with respect to CFCs

A total stop to production and use of CFCs and halons is not only the most effective means of protecting the ozone layer, but it is also the most economic short-term measure for reducing the additional anthropogenic greenhouse effect. In order to ensure comprehensive protection of the Earth's atmosphere, therefore, the Commission regards it as essential for the stipulations of the Montreal Protocol to be strengthened even further than was done by the agreements of the 2nd conference of the signatory states in London. Specifically, this means:

- To move up from 2010/2015 to 1997 and to completely stop production and use of all fully halogenated CFCs and halons as well as carbon tetrachloride and methyl chloroform;

- to stop by 2005 at the latest production and use of partially halogenated chlorofluorocarbons, which had not yet been included in the Montreal Protocol;

- to permit a time-limited and restricted use of radiatively-active fluorocarbons only if it can be proven that no substitutes are available;

- to eliminate all currently existing exemptions for the industrialized and developing countries; for the latter substitutes and technologies must be made available and financial assistance must be provided;

- to supply funds derived from a tax of 10 deutschmarks on each kilogram of CFCs and halons produced and 5 deutschmarks on each kilogram of H-CFCs;

- to identify world-wide all raw materials, intermediate and finished products that are made of ozone-depleting and climate-influencing substances and that should be subject to an import ban under the provisions of the Montreal Protocol;

- to provide government monitoring on production and use as well as the achieved reduction rates in a manner that is transparent to the public.

4.2 Measures for the protection of tropical forests

For the preservation of the tropical forests, the Commission has developed the following three-stage plan:

- Firstly, reducing the rate of forest destruction between 1990 and 2000 to that of 1980;

- secondly, stopping the destruction of tropical forests by 2010 at the latest, so that the total forested area in absolute terms no longer declines;

- thirdly, between 2010 and 2030 restoring the forest stand to the level of 1990.

This plan is to be financed by establishing an International Trust Fund of the order of 10 billion DM to be paid annually for the most part by the industrialized nations.

The Commission proposes that an International Convention for the Protection of Tropical Forests be adopted at the 1992 UN Conference on Environment and Development in Brazil separately from an International Convention on Climate and Energy. This is reasonable, because the global climate is just one aspect of many and because here the focus should be mainly on conservation of species diversity, protection of indigenous peoples, social, ecological and economic issues, and on factors that affect the climate more on local and regional scales.

4.3 Measures and reduction targets in the areas of energy and climate

The reduction of the additional anthropogenically-induced greenhouse effect requires drastic changes in the global energy systems. These changes must be implemented as soon as possible on the basis of carefully worked-out long-term strategies. The magnitude of the reductions depends on the reduction potential that can be mobilized in each individual country. The allocation of the reduction shares must take into account both the different economic and technological situations in the various countries and criteria that ensure equitable treatment of all. It is clear that the developing countries need to catch up. A binding global plan for reducing emissions can only succeed if fair reduction shares are allocated based on a catalog of criteria that takes into consideration all important aspects.

The Commission has formulated the following criteria for identifying reduction targets for individual countries:

- Economic strength
- The existing energy supply structure and the efficiency of the energy system
- Total accumulated emissions and the current per capita emission rate
- Population development and refugee flows
- The export/import balance of energy-intensive products and processes
- Emission per unit land area and the influence of the climate on energy use (heating, refrigeration)
- The anticipated extent of climate impacts and the level of environmental awareness.

Based on present knowledge, the Commission considers the following region-specific reduction targets for energy-related CO_2 emissions, using 1987 as the base date, both necessary and feasible:

- For the Federal Republic of Germany and other economically-strong industrialized countries with very high per capita CO_2 emissions, reductions of at least:
  - 30 % by 2005,
  - 50 % by 2020, and
  - 80 % by 2050.
- For economically-weaker industrialized countries, reductions of at least:
  - 20 % by 2005,
  - 40 % by 2020, and
  - 80 % by 2050.

- For Eastern Europe Countries including the USSR initially an even slower procedure is indicated for economical and technological reasons. leading to reductions of:
  - 10 % by 2005,

- 30 % by 2020, and
- 80 % by 2050.

- Equity considerations lead to a greater share in global CO_2-emissions in the developing countries as compared to the industrialized countries in the first half of the 21st century; at the same time the rate of increase of fossil fuel use per year must be slowed down so that the transition to a more efficient energy economy based on renewables is not impeded, this results in an increase of:
  - 50 % by 2005,
  - 60 % by 2020, and
  - 70 % by 2020.

- For industrial countries taken together reductions of at least:
  - 20 %by 2005,
  - 40 % by 2020, and
  - 80 % by 2050.

- For the world as a whole, reductions of at least:
  - 5 % by 2005,
  - 20 % by 2020, and
  - 50 % by 2050.

The target formulated at the 1988 conference "The Changing Atmosphere: Implications for Global Security", namely a 20 % reduction in global CO_2 by the year 2005, is, as the foregoing discussion shows, not practicable. The reason is that the economically-strong industrialized countries would be required to implement reductions of considerably more than 30 % in order to offset the increases in the developing countries.

In order to mobilize the existing emission reduction potential, the Commission proposes that, together with a Reduction Protocol an International Trust Fund be established, initially having DM 20 billion at its disposal. In addition, an International Environment Council should be set up under the auspices of the United Nations for the purposes of strenghtening global environmental policies and supervising the measures to be implemented.

4.4 Overall strategy for the protection of the Earth's atmosphere

The different sectoral and international agreements must be integrated into an overall strategy. The Commission recommends that the timetable shown in Table 3 be adopted, culminating in the ratification of an overall strategy on the protection of the Earth' atmosphere.

Table 3: The 1990s, a Decade of Critical Decisions

Time-table for the development of an overall strategy for the protection of the Earth's atmosphere as prepared by the Enquete-Commission of the German Parliament

1990	Recommendation by the Second World Climate Conference in Geneva to negotiate international conventions on • Climate and Energy and • Protection of Tropical Rainforests to be completed at the UN Conference on Enviroment and Development (UNCED) in 1992 in Brasilia.
1991	Two preparation conferences held to negotiate the draft conventions.
1992	Acceptance of the two international conventions in Brasilia and strengthening of the Montreal Protocol according to the version agreed upon at the conference of contracting nations in London.
1994	Acceptance of concrete implementation protocols to both conventions.
1995	Ratification and implementation of the protocols and the international funding mechanism.
1998	Preparatory Conference for combining the sectoral international conventions.
2000	Adaption of a complete strategy for the protection of the Earth's atmosphere.

After: EK (1990)

Because of the lack of reliable estimates on the emission reduction potential, it is not yet possible at this time to establish concrete reduction targets for individual countries. In the next section, the Federal Republic of Germany is taken as an example to demonstrate what detailed studies are needed and which measures can be applied to attain the required reduction goals.

5. Recommended action for reducing energy-related climatically-relevant trace gas emissions in the Federal Republic of Germany

In order to prevent further aggravation of the additional anthropogenic greenhouse effect and in the light of what is already known, the Commission considers it essential to take appropriate steps without delay to reduce emissions in the Federal Republic of Germany. Besides a coordinated approach in the EC, in the opinion of the Commission it is also urgently necessary for the Parliament and the Federal Government to initiate action at the national level that will permit the Federal Republic of Germany to assume a leading role. It is imperative to act now and not to wait until all the respective international agreements have been made.

Due to lack of time, so far the Commission has primarily concerned itself with energy-related climatically-relevant trace gas emissions. In the next study phase, attention will focus on such interrelated topics as "ecosystems - climate - agriculture", which have an enormous importance for food security.

5.1 Reduction targets for energy-related climatically-relevant trace gas emissions

An initial evaluation of the potential for reducing emissions, carried out for the Commission by some 50 different institutes and laid down in a total of around 150 energy studies, has yielded the reduction plan shown in Table 4 for CO_2 and other directly or indirectly climatically relevant trace gases. Because of the different energy structure in the Eastern part of Germany, altered reduction contributions could result in some of the sectors as a consequence of the former GDR joining the Federal Republic of Germany. Taken as a whole, however, as already shown for the case of CO_2, the recommended reduction targets (expressed as percentages) will not be changed because of German unity.

Table 4: Reduction plan for the energy-related and climate-relevant trace gas emissions of the F.R. Germany plus the GDR for 2005 (binding reduction target) and the years 2020 and 2050 (goals) related to base year 1987

Trace gas	Base year 1987 Mio t	Reduction target by 2005 %	Goals by 2020 %	Goals by 2050 %
Carbon dioxide (CO_2)				
FRG	715	-30	-50	-80
FRG + GDR	1067	-30	-50	-80
Methane (CH_4)	1.8	-30	-50	-80
Nitrogen oxides (NO_x)	2.6	-50	-70	-90
Carbon monoxide (CO)	8	-60	-75	-90
Non-methane volatile organic compounds (NMVOC)	1.5	-80	-90	-95

After: EK (1990)

5.2 Assessment of the CO_2 emission reduction potential

Within the scope of the Commission's energy study program the CO_2 emission reduction potential for the year 2005 as compared to the reference year 1987 was estimated for the following three scenarios:

- A reduction scenario "energy policy" in which the share of electricity from nuclear energy of 31.2 % in 1987 is maintained.
- A reduction scenario "nuclear energy phase - out".
- A reduction scenario "nuclear energy expansion" (in which the share of electricity from nuclear reactors is almost doubled to 60 % and the share in primary energy consumption is more than doubled from 11.4 % in 1987 to 24.6 % in 2005).

The most significant result shown in Table 5 is that considerable increase in the use of atomic energy would not cut down on CO_2 emissions as compared to a phase-out of nuclear power. In all three scenarios the CO_2 emission reduction potential is almost the same, at 34-36 %. There are, however, appreciable differences among the various sectors, with the greatest emission reductions being achieved in the end-use energy sector for a nuclear phase-out, while the energy conversion sector would account for the largest reductions with a nuclear energy expansion. However, these reductions are not independent of one another, since energy savings in the end-use energy sector (particularly with respect to electricity) would also lead to reduced energy consumption and emissions in the conversion sector. It is also conspicuous that, in the end-use energy sector, the greatest potential for reduction exists in private households and - to a lesser extent - in the category "commercial". An enormous, still virtually untapped savings potential exists here in connection with electricity and heating, and thus measures to reduce all forms of electricity use would be particularly worthwhile. By contrast, only smaller reductions or even increases in emissions are obtained in the sectors of industry and transportation. This could be partially attributable to the rate of economic growth of 2.4 % per annum which was assumed in all scenarios, as well as to significant increases in traffic. If appropriate political action were taken, it would be possible to mobilize a considerable reduction potential in the industry and transportation sectors as well. Of importance is the fact that in the end-use energy sector an expansion of nuclear power would lead to an emission reduction of only 9.8 % if nuclear power were phased-out, thereby hindering rather than facilitating the mobilization of the CO_2 reduction potential.

Table 6 shows the changes in use of the various energy carriers on which the scenarios was based. Compared to nuclear phase-out, with nuclear expansion more hard and brown coal but less mineral oil would be used. Substitution of the other energy carriers by the more CO_2-benign natural gas is nearly three times as high in the phase-out scenario as in the expansion scenario. Expansion of nuclear power would tend to inhibit rather than to promote the use of renewable energy sources. In total, the phase-out of nuclear power would lead to more than twice as much energy saved as compared to expansion; and, moreover, less would be consumed than in the "energy policy" reduction scenario.

Additionally, the Commission requested the "Öko-Institut" to assess the consequences of a nuclear phase-out as early as 1995. Interestingly, this scenario revealed that, compared to the "energy policy" scenario, the accumulated CO_2 emissions from 1987 to 2005 would only be 1.7 % higher in the "1995 nuclear phase-out" scenario. This is due to the greater mobilization of the efficiency potential and the greater utilization of renewable energy as well as to the assumption that ambitious measures will be implemented in the transportation sector. The reliable supply of electricity is also ensured. In addition, the "1995 nuclear phase-out" scenario would not only reduce the risk of nuclear accidents to zero within just 5 years, but would also minimize both the danger of proliferation of atomic weapons and the amount of nuclear waste.

Table 5: Reduction potential of CO_2-emissions[1)] by sector and scenario for the F.R. Germany in 2005 as related to 1987

Sector	Reduction potential (%) until 2005 for scenarios		
	Energy policy	Phase-out of nuclear energy	Expansion of nuclear energy
Residential	-8.5	-10.1	-7.1
Commercial	-3.3	-5.0	-2.8
Industry[2)]	-0.6	-2.5	+1.8
Transport	-1.9	-3.2	+0.1
Sectors not considered	+0.4	-0.1	-0.1
Substitution by cogeneration	-2.8	-4.1	-1.7
Sum of end-use energy sector	-16.8	-25.0	-9.8
Cogeneration and heating plants	+3.6	+5.2	-4.1
Other electricity production	-13.3	-6.9	-15.0
Other contributions and statistical differences	-2.2	-2.0	-1.9
Sum of conversion sector	-11.9	-3.7	-21.0
Sum of end-use energy and conversion sectors	-28.7	-28.7	-30.8
Energy-conscious behavior	-5.0	-5.0	-5.0
Total reduction	-33.7	-33.7	-35.8

1) Referring to end-use energy including the German share in international air traffic but excluding non-energy use.
2) Including fossil fuel emissions from internally produced and used electricity.

After: EK (1990)

Table 6: Change in energy use (%) by energy carrier[1] in the reduction scenarios for the F.R. Germany in 2005 as related to 1987

Energy carrier	Change (%) in the scenarios until 2005		
	Energy policy	Phase-out of nuclear energy	Expansion of nuclear energy
Hard coal	-6.7	-4.4	-7.3
Lignite	-3.7	-4.3	-5.2
Mineral oil	-13.9	-20.4	-12.3
Natural gas	+4.2	+10.0	+3.4
Sum of fossil fuels	-20.1	-19.1	-21.4
Hydrogen	+0.3	+0.3	+0.3
Wind energy	+0.4	+0.8	+0.1
Photovoltaics	0	+0.1	0
Renewables in the end-use energy sector	+3.6	+5.6	+1.7
Net electricity import	0	0	0
Nuclear energy	+1.8	- 11.4	+11.1
Sum	-13.9	- 23.7	-8.2
Energy-conscious behavior	-5.0	- 5.0	-5.0
Total reduction	-18.9	- 28.7	-13.2

[1] Deviations from primary energy use because non-energy use and other energy trade etc. are not taken into consideration.

5.3 Strategy for attaining the reduction targets

In order to achieve emission reductions in Germany of 30 % each for CO_2 and CH_4, 50 % for NO_x, 60 % for CO, and 80 % for non-methane volatile organic compounds by the year 2005 as compared to 1987, the Commission recommends that

- the Federal Government initiate a crash program of far-reaching measures on the basis of existing legislation;
- the Federal Government submit the drafts of additional amendments by 1 September 1991;

- the Federal Government submit a report on the required legislative amendments to the German Bundestag by 1 December 1991, elaborate a plan for the financial policy measures and present the corresponding draft bills by 1 July 1992;
- the Federal Government submit progress reports on the required environmental, energy, and transportation legislation to Parliament every 2 years, beginning on 1 December 1993.

Within this process of dynamic adjustment and reassessment, the focus is as follows:

- **In the energy sector:**

The Electric Utilities Act, the Energy Conservation Act, the Thermal Insulation Regulation, the Federal Tariff Code, the Waste Heat Recovery Ordinance, the Heating Systems Ordinance, the 3rd Electricity Generation Act, the Federal Pollution Control Act, the Low-Income Housing Act, etc. These must be supplemented by appropriate new legislation, e.g. regulation of refunds on electricity sold to utilities as well as energy prices that will encourage more efficient use.

- **In the transportation sector:**

Integrated and environmentally-benign transportation systems are required shifting transportation services to environmentally compatible types of transport, minimizing increases in road traffic, improving transportation efficiency, and moderating traffic loads. In addition - among other things - the Right of Way Act and the Urban Transportation Funding Act must be changed.

5.4 Priorities for the reduction measures

In order to reach the set reduction targets, the following basic principles must be observed:

- Energy conservation by more efficient use and conversion, as well as by more energy-conscious behavior, must be given top priority.
- Renewable energy use must be extended as a matter of greatest urgency so that it will account for a growing proportion, and ultimately the largest proportion, of the energy supply.
- During the transition period toward a low-pollution energy economy, fossil fuels (coal, oil and gas) must be used as environmentally and climatically benign as possible.
- The views vary widely on whether nuclear power should be used in the future. The serious problems involved should be further discussed and the justification of continued use should be clarified.

If the goal is to prevent continued increases in energy use and emissions in spite of future economic growth and instead achieve substantial decreases, then full advantage must be taken of the technological possibilities. Space does not permit a detailed discussion of the technological potential for reducing CO_2 emissions by means of more efficient energy use as determined by the energy studies carried out for the Commission; however, a brief overview is given in Table 7. The potentially achievable reductions range from about 10 % through cogeneration to approx. 90 % by upgrading the existing housing stock.

Table 7: Current (1987) technical CO_2-reduction potential through rational energy use in the FR Germany

Sector	Energy use PJ	Reduction potential %
Old housing stock	1600	70-90
New housing	300[1)]	70-80
Commercial	1290	40-70
Electric appliances	270	30-70
Cars and aircraft	1420	50-60
Warm water use	200	10-50
Buses, trucks, fuels in industry, powerplants and refineries	6400	15-25
Cogeneration and electricity in industry	950	10-15
Total	11350	35-45

1) in 2005

After: EK (1990)

Finally, it is shown how effective the global reduction measures recommended by the Commission will be.

6. Effectiveness of the measures recommended by the Enquête-Commission for mitigating the additional man-made greenhouse effect

The Commission has developed four scenarios on CFC reduction and three scenarios on the reduction of CO_2 emissions. Due to the tight time schedule and the precarious data situation, the development of reduction scenarios for CH_4 and N_2O was postponed for the time being. The scenarios include 14 of the most important trace gases contributing to the additional anthropogenic greenhouse effect, namely CO_2, CH_4, N_2O, CFCs 11, 12, 113, 114 and 115, H-CFC 22, halon 1301, carbon tetrachloride, methyl chloroform, tropospheric ozone, and stratospheric water vapor. The time frame extends from the preindustrial conditions in 1860 to the year 2100. The scenario results are expressed relative to 1860 in order to include the climatic burden of past years. An average rate of temperature change of 0.1 °C per decade critical for many ecosystems, was used here for a preliminary evaluation of the results. This rate of change adds up to a mean global warming ceiling of about 2 °C in 2100 as compared to the preindustrial level. All calculations were performed with the one dimaensional Muenster Climate Model. As with all replicas of reality, the results from this modelling effort have a certain degree of uncertainty. In the following this so-called climate sensitivity of the model calculations is taken into account by giving the average values, as well as (in brackets) the uncertainty range.

6.1 Results of the scenario calculations

The scenario analyses allow the following conclusions to be drawn on the effectiveness of different reduction measures (see Figure 2).

- **Without measures**

If the current trend continues (business-as-usual), then a mean global warming of 2.95 °C (2.03 to 4.19) is to be expected by the year 2100, thus exceeding the upper warming ceiling by more than 100 %.

- **Strenghtening of the Montreal Protocol in the version adopted by the 2nd conference of its signatory states in London**

The Montreal Protocol in the London version with exemptions will phase-out the production of CFCs and methyl chloroform as late as 2010 and 2015, respectively, and it does not contain any provisions for partially-halogenated substitutes such as H-CFC-22. The recommendations of the Commission would, if adopted, remove all exemptions and lead to a complete production stop by 1995 in the Federal Republic of Germany, by 1997 for the rest of the EC, and by 2000 for the entire world. H-CFC 22 would be phased-out by 2005.

Figure 2: Effectiveness of the measures of the Enquête-Commission to reduce the additional greenhouse effect. A-B: Strengthening of the Montreal Protocol as revised by the second conference of contracting nations in London; B-C: Tropical forest conservation plan; C-D: CO_2-emission reduction targets for fossil fuel use. the temperature change as related to the preindustrial level of 1860 was calculated with the Muenster Climate Model.
Source: Bach a. Jain

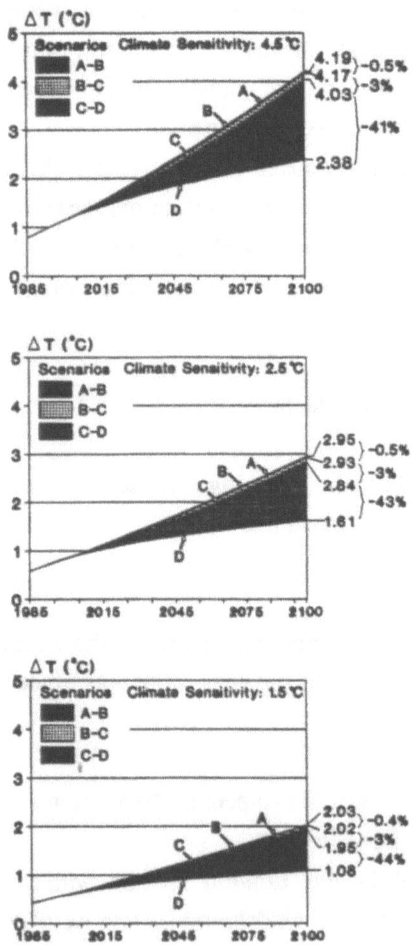

Conclusion: The strenghtening of the current Protocol in its London version results only in a small reduction of the global warming from 2.95 °C (2.03 to 4.19) to 2.93 °C (2.02 to 4.17). The strenghtening of the regulations is nevertheless important and necessary in order to protect the stratospheric ozone layer. The upper warming ceiling of 2 °C is still exceeded even at the lowest climate sensitivity.

- **Plan to save the tropical forests**

The three-phase plan developed by the Commission is aimed at lowering the deforestation rate in the tropics to below the level of 1980 by the year 2000 in the first phase, to halt the destruction of tropical forests completely by 2010 at the latest in the second phase, and in the third phase (2010 to 2030) to restore the tropical forest stand to the level of 1990.

Conclusion: Through these measures global warming will be reduced from 2.93 °C (2.02 to 4.17) to 2.84 °C (1.95 to 4.03). As a consequence, the temperature rise will drop slightly below the 2 °C ceiling at the low climate sensitivity. This calculation only takes CO_2 into account, and not the many other greenhouse gases that are also emitted when forests are burned. Of importance is the fact that the burning of a primary forest releases enormous amounts of CO_2, while replanting a secondary forest on the same land area is only capable, over a period of many decades, of sequestering only about half of the originally released CO_2 from the atmosphere and fixing it in the biomass. The preservation of tropical forests has high priority not only on ecological and climatic grounds, but also for reasons to preserve the living space for indigenous peoples.

- **Reduction targets for energy-related CO_2 emissions**

The Commission is convinced that its reduction plan of energy-related CO_2 emissions for the Federal Republic of Germany, calling for cuts of 30 % by 2005, 50 % by 2020 and 80 % by the year 2050 with respect to 1987 levels, can also be applied to other economically strong industrialized countries with particularly high per capita emission rates. For economically weaker industrialized countries, a slower reduction rate, namely 20 % by 2005, 40 % by 2020, and 80 % by 2050 is proposed. The industrialized countries of Eastern Europe, including the USSR, for technical and financial reasons initially can be expected to follow smaller reduction rates, namely 10 % by 2005 and 30 % by 2020, while by 2050 a reduction of 80 % should then also be attainable. Equity considerations lead to a greater share in global CO_2 emissions in the developing countries as compared to the industrialized nations, namely an increase of 50 % by 2005, 60 % by 2020, and 70 % by 2050. At the same time, the rate of increase is slowed down so that the transition to a more efficient energy economy based on renewables is not impeded. For the industrialized and developing countries taken together, CO_2 emission reductions of 5 % by 2005, 20 % by

2020 and 50 % by 2050 will result. But for operability it is clear that reduction targets for individual countries must be allocated in a differentiated manner.

Conclusion: The additional anthropogenic greenhouse effect can be substantially mitigated by energy-related CO_2 emission reductions, with the result that the degree of warming will drop from 2.84 °C (1.95 to 4.03) to °C (1.08 to 2.38), or by ca. 40 %.

6.2 Evaluation of the results

If the high climate sensitivity should turn out to be the most realistic, the global warming ceiling of 2 °C would still be exceeded by 0.4 °C or about 20 % despite the measures recommended by the Commission. Additional action would be necessary. This could be directed at reducing emissions of CH_4 and N_2O. In the above scenarios, from 1985 to 2100 the CH_4 and N_2O emissions rise by 96 % and 40 %, respectively. Additional model calculations could reveal which combinations of CH_4 and N_2O reductions would be required to keep global warming below the upper limit of 2 °C.

The present results give a first indication of the priorities of the measures that have to be taken.

- If steps are taken to ensure that partially halogenated H-CFCs, as is assumed here, are no longer in use by the year 2040, and if production does not increase beyond the projected levels, then additional agreements going beyond the stipulations of the Montreal Protocol as formulated in the version of the 2nd conference of the signatory states in London will only have a small additional impact on global warming. They are nonetheless important and necessary in order to protect the stratospheric ozone layer.

- The plan to save the tropical forests will make a significant contribution to CO_2 emission reduction by lowering the deforestation rates and by initiating systematic reforestation during the first half of the 21st century. By year 2100, the reference year in the present model calculations, the mitigating effect has, however, become relatively small. Nevertheless, preservation of the tropical forests deserves highest priority not only for ecological reasons, but also, to preserve the living space of indigenous peoples.

- The reduction of energy-related CO_2 emissions is by far the most effective way of combatting the additional man-made greenhouse effect. The greatest results can be expected in industrialized countries that have the highest technological know-how and the greatest economic strength. Transfer of technical and economic assistance to other countries could also mobilize relatively quickly the large existing reduction potential there.

- Further measures are necessary including those for CH_4 and N_2O.

7. Climate protection policy

The scientific facts convey a message that is unequivocal: The impact of human activities on the Earth's climate has reached a critical level. This is because of the unimpeded growth dynamic that underpins our society. For Man to survive, the suicidal logic of growth must be abandoned and our systems must be restructured in a socially, environmentally, and climatically benign way.

A comprehensive climate protection policy is therefore urgently needed. The industrialized nations must assume a leading role in implementing the necessary far-reaching reforms to restructure the present systems, since it is they who have the technological know-how and the financial power. Moreover, in their own best interest, the industrialized countries must support the developing countries in an unselfish manner.

A tractable climate protection policy requires a change in the outlook of the principal actors:

- The **energy companies** must be more open to efficient energy use, decentralized energy production, cogeneration, and an increasing utilization of renewable energy sources. They must reduce the wasteful use of energy. Moreover, they must take more notice of the energy services needed by consumers.

- The **chemical industry** must stop production of ever new substances that have unpredictable impacts on the environment and the climate.

- **Agrobusiness**, which is destroying the environment, the landscapes, rural farming and forestry, notably in the developing countries, must be replaced by ecologically-compatible forms of agriculture and forestry that preserve the environment.

- Conventional **traffic planning** is deleterious to the environment, landscapes, forests, health, and climate. To avoid a catastrophic collapse of the entire transportation system, individual motorized traffic must be speedily replaced by an integrated transportation concept that emphasizes railways and local forms of public transport together with urban planning that reduces the need for traveling.

The success of a climate protection policy will hinge on whether or not, and if so, how fast the numerous possibilities for action can be put into practice. The creativity and commitment of each and every one of us are called for. We owe it to our descendants to maintain our planet's ability to support life.

References

Bach, W. (1990): Beitrag zu Mensch-Natur-Technik, Expo '92, Sevilla; unveröffentlichtes Manuskript.

Bach, W. und A. Jain (1990): Die Wirksamkeit der von der Enquête-Kommission empfohlenen Maßnahmen. Beitrag zum Bericht der Enquête-Kommission "Vorsorge zum Schutz der Erdatmosphäre" des Deutschen Bundestages.

Enquête-Kommission (EK) (1990): Schutz der Erde. Eine Bestandsaufnahme mit Vorschlägen zu einer neuen Energiepolitik. Economica Verlag/C.F. Müller Verlag, Bonn.

UNEP/WMO (1990): Intergovernmental Panel on Climatic Change (IPCC), Policy-makers Summary. Working Group Reports I-III.

3.4 The Externalities of Global Warming

Professor David R. Hodas
Widener University School of Law
P.O. Box 7474, Concord Pike
Wilmington, DE 19803

Measuring The Cost of CO_2 Induced Global Warming

The societal cost of CO_2 emissions from fossil fuel combustion can be measured as (1) the cost of enduring or repairing the *damage* caused by global warming; (2) the cost of *controlling* the harm by reducing greenhouse gas emissions; or, (3) the cost of *mitigating* the harm by sequestering emitted CO_2 in trees.

If the cost of enduring or repairing the damage caused by global warming could be calculated, then an emitter's proportional role in global warming could be determined and the environmental costs charged against (internalized by) that emitter would be the product of the total damage times that plant's proportional contribution.[1] However, the potential extent, timing and effects of global warming are based on computer-based general circulation models, which predict only large scale weather phenomena such as seasonal temperature changes and broad rainfall patterns.[2] Because the process of climate change varies regionally, is not necessarily linear - an area may become wetter before it becomes drier or colder before warmer - and because the data are not detailed enough to describe regional results, the models are unreliable for regional analysis of the impacts of climate change. Even if regional climate change could be predicted accurately, the impacts of those changes are so speculative and unmeasurable that it is not currently possible to predict accurately the damage costs associated with these effects in the United States, let alone the world.

Controlling emissions by scrubbing exhaust gas to remove CO_2 has been almost universally dismissed as too costly, at $240 per ton of carbon, and too technically difficult to be practical.[3] One idea for disposing of significant amounts of CO_2 is to pump it in liquid form under high pressure in a pipe 100 miles out to sea, where it would be discharged at least 500 feet below sea level, to dissolve in sea water.[4] Although neither the technical feasibility nor the reliability of the cost estimates for an undertaking of this size have been demonstrated, for electric generators the capital costs of such a project are estimated to be between \$890 and \$1520/KW for the CO_2 scrubbing and pipeline disposal, in addition to a 25 % energy penalty, a 22 % capacity penalty and operational and maintenance costs of at least \$10 per kW year.[5] Other suggestions include dumping a high density solution of sea water and CO_2 into the ocean floor, dumping blocks of CO_2-hydrate or solidified CO_2 (dry ice) into the ocean, where they would sink to the ocean floor.[6] The uncertainty about the ocean's processes, the immense energy required to run such an operation (30 to 45 % of the total electric energy produced would be required to operate this disposal system), and the unknown ecological effects (such as shifts in plankton populations) of rapidly increasing the CO_2 content of the ocean make these suggestions infeasible.

Alternatively, the cost of supply side[7] and demand side energy efficiency improvements could be used as a measure of CO_2 control costs. However, the data is sparse and the models used are fraught with uncertainty and questionable assumptions.[8] More importantly, even if all the available supply and demand-side improvements were implemented, which assumes that major institutional obstacles to energy efficiency improvements are removed, further measures would still be necessary to achieve even a 20 % reduction in CO_2 emissions from the level of 1990.[9] Examples of cost estimates \$23,17 per ton of carbon[10] for reduced energy consumption in buildings, \$18,11 per ton of carbon for fuel switching, and \$176 per ton of carbon saved to increase automobile efficiency to 44 miles per gallon, assuming \$0.53 per gallon to achieve those improvements which would save 22 pounds of CO_2 per gallon."

The approach for valuing the effects of global warming adopted here is to price the cost of mitigating the effects of CO_2 emissions by sequestering carbon in trees. Tree cultivation, a long-known art, is an approach far more susceptible to concrete cost analysis than either CO_2 scrubbing or measuring environmental harm from global warming.[12] Moreover, the cost of planting trees to sequester carbon is a measure that is independent of fuel and power use, so it provides a yardstick against which to measure the seemingly infinite variety of efficiency and offset options available in the market. Although forestry is not the sole or complete solution to the problem of global warming, it can significantly delay its effects.[13]

General Forestation Issues

There is no world-wide tree planting market that establishes a price for the cost per ton of sequestered carbon in trees. Rather, the cost of carbon fixation and sequestration[14] by cultivating new trees and protecting existing forests is as complex and diverse as the world-wide variety of ecosystems. For instance, what happens after all available land is used to plant all possible trees? Trees could be left standing in mature forests or harvested for lumber and paper, or could be burned for fuel, releasing sequestered carbon that will not be removed from the atmosphere unless replacement trees are planted. A tree planting/burning/planting cycle would not reduce emissions, but would result in a steady state of emissions sustained by a renewable fuel. However, in areas already short of fuel sources, as most of the third world, the availability of wood fuel may not displace existing fossil fuel use, but supplement it. In that case, unless fast-rotation energy forests protect existing natural forests by displacing fossil fuel use, the fast rotation forests will not reduce CO_2 emissions but will merely delay their build-up in the atmosphere for the few days during which the trees are growing.[15]

To determine cost, one must choose whether to reduce demand for forest land and destructive products (especially in the tropics) and/or increase supply of forestry land and responsibly harvested forest products.[16] Although these strategies can be pursued simultaneously, some are better suited to undeveloped countries and some are more appropriate for developing countries. For instance, high volume branch trees such as Douglas fir in the Pacific Northwest grow slowly (80 to 100 years to mature fully) and provide wood most useful in industrial countries whose relatively stable political, social and economic institutions favour long-term forest protection and management. However, in tropical, developing countries that need to maximize growth rates, restore deforested land and produce firewood and timber quickly, and which have limited ability to achieve long-term forest protection, short rotation forest crops may be appropriate.

Cost will depend on the scope of the sequestration necessary to offset CO_2 emissions. For instance, the United States would have to increase its forests some 7.5 times, to about six hectares of forest per person, the U.S. would need 1,500 x 10^6 hectares of average forest to absorb the carbon it emits annually. Unfortunately, the total land area of the U.S. is only 913 x 10^6 hectares, and not all of this will support forests.[17] More intensive growth could reduce the land required. Increased CO_2 concentrations may stimulate tree growth rates and lower trees' water requirements, but temperature changes and other factors may cause forest migration and loss, and temperature extremes associated with climate change may cause trees to lose vigour and become more susceptible to insects and disease.[18] Additional growth stimulation would have to come from intensive management practices such as species selection, fertilization, spacing control, genetically improving trees, and full stocking of forests.[19]

However, these ideas may be expensive and ecologically dangerous. The cost of fertilization may equal or exceed the value of tree growth that fertilization provides.[20] Moreover, nitrogen fertilizers release the potent greenhouse gases methane and nitrous oxide from forest soil. Although genetic research may create stranger, faster growing trees, tropical deforestation may reduce genetic diversity among trees, thereby limiting options for genetic improvement.[21] Large-scale tree plantations may strain surface and groundwater supplies in areas already short of water, such as the Southern Great Plains of the United States. Other plantation problems include rapid depletion of soil nutrients by fast-growing species; reduced growth rates on hillsides at elevations above 450 meters; frost sensitivity; erosion; lack of large, contiguous land areas; poor transportation; poaching; lack of capital for establishment; and lack of educated, skilled persons to operate the plantations.[22] To reforest pastures, resilient monoculture grasses must be suppressed. Few tree seeds are naturally dispersed into large, highly degraded Amazon pastures.[23] Severe economic, population and social pressures in developing countries may inhibit the success of forestry projects. Most importantly, do we want to replace natural forest with sterile, mono-culture stands of trees?

As climate changes, genetically "improved" trees will also be vulnerable to disease, pests, toxins, or environmental conditions for which the trees are not specifically adapted, and monoculture forests would have little genetic diversity, reducing their resistance to infestation by pests that can cause forest decline. Moreover, as the climate becomes warmer, fungi and insects will increase; and in those areas where climate change results in reduced precipitation, fungi, bacteria and weeds will be advantaged.[24] Moreover, the natural rate of forest migration, will probably not keep up with the rate of forest decline from climate change.[25] Furthermore, unless acid precipitation and ozone are reduced, even if global boreal forests increase in total area and tree density as a result of global warming, the forests will probably contain less biomass and fix carbon at lower rates than current boreal forests.[26]

Nevertheless, forestation has significant potential to mitigate CO_2 emissions. But it will only occur on a project by project basis, each with its own approach to carbon sequestration and correspondingly unique costs. Once the CO_2 emission offset requirements are set,[27] then the marginal cost of CO_2 sequestration becomes a least-cost option against which other alternatives can be measured.[28] The next section will review cost calculations of various projects in an attempt to set a range, to the extent current data can support it, for the cost of sequestering carbon.

Cost of Forestry Based Carbon Sequestration
Applied Energy System's Guatemala Carbon Sequestration Project

The first, and currently only, forestry project designed to offset carbon emissions from an electric generation power plant was the result of Applied Energy Service's ("AES") desire to mitigate the carbon emissions of a new coal fuel power plant it was building in Connecticut. The project, designed to sequester 387,000 metric tons of carbon per year for forty years, has an estimated cost of $14 million (shared between AES, CARE, Guatemala, USAID, and Peace Corps), and will involve 40,000 farm families planting 24.4×10^6 trees in wood lots for poles and timber, 26.5 in agroforestry for fuel wood, fodder, and soil nitrogen fixation, in fruit and nut production, in alley cropping, and in live barriers for soil conservation and line fencing, 2,400 hectares of forest saved by fire protection and carbon added to the soils as a result of these activities.[29]

For carbon posting purposes, the major question is how much carbon will be sequestered- a question not easily answered. World Resources Institute (WRI), which analyzed the proposals for AES, projected that 85 % of the trees would be consumed as fuel over forty years, leaving only 15 % of the planted trees as a net addition to the earth's biomass, or 0.6×10^6 tons of carbon.[30] However, over the forty years, the woodlots' carbon would displace wild forest exploitation. WRI assumed that of the 85 % of the woodlot harvest for fuel, only two-thirds of that amount contained usable wood, leaving 2.4×10^6 tons of carbon that would displace forest exploitation. WRI then assumed that if the farmers had collected wood from the forest, additional carbon would be released from the soil, roots, leaves, and destruction of other trees so that every "usable" ton of woodlot carbon would save 1.5 tons of natural forest carbon. Thus, on the assumption that the woodlots will save 3.6×10^6 tons of carbon in the forests, WRI credited the woodlots with a total of 4.2×10^6 tons of carbon sequestered.

A similar analysis was performed on the agroforestry component, resulting in a credit allowed by WRI of 11.2×10^6 tons. Since WRI assumed that wild forest was already credited with being protected by the woodlot and agroforestry programs, it credited fire fighting efforts with saving only 2,400 hectares. Finally, the forestry would increase the carbon content of affected soils by 0.3×10^6 tons.

Thus, WRI's final estimate of the total carbon to be sequestrated by the project is 16.3×10^6 tons, obtained at a total cost, in cash or kind, of $14.3 million to offset the forty year CO_2 emissions of a 183 megawatt ("MW") coal-fired plant. If the project is successful, the installed cost of the sequestration will be $0.78 per kilowatt (KW) which is 5.1 % of the total project cost of $275 million, or is $4.21 per ton of carbon sequestered.[31] On a kilowatt hour basis, assuming the power plant operates at a 90 % capacity factor, the cost is .12 cents/kWh over forty years.

These cost estimates must be used cautiously because they are based on assumptions made in the absence of hard data.[32] For instance, the amount of hectares and carbon to be saved by forest protection was, ultimately, a guess not supported by any meaningful data from past projects. Similarly, the woodlot and agroforestry numbers are only best-guess estimates, since there are no data on the carbon sequestration results of comparable projects. The land on which the trees will be planted may be too depleted to support the growth rates and carbon fixation rates assumed in the calculations.

Moreover, the costs of the project are understated.[33] For instance, there are no direct labor costs included in the project; rather, WRI assumed, that the farmers' labor could be compensated with food aid. The calculations for this project do not include costs associated with pests, natural disasters, social and political unrest, institutional barriers, significant fire protection, prevention of human forest destruction, or the otherwise unherently risky nature of tropical forestry projects. The WRI calculations neither consider possible detrimental effects on carbon fixation rates of increased temperature such as changed precipitation patterns, or increased severity of storms that might occur as a result of global warming, nor evaluate the risks of failure of the forestry program due to desperate political or social conditions caused by rapid population growth. Moreover, the costs of the first few projects - which will tap the least expensive forestry opportunities, although at the most expensive point on the learning curve - will probably be less than the cost of later projects designed to offset additional carbon releases. Additionally, the cost of project development, proposed evaluation, and project follow-up evaluation were not included in the above figures, cor were the costs for detailed verification, quality control oversight, modification or replacement of projects that break down, or implementation of other enforcement mechanisms. Since the AES project is voluntary, there is no requirement for independent verification or enforcement of the carbon sequestration results,[34] a requirement that would be costly but crucial if a project were to be given a carbon credit by a government agency as part of an emissions-trading or offset program.

The WRI analysis assumes that "social forestry" - forestry pursued to meet social needs such as carbon reduction - will save natural forests from destruction by reducing pressure to clear forests for building materials, firewood, and agriculture, but does not propose how the success of a project should be measured or evaluated. If the woodlots and agroforestry do not offset demand, then the project's carbon sequestration will not be realized.

Taken together, uncertainties about WRI's carbon-offset calculations limit the value of any results derived from them. Generally, the cost of marginal carbon sequestration will be considerably higher than in the AES project. The complete results of the AES project evaluation will not be available for decades, at which point decades of additional CO_2 emissions will have occurred and many decisions made in the interim will be expensive or impossible to correct. To the extent that a project like this is used to establish a surrogate measure for global warming costs, the cost evaluation should allow considerable latitude for increased cost, decreased effectiveness, social, political and scientific uncertainties and for the effect of historic emissions. Although the AES project has shortcomings, AES nevertheless deserves much credit for attempting, on a voluntary and experimental basis, the first forestry carbon sequestration project to offset power plant carbon emissions.

Conservation Foundation/World Wildlife Fund Project

In response to the AES/WRI request for proposals, the Conservation Foundation/World Wildlife Fund (CF/WWF) submitted a proposal designed to sequester 10.9×10^6 tons of carbon in ten years and 21.8×10^6 tons in forty years by halting rampant deforestation in Costa Rica through forest management and reforestation. As with the CARE proposal, CF/WWF intended to leverage AES's \$2 million grant; WWF planned to create a debt-swap endowment that would provide loans and technical assistance to small landowners.

The CF/WWF proposed to sequester carbon by slowing deforestation by actively protecting and managing forests, by reducing logging and clear cutting on private land and by stabilizing and reforesting land converted to pasture by clear cutting. In Costa Rica, loggers pay little or nothing for the right to harvest trees, which are then sold to the mills. The mills neither own nor manage the forests, nor do the loggers, who do not live in the area and have no interest in a sustained yield from a forest; thus a "tragedy of the commons" is occurring in the form of unbridled destruction of a common resource.[35] Moreover, the logging occurs in previously inaccessible areas, made accessible by the logging roads, allowing migration of agricultural squatters who use slash-and-burn techniques, furthering the increase of a national park and a forest reserve by strengthening forest protection institutions and coordinating state and national Costa Rican forest protection services and to stop destruction of privately held forests by providing local organizations, cooperatives and businesses with technical assistance in selective timber extraction, forest regeneration, waste reduction, and improvement of extraction efficiency.

The CF/WWf endowment fund would lend money to farms that plant trees to cover lost opportunity costs incurred by taking agriculture or pasture land out of use; the future value of the planted trees would provide the collateral. The plots would have a mixture of fast and slow growing trees, so that wood and crop harvests would continue from year seven to fifty, and into perpetuity as trees were replanted, with payments on the loans beginning after year ten, and then at a low interest rate. CF/WWF believes that, if this approach is successful, peasant farmers will no longer need to create farms by clear-cuting forests to create farms every five to ten years.

In calculating the carbon-sequestration effect of this plan, CF/WWF compared the ecosystem carbon content of various land uses before and after the project. CF/WWF calculated the carbon that would be sequestered by reforesting 1,200 hectares per year for ten years as the difference in carbon content between a thirty to forty year old mature plantation (256 tons of carbon per hectare), for a net carbon gain of 181 tons per hectare. For 12,000 hectares this is 2.2×10^6 tons. CF/WWF did not address what would happen to the carbon in the wood after the plantation reached maturity, i.e., whether it would be burned as biomass fuel, thereby releasing carbon, or used as durable construction materials, thereby remaining sequestered. The forest management and protection aspects of the project were analysed by comparing the estimated carbon content of the forest reserve and park with and without the project; the difference projected as a result of the program would be a savings of 6.2×10^6 tons of carbon from release by deforestation.

In the national park area, which was 100 % natural forest, CF/WWF estimated that, without protection, the area would become 30 % natural fores, 50 % logged forest and 20 % pasture. The carbon difference between 100 % forest and the hypothetically lower future deforestation because of the protection of the program compared to the projected deforestation without the program's protection, produced projected carbon savings of 2.5×10^6 tons.

The cost of these savings, including estimated debt-swap leverage, would be about $6 million; on an annual basis the cost would be about $720,000 for 273,000 tons, or $2.64 per ton. CF/WWF proposed to repeat the entire process during years ten to forty in a different area of the country, to double the eventual carbon savings. Thus, the total projected carbon savings would be 21.8×10^6 tons for an initial investment of $6 million. assuming the second round of reforestation starts twenty years after the first, and a 6 % discount rate, the effective cost of carbon sequestration would be reduced to about $2 per ton. As with the CARE project, the costs of sequestering carbon are astoundingly low.

CF/WWF does not account for what happens to the trees that will be ready for harvest after seven to ten years, but assumes they will be harvested to provide farm income. Once harvested, if used for biomass fuels, all the carbon will be released, and no long-term sequestration will result. Carbon will be sequestered in the wood used for construction materials, but even so, some carbons in the leaves, bark, and branches will be released. CF/WWF provides no analysis of the steady-state level of carbon on the reforested pasture, other than to indicate that these trees will be used to reduce pressure on forests. Using WRI's analysis of the disposition of woodlot carbon, CF/WWF's 2.2×10^6 tons would be reduced by 85 % to 330,000 tons sequestered, which would dramatically increase the cost per ton. CF/WWF separately accounts for the carbon saved through protection of existing forests. Using WRI's analysis of natural forest saved by reforestation, CF/WWF's 2.2×10^6 tons of carbon is multiplied by 0.66 (usable branch wood multiplier) and by 0.85 (proportion of woodlot harvested) to make only 1.12×10^6 tons of woodlot carbon, or $1.68\ 10^6$ tons of forest carbon sequestered.

CF/WWF's analysis assumes that its project will be completely successful. CF/WWF does not include any money for fire protection, nor does it demonstrate that forestry protection and management efforts will be completely successful. As with WRI/CARE, CF/WWF does not provide a margin for risk of failure. Nevertheless, even if burdened with add-ons to account for risk and historic emissions, the CF/WWF project, if it works, appears to be a low-cost approach to mitigating the global warming problem.

As a fall-back from its social reforestation, CF/WWF assumes a cost of $700 per hectare to reforest the land areas included in its proposal. At a 3 % real discount rate (4.2 % carrying charge) $700 per hectare is the equivalent of $29 per hectare-year; at a 12 % real carrying charge the cost would be $84 per hectare per year. If all the carbon is sequestered (181 net tons per hectare about 4.5 tons per year over forty years), the cost would range between $6.50 and $18.70 per ton of carbon, depending on the carrying charge.

Although it is impossible to quantify the risk of failure associated with such forestry projects, there is an intuitively inverse relationship between cost and risk.[36] When the projected carbon sequestration is based exclusively on halting deforestation in the future, the uncertainty is increased by both the risks that the replanting will not be as successful as envisioned, and the risks that the replanted and "protected" areas will be deforested anyway.

Costa Rican Government Project

Costa Rica has its own reforestation program, which in 1989 will reforest 1375 hectares divided among 435 farms.[37] This new, although modest, program is funded by the Foundation for Forest Development (FFD), which will give local farm-based forestry groups technical assistance, capital, and administrative help to encourage planting trees for windbreaks , firewood and agricultural forestry. According to FFD figures, the cost of establishing and maintaining a plantation will be 86,000 Colones per hectare,[38] broken down as follows.

Supplies (plants, medical, etc.)	21,000
Manual labor	53,000
Administration	12,000
	86,000 or $1,079 (U.S.)

The project description does not discuss carbon sequestration. However, it is assumed, that each hectare planted for firewood will displace demand for deforestation standing forest, thereby reducing carbon release. Each hectare for windbreaks and agroforestry will sequester carbon directly, since these uses envision long-lived trees. Once the woodlots reach steady state, each hectare will contain about 90 tons of extra carbon. Assuming that the average tree is half-grown, the woodlots would directly add 2.25 tons per hectare per year to the standing carbon inventory over a 40-year project term. Using WRI's method, woodlots that displace forest exploitation will produce 181 x 0.66 x 0.85/20 = 5.1 tons of carbon per hectare annually. This displacement of forest exploitation would start when the woodlots are reasonably mature, perhaps about year fifteen, resulting in 5.1 x 1.5 = 7.7 tons of carbon saved per hectare per year. The cost of the net carbon sequestered would be $6.30 per ton at a 4.2 % carrying charge of $23.60 at a 12 % carrying charge.

Macro Cost Estimates

EPA and the New York Energy Plan both summarize several generic estimates.[39] Chernick and Caverhill also review macro-sequestration estimates, calculate the cost to be between $40 and $ 200 per ton of carbon, suggest the middle range of $120 per ton is reasonable, but adopt the more optimistic figure of $80 per ton of carbon.[40] The Tellus Institute reviewed Chernick and Caverhill and adopted $80 per ton.[41] Koomey[42] compares several generic estimates and reports Chernick and Caverhill at $84 per ton (1989 $), Schilberg et al.[43] At $54 per ton, and the California Energy Commission at $26 per ton.[44] The Schilberg estimate was based upon the estimated average cost of establishing and maintaining large stands of trees in California and the Pacific Northwest. More recently, in evaluating the air quality benefits of proposed new units, the California Energy Commission adopted $54 per ton of carbon as the appropriate marginal cost.[45] By comparison, Marland reports that the cost of establishing and tending large scale plantations in the Sahel in Africa is between $67 and $120 per ton of carbon.[46]

Pacific Northwest Forestry

The cost of sequestering carbon in the United States has been analyzed for the Pacific Northwest, for existing agricultural lands, and for urban areas. Two sources have estimated the costs of planting forests in the Pacific Northwest to offset carbon emissions. One was performed by Shepard Buchanan of the Bonneville Power Administration[47], and the other by Howard Reichmuth of the Pacific Power and Light (PPL), and David Robison of Lambert Engineering for a West coast utility.[48]

Buchanan estimated the cost of growing trees to store carbon without harvesting to obtain a "reasonable first approximation" of the environmental costs associated with the CO_2 emissions from fossil-fueled power plants. He estimated that an average mixed-age stand of Douglas Fir sequesters 1.14 tons of carbon per acre per year and that the cost of planting commercial stands of Douglas Fir, as estimated by the Forest Service, was $450 per acre. Buchanan then annualized $450 per acre using a capital recovery factor of .043 to arrive at a cost of $19.50 per year for planting the trees, assuming a 3 % real discount rate over forty years and no taxes. Dividing the amortized cost of tree planting by the sequestration rate of 1.14 tons of carbon per acre per year he calculated a sequestration cost of $17.08 per ton. In comparison to Buchanan's 3 % real discount rate, using a 5 % real discount rate the cost would be $29.91 per acre per year and $26.23 per ton of carbon; at a 12 % real discount rate, the cost would be $54 per acre per year and $47,40 per ton of carbon.

These costs are based on land which is currently replanted with trees as it is clear-cut. The actual costs for marginal sites would be higher, since only higher-priced, less accessible, or less productive land would be available for planting.[49]

The Reichmuth/Robison analysis arrives at lower costs for carbon offset, because it uses different assumptions than Buchanan used. Reichmuth and Robison assume that a temperate forest can fix carbon at the average rate of 1.7 tons of carbon per acre per year. They also assume that \$250 per acre (\$618 per hectare) will purchase a "stewardship" forestation contract. Amortizing the \$250 per acre forestry cost over forty years at 3 %, the cost per ton of carbon sequestered would be about \$6.30 per ton. Adding in land costs at \$100 per acre would increase the total cost to \$8.85 per ton at a 3 % real discount rate, \$13.60 at a 6 % real discount rate, or \$24.70 at a 12 % real discount rate.

Conservation Reserve Program

Another approach to offsetting carbon emissions is the proposal of Dudek and LeBlanc[50] to use the Conservation Reserve Program (CRP), created by the Food Security Act of 1985,[51] to subsidize the planting of trees which will reduce erosion and surplus agricultural production and sequester carbon to mitigate global warming. CRP contracts with owners of erodible cropland to plant trees or other vegetation cover for ten years; the CRP pays half the planting costs and an annual rental fee to the landowners. As of May, 1989, of about 5.6 million acres (2.3 million hectares) designated as eligible to be forested within the program, only 1.6 million acres (0.6 million hectares) have been enrolled.[52]

Dudek and LeBlanc proposed that the CRP be used to offset the added CO_2 emissions from oil- and coal-fired power plants planned to go on-line between 1987 and 1996 and they assumed that the plants would operate at 65 % capacity and would be 33 % efficient. Since different tree species would be planted in different regions of the country, and because of scientific uncertainty as to carbon fixation and sequestration rates, Dudek estimated the necessary acreage by using fixation rates ranging between five and fifteen tons of CO_2 per acre year, based on silver maples, the least efficient tree reviewed.[53] This assumption led to the conclusion that 8.5 to 25.4 million acres (3.4 to 10.3 million hectares) could sequester the 126.8 million of annual CO_2 emissions (or 34.6 million tons of carbon) from the new coal-and oil-power plants planned to go on line through 1996.

They assume planting costs of \$150 per acre, annual rental costs of \$100 per acre and fertilization costs of \$25 per acre, all in 1985 dollars;[54] for 16.95 million acres (6.9 million hectares) the total costs are \$3.5 billion to establish the trees plus \$2 billion per year for rental in 1989 dollars.[55] The total cost of carbon fixation would be about \$53 per ton of carbon for a 3 % real discount rate (a 4.2 % carrying charge over forty years), and \$58 per ton for a 12 % carrying charge.

Foresting Urban Areas

Developed urban areas create "heat islands" which in warm, mid- and low-latitude cities can increase temperature an average of 3 °C to 5 °C, thereby adding to discomfort and increasing air conditioning loads.[56] Planting one shade tree on the west side of a house and two on the south side, can reduce a house's cooling energy needs by 47 % and peak load cooling by 34.2 %.[57] Because it both absorbs carbon and reduces the need for electricity, a well placed urban tree is potentially fifteen times more valuable than a forest tree in mitigating global warming and the estimate cost of planting and protecting these trees will be \$26 per ton of carbon conserved.[58]

Conclusion

Since the damages from global warming cannot be calculated, the external environmental cost of global warming should represent the cost of the marginal measures necessary to mitigate climate change by sequestering carbon in trees. The most reliable estimates of sequestration costs are from temperate industrial countries, because they are more politically and socially stable then developing countries and have an established commercial forestry industry. But more importantly, in the temperate zones the rate of sequestration is verifiable by direct observation of tree growth rates in new forests.

In contrast, in projects proposed for tropical countries, the sequestration rate is not be verifiable by direct measurement, since the rate is based upon estimated reductions in off-site deforestation. Until it can be verified that tropical social forestry sequesters carbon at identifiable rates, the cost of social forestry, which is currently estimated to be relatively low (at least for small increments), should not be used as an estimate for marginal global-warming mitigation costs.

Given the potentially catastrophic consequences of global warming, any estimate of mitigation costs must be adopted with an adequate margin of safety; that is, any error should be in favor of higher costs to reflect the risk of currently unforeseen catastrophe. The higher the internalized cost, the stronger will be the signal directing decisions away from further global warming. Although the AES project, CF/WWF proposal and Costa Rican project are commendable, their unrepresentatively low costs, unverifiable sequestration rates, and failures to include risk of failure, all constrain their use as a sequestration models to establish a global warming externality value. Similarly, a margin of safety requires that the cost not be based upon high intensity plantations, exotic genetic fencing, or fertilization since these measures raise the specter of new environmental problems. However, less exotic African Sahel plantations were reported by Marland to sequester carbon at a cost of $67 to $120 per ton.

Of the American, region-specific estimates reviewed, the best appears to be the analysis of Buchanan, $17.08 per ton of carbon for 3 % financing or $47.45 per ton for a 12 % real discount rate, which is similar to Schilberg at $54 per ton. Reichmuth and Robison's assumptions are speculative and undocumented, and their $19-25 per ton results are unreviewable. Dudek's estimated sequestration rates are optimistically high, and the $53-58 per ton values resulting from those estimates are correspondingly low, although the result is similar to those obtained by Buchanan (at a 12 % real discount rate) and Schilberg. The Akbari figures for the cost of energy conservation through urban heat island forestry represent an attractive, low cost, conservation option, not a marginal carbon-sequestration cost.

References

[1] Buchanan, S., *Costs of Mitigating Greenhouse Effect for Generic Coal, Oil, and Gas-Fired Plants* at p. 1 (draft of January 24, 1989). (Unpublished report).

[2] Gibbs and Hoffman, *An Approach for Generating Climate Change Hypothetical Given Limitations in Current Climate Models* in Shands and Hoffman (ed.), *The Greenhouse Effect, Climate Change and United States Forests* (1987 at 95). The models only crudely describe ocean/air interaction, evolution and migration of vegetation, cloud and ice formation and melting and do not permit regional climate forecasts or identify specific global warming impacts. Moreover, the models provide little insight into possible non-linear, chaotic bio-chemical planetary feedback systems that may affect the climate; scientists can only speculate as to the surprises nature has in store for the world. R. Coppock (National Academy of Sciences), in letter to Richard L. Ottinger, April 6, 1990.

[3] *See, e.g.*, New York State Energy Office, Staff Report, *Draft New York State Energy Plan*, Vol. VI, p. 39 (1989). Control of other greenhouse gases may be feasible, but the costs have not generally been determined.

[4] Steinberg M. and H. Cheng, *A Systems Study for the Removal, Recovery, and Disposal of Carbon Dioxide from Fossil Fuel Power Plants in the U.S.*, Brookhaven National Laboratory, Department of Applied Science, Process Sciences Division, BNL 35666 (February, 1985).

[5] U.S. Environmental Protection Agency, *Policy Options for Stabilizing Climate Change*, Draft Report to Congress (Feb. 1989). ("*EPA Report*") at p. VII-135, Table 7-9. Chernick, P. and E. Caverhill, *The Valuation of Externalities From Energy Production, Delivery, and Use*, Boston, MA: PLC, Inc., December, 1989. The cost of scrubbing is estimated to be $200-500 per ton of carbon.

[6] de Baar, H. and M. Stoll, "Storage of Carbon Dioxide in the Oceans" in Okken, P.A., R. J. Swart and S. Zwerver (ed.), *Climate and Energy* (1989).

[7] Supply-side improvements include renewable energy sources (such as hydropower and pumped storage, biomass/biofuels, geothermal, ocean energy, solar thermal electric, wind energy, photovoltaics, solar heat, and associated technologies such as energy storage, hydrogen, fuel cells, and transmission and distribution improvements), advanced coal and gas-fired electricity generation technologies and fuel switching. *See EPA Report, supra* note 2 and U.S. Dept. of Energy, *The Potential of Renewable Energy: An Interlaboratory White Paper*, DE90000322, SERI/TP-260-3674, March, 1990.

[8] Zimmerman, M., *Assessing the Costs of Climate Policies: The Uses and Limits of Models*, The Alliance to Save Energy, (April 10, 1990).

[9] *EPA Report, supra* note 5.

[10] Dudek, D., A. LeBlanc, P. Miller, *CO2 and SO2: Consistent Policy Making in a Greenhouse*, Environmental Defense Fund (January, 1990) at p. 9.

[11] Telephone conversation with Marc Ledbetter, ACEEE, June 7, 1990; Ledbetter and Ross, "Supply curves of Conserved Energy from Automobiles", Lawrence Berkely Laboratory Applied Science Division, Berkely, CA (March 1990).

[12] However, there is a danger that if tree planting is concentrated in monoculture plantations supported by pesticides, herbicides and fertilizers, species diversity may be

diminished and ecosystem stability jeopardized. We must act carefully to avoid creating new environmental problems from the technological fixes to current environmental problems, i.e., the environmental equivalent of iatrogenic disease - disease caused by medical treatment.

13 Marland, *The Prospect of Solving the CO2 Problem Through Global Reforestation*, U.S. Dept. of Energy, DDE/NBB-0082, 1988.

14 "Fixation" refers to the ability of a tree to absorb carbon, whereas "sequestration" refers to the ability of an ecosystem to retain the carbon over time. For instance, a rapidly growing tree may fix carbon at a high rate, but release it quickly when it is burned for firewood, perhaps resulting in no sequestration, but still displacing fossil fuel and reducing atmospheric carbon levels. In comparison, a slow growing hardwood may fix carbon at a low rate each year, but hold it for 100 years in a forest, and for decades more in the form of furniture and construction lumber, resulting in a high, long-term sequestration rate.

15 Trexler et al., *Forestry as a Response to Global Warming: An Analysis of the Guatemala Agroforestry and Carbon Sequestration Project*, World Resources Institute, June, 1989, at 63, note 35. See Chernick and Caverhill, *The Valuation of Externalities from Energy Production, Delivery and Use*, PLC, Inc., Boston, MA, December, 1989 at pp. 65-66 for a discussion of the fossil fuel vs. wood fuel problem.

16 *EPA Report*, *supra* note 5 at p. VII-203.

17 Marland, *supra* note 13 at p. 6.

18 Sandenburg et al., "Rising CO_2 Climate Change and Forest Management: an Overview", in Shands and Hoffman, *The Greenhouse Effect*, *supra* note 2, at 114.

19 Marland, *supra* note 13 at p. 27.

20 *Id.* at p. 43.

21 *Id.* at p. 47. Marland notes that while there are thousands of tropical tree species, 87 % of tropical forestry depends on three types (pine, eucalyptus and teak). In temperature forests such as in the USSR, with the worlds largest forest, there are 570 species, only thirty of which have major economic significance.

22 *Id.*

[23] 85 % of the seeds are dispersed by animals that do not live in pastures, most seeds that do reach the pastures are eaten and water and light conditions at the soil level are different in pastures than in forests.

[24] Sandenburgh et al., *supra* note 18 at p. 115.

[25] *EPA Report*, *supra* note 5 at pp. VII-242-243. If atmospheric CO_2 doubles, the decline in boreal forest biomass is estimated to produce a global net loss of 24 billion tons of carbon.

[26] *Id.* at p. VII-243.

[27] Because this is an environmental externality, the offset requirement must be imposed as a legal requirement if the remedial costs are to be internalized; the offset level will be based on value judgements that depend on whether emitters should offset past and future emissions or only a portion of them. The larger the offset ratio, the greater the cost of mitigation; as the internalized costs increase, alternative mitigation measures become marginally more beneficial, thus increasing CO_2 reductions in the atmosphere.

[28] Forestry may also provide important social, environmental and economic benefits that may justify the projects irrespective of carbon sequestration. However, those benefits should not be subtracted from the project costs, since they are unrelated to carbon sequestration (i.e., the prevention of global warming). Because forestry is being used as a surrogate for the external costs of global warming, only carbon sequestration-related costs and benefits should be considered in calculating the costs to be internalized by the electric power industry. This is consistent with the practice that benefits of control technology are not counted when estimating the cost of the control. Just as benefits such as enhanced employment from SO_2 scrubber construction are not subtracted from the cost of the scrubber, so the non-global warming benefits of using forests as carbon scrubbers should not be subtracted from the cost of sequestering carbon in trees.

[29] Trexler et al., *supra* note 15 at pp. 18, 58-59.

[30] *Id.* at p. 49.

[31] $14.3 million multiplied by a typical utility's real carrying charge of 12 % equals $1.72 million annually. Dividing the annualized cost by the annual carbon savings (0.41 million tons) gives $4.21/ton.

[32] Conversation between Trexler and David R. Hodas, October 24, 1989.

[33] *Id.*

[34] However, CARE is to provide AES with information about the pre-project status of selected evaluation sites, quarterly monitoring reports on seedling outplanting rates, soil conservation treatment and fire brigade activities, annual reports on seedling survival rates, tree growth measurements on selected plots, soil carbon changes on selected soil conservation plots and estimated carbon savings through fire protection activities, and provide three external evaluations that estimate the carbon sequestration of the project. Applied Energy Services, "Monitoring and Evaluation Plan For Carbon Flux CARE-Guatemala Agroforestry and Resource Conservation Project" (undated).

[35] CF/WWF, Proposal: *The Forestry Fund: Endowment for Forest Protection, Management & Reforestation in Costa Rica*, May 1988 at 1, 7-8.

[36] Trexler, *supra* note 15 at p. 43.

[37] 4 *Boletin Forestal* 1, Costa Rica National Forestry Service (March, 1989).

[38] About 80 Costa Rican Colones are equivalent to one United States dollar. (Bank of America Foreign Currency Indication Rates, November 1989).

[39] *EPA Report*, Table 7-21, supra note 5; *Draft New York State Energy Plan, supra* note 3, Vol. VI, Tables 11 and A5 at App. 40.

[40] Chernick and Caverhill, *supra* note 15 at p.73.

[41] Bernow, S., & D. Marron, *Valuation of Environmental Externalities For Energy Planning and Operations: May 1990 Update*, May 18, 1990 at p. 26.

[42] Koomey, *Comparative Analysis of Monetary Estimates of External Environmental Costs Associated with Combustion of Fossil Fuels*, Energy Analysis Program, Applied Sciences Division, Lawrence Berkeley Laboratory, Berkeley, CA, April, 1990, at p. 27.

[43] Schilberg, G., J. Nahigan & W. Marcus, *Valuing Reductions in Air Emissions and Incorporation into Electric Resource Planning: Theoretical and Quantitative Aspects* (re. CEC Docket 88-ER-8), JBS Energy, Inc. for the Independent Power Producers, August 25, 1989, at p. 9.

[44] California Energy Commission Staff, Energy Siting and Environmental Protection Division, *Valuing Emission Reductions for Electricity Report 90*, Staff Issue Paper #3R, Docket #88-ER-8, November 21, 1989.

[45] Tooker, C. & R. Buell, "LUZ SEGS Units IX and X Air Quality Benefits As Determined Under ER-7, "California Energy Commission, written testimony (May 30, 1990).

[46] Marland, *supra* note 13, at p. 55.

[47] Buchanan, *supra* note 1.

[48] Reichmuth and Robison, *Carbon Dioxide Offsets* (March 2, 1989). (unpublished report, available from reichmuth at Pacific Power and Light).

[49] *EPA Report*, *supra* note 5 at p. VII-235.

[50] Dudek, D. & A. LeBlanc, *Offsetting New CO2 Emissions: A Rational First Greenhouse Policy Step*, Environmental Defense Fund, April 2, 1990, at p. 4.

[51] 16 U.S.C. §§ 3831-3836; Department of Agriculture Regulations Governing the Conservation Reserve Program are located at 7 CFR 704.1 *et seq*.

[52] Dudek & LeBlanc, *supra*, note 50.

[53] These carbon-fixation rates (up to ten tons of carbon per hectare-year) require intensive plantation management (*EPA Report*, *supra* note 5 at p. VII-227) and would also result in only a few years of growth before harvesting. If the wood is burned, there would be little long-term sequestration.

[54] The costs do not include the cost of additional trees to offset the effect of nitrogen fertilizers, which cause forest soils to release methane, an important greenhouse gas.

[55] *Id*. This calculation assumes 4 % inflation from 1985 to 1989.

[56] Akbari et al., *The Impact of Summer Heat Islands on Residential Cooling Energy Consumption*, presented at the 8th Miami International Conference on Alternate Energy Sources (December 14-16, 1987) [hereinafter cited as *Impact*]; and Taha, et al., "Residential Cooling Loads and the Urban Heat Island - the Effects of Albedo", 23 *Building and Environment* 271-283 (1988).

[57] Akbari et al., *Strategies for Reducing Urban Islands: Savings, Conflicts and City's Role*, presented at 1st National Conference on Energy Efficient Cooling, San José, CA, Oct. 21-22, 1987, [hereafter cited as "Strategies"].

[58] Akbari et al., "*The Impact of Summer Heat Islands on Cooling Energy Consumption and CO2 Emissions*", presented at ACEEE Summer Study on Energy Efficient buildings, Asilomar, CA, July 12, 1988, [hereafter cited as "ACEEE"]; *EPA Report*, *supra* note 5 at p. VII-235. Heat island effects can also be reduced by planting trees to shade streets with dark surfaces. The American Forestry association estimates that there is room to plant one million trees along urban streets, although it might cost up to $200 to plant each urban street tree. N. Sampson, American Forestry Association, speaking at Global Warming Dialogue, Center for Clean Air Policy, Washington, DC, (December 8, 1989).

4. SUBJECT AREA 2:

MONETIZATION OF ENVIRONMENTAL DAMAGES

4.1 Environmental Externalities Measurement: Quantification, Valuation and Monetization

Stephen Bernow, Bruce Biewald and Donald Marron
Tellus Institute
Boston
Massachusetts
USA

Introduction

Energy Resources impose a variety of costs on society. Some of these costs, such as the direct costs of facility construction, operation and fuel consumption or the cost of end-use equipment, are borne directly by the producers and consumers of energy. Other costs, however, are borne by society and the environment at large, not just by the producers and consumers of energy. Most notable among these *external* costs are the damages to the environment and human health that result from the construction of energy facilities and from the extraction, processing, transport, combustion and disposal of waste and spent fuels. Some of these impacts can become reflected (or "internalized") in direct costs, to the extent that regulations and engineering practice respond, however belatedly, to scientific research and public concern. Others, however, remain as the deleterious consequences of satisfying society's identified energy needs. Taking account of these impacts in energy policy, planning and operation could alter the magnitude and mix of resources used to meet our energy demands, with salutary effects for the environment, public health and long-term economic and ecological sustainability.

These environmental impacts from energy resource use can have significant, often severe and sometimes irreversible impacts on air, water, soils, biota, habitats and human health. Such impacts occur on local , regional and global scales, and are often quite complex and difficult to predict. The release of pollutants and the use of local resources by energy related activities can directly affect human health and amenity and the integrity of natural systems in the immediate vicinity of the facilities, thereby raising important siting issues. At the regional impact level, discussions over the last decade in North America and Europe have sought the best means of reducing the release of acid gases that could seriously damage habitants, agricultural production and a variety of other human and ecological systems. Pending national legislation in the United States will address this issue with acid gas release constraints and associated additional energy costs. At the global level, attention has been drawn to the accumulation of those greenhouse gases in the upper atmosphere which are released from anthropogenic sources, especially from energy and land use. This accumulation threatens global climate change, with potentially severe environmental, social and economic consequences by the middle of the next century. Thus, resource development decisions must be taken under conditions of great uncertainty, where the risk of deleterious and potentially catastrophic outcomes must be taken into account.

Utility planning in the United States appears to be at the beginning of a major transition, from traditional least-cost planning focussing on supply options to an integrated approach where both supply and demand options are evaluated in light of both and economic and environmental factors. To the extent that environmental costs or sustainability goals are *not* included in electric utility decisions, the outcome will be inefficient from a societal perspective. The "wrong" types of resources will be pursued, as evaluated from a standpoint that considers environmental preservation, human health and long-term sustainability along with traditional monetary costs and benefits. In order to improve energy decisions generally, and electric utility resource decisions in particular, economists and environmentalists have argued that these external costs should be "internalized" in resource planning, acquisition and operating decisions. In this way, more appropriate price signals to planners would be established, thereby leading to the satisfaction of energy needs at the lowest overall cost to society.

Quantification, Valuation and Implementation

The art and science of attributing costs to the environmental impacts of energy production and use are still in their developing stages. For the costs of these impacts to be fairly and accurately considered in energy planning analyses and decisions, an appropriate framework for their incorporation, consistent with their character, is important. By clearly defining terminology and proposing a conceptual framework for environmental externalities measurement, we hope to contribute to the more precise articulation of the issues involved in integrating environmental externalities, as well as the implementation of a procedure for their inclusion in an integrated resource planning process.

Quantification

Quantification of environmental externalities refers to the estimation of the physical magnitude of environmental impacts. Calculation of the tons of oxides of sulfur (SO_x) emitted from a power plant per megawatt-hour of electricity generation is a relatively straightforward example of quantification. While environmental impacts can be expressed in terms of an ordinal ranking, quantification of environmental externalities can, in principle, provide more information than can be embodied in the ordinal ranking of impacts. Quantification makes possible statements such as "the X impacts of project A are twice as great as those of project B". Many environmental externalities can not be so readily quantified, however. Calculation of environmental damages, such as crop or forest loss, public health consequences or habitat degradation, is very complex.

Quantification of environmental impacts begins with the task of identifying and defining the externalities themselves. We can distinguish between three overall stages or levels of environmental externality impacts: *initial loadings, intermediate effects*, and *ultimate impacts*. In some analyses, the initial pollutant loading themselves, e.g., tons of SO_x emissions or tons per MWH generated, are viewed as the externality. In others, the externality could be the intermediate effects of those loadings, such as ambient SO_x concentrations in air or the acidification of a lake. In still others, the externality could be the ultimate impacts on human health, the environment or the economy, such as increased mortality, destroyed fish populations or reduced crop yields. These physical, biological, economic, aesthetic, demographic and other impacts can be represented both quantitatively and qualitatively.

Each of the stages of the causal chain have temporal, spatial and sectoral aspects. Each can itself be further disaggregated. Initial loadings can be disaggregated into pollutant type categories. Intermediate efforts can be disaggregated into sub-stages such as air concentration and acidification. Finally, the ultimate impacts can be disaggregated by receptor/impact type such as human health, wildlife and so on. Each can be analyzed using technical, economic, physical and ecological parameters and relationships.

Estimating the direct *physical* impacts or damages of environmental loadings (e.g., air emissions, water emissions) can be a large and exceedingly complex task, fraught with uncertainty. These impacts depend on: the *emissions loadings* themselves into various media (e.g., air, water, soils); the *transport* of pollutants through those media, based on the physical characteristics of the emissions (e.g., velocity) and on climatological (e.g., wind speed, wind direction) and topographical conditions; the *exposure* of receptor area or populations; and, the *dose/response* relationship of those populations. Moreover, there can be a series of subsequent physical and non-physical (e.g., socio-economic) impacts that ensue in both the near and long term. The local, regional and global consequences of air emissions, for example, depend upon atmospheric, biological, chemical, geophysical, ecological and physiological relationships across time, space, and socio-economic and cultural conditions. In modelling such relationships, elements of scientific uncertainty, availability of accurate data and the ecological response to perturbation (including non-linearities and feedback effects, among other factors) must be taken into account. The net result is that the prediction of the direct impacts of environmental perturbations are site-specific and often may be very difficult.

Valuation

Various approaches can be used for the valuation of the environmental externalities of energy resource use. Among these are ordinal ranking, scoring and monetization. Within each of these general approaches, one can further distinguish between methods for establishing values.

Many environmental impacts can be directly valued only in qualitative terms. For example, the character of land might be expressible only in terms such as whether the land is pristine, residential, commercial or industrial. Such qualitative descriptions can serve as a basis for the ordinal ranking of environmental impacts, the most basic level of environmental evaluation, or for a scoring system that assigns points between some minimum and maximum. They may also, as we shall see, be monetized under certain conditions.

The valuation of environmental externalities in money terms is referred to as monetization. The term has been used to mean the estimation of the monetary costs to society of environmental damage. A more modest notion of monetization might involve the judicious use of abatement costs to represent the monetary value to society of externalities. Both of these cases of monetization represent attempts to objectively develop the cost associated with environmental loading from damage or abatement costs associated with that loading.

Explicit monetization places a monetary value environmental externalities based on some criteria or measurement. Explicit monetization can be based on damage costs, including the costs of lost product, remediation or adaptation. This kind of monetization is referred to as *direct monetization*. Explicit monetization can also be determined from control costs (e.g., abatement, offset), regulators' "revealed preferences", contingent valuation or the price of marketable permits, in which case it is referred to as *indirect monetization*.

Monetization can also be developed by establishing a scoring scheme for both the environmental impacts and price attributes of resources and then calculating the value implicitly placed on various externalities. Scoring of environmental impacts can be based, for example, on expert or public opinion surveys or on regulators' subjective estimates. When these kinds of scoring systems are combined with monetary costs they effectively embody monetary values. Thus, these resulting monetary values represent a process of *implicit monetization*.

The common element in all monetization methods is the use of dollars (or other currency units) to value environmental externalities, rather than "points" or "credits" that have no inherent meaning. Given current scientific and economic knowledge, it is possible to explicitly monetize some, but by no means all, environmental impacts. Various methods exist for estimating costs based on the damages caused by environmental loadings and/or the cost of controlling such loadings.

While we may aspire toward direct monetization of all environmental externalities, this aim is not likely to be achieve. For some environmental impacts, objective costs are not available for a number of reasons. In some cases, only lack of study and knowledge prevent any reasonable determination of costs at this time. In other cases, however, such as for loss of a species or even of a specific population of biota, it may be impossible, in principle, to develop direct monetary costs. Here, preservation or sustainability goals could be imposed. These could be used as constraints on resource planning, from which a shadow price (i.e., the incremental cost of meeting these constraints) can be established.

Implementation

The third dimension of our framework is *implementation*, whether in a resource bidding system, a least cost planning protocol or a facility evaluation. When implementing as externality valuation system, the externality values and other attributes of resources are combined in order to compare and select among options. We here argue that monetization of externalities is desirable to the extent feasible, when the objective is to combine externalities costs with direct economic costs using a common *numeraire*. In fact, any attempt to evaluate externalities and direct economic costs on such a common basis (e.g., a scoring system for utility bidding) can be used to derive implicit monetized values. We conclude that monetized estimates should be made explicit whenever this is possible.

However, there are instances where direct valuation of certain impacts is not possible because information on, for example, damages or revealed preferences, is lacking or because policy dictates constraints. It may be sometimes desirable to compare resource plans by estimating and considering their economic and environmental attributes separately without reduction to a single factor. Thus, the cost of avoiding certain environmental or public health risks would be derived by comparing the costs of alternative resources plans with different environmental attributes. Least cost planning in the context of proposed new acid gas targets or, potentially, a protocol on CO_2 emissions would be examples of this.

The Argument for Monetization of Environmental Externalities

We have two basic reasons for arguing that environmental externalities should be monetized whenever possible for evaluation alongside economic costs. First, the use of monetized values makes clear and understandable comparisons possible between direct economic costs (such as fuel costs) and environmental costs. If these costs of energy planning are not expressed in common units, comparisons become confused and the trade-offs between economic factors and environmental factors may become less comprehensible. Second, and more importantly, monetization allows for the consistent treatment and evaluation of environmental issues in a manner that other methods do not. Once costs are developed for air emissions, for example, those costs can be easily applied in many contexts within the energy sector, such as planning, acquisition of resources and operation of electric generating facilities, as well as in other sectors of the economy.

Even if the direct costing of the externality is not possible, monetization can be used to greatly facilitate comparisons between economic and environmental costs. The use of a monetized system for environmental externalities provides for rational and consistent decision-making. This does not require that objective cost figures be developed. Even where the value placed upon a particular impact is inherently subjective, it is preferable to state the value in cost terms so that it may be readily compared with or added to direct costs. It is possible to translate a scoring system, so that price and environmental values can be compared and, if desired, combined. It would be clearer if this were done explicitly, without recourse to the intermediate step of scoring.

With a valuation system that uses "points" for environmental impacts, the scoring system for environmental externalities has little or no meaning unless the treatment of price is defined. In evaluating the externalities of various proposals, the tradeoff between bid price and environmental points is crucial. The value of each environmental point is determined by the treatment of price. For example, in a system with 25 points for externalities and 50 points for price, if the 50 price points are assigned across a range from 0 to 10 cents per kwh, then each rating point would be worth 0.2 cents per kwh.

The use of monetization of environmental externalities ensures that the value placed upon the environment does not scale with the prices bid. If the same 50 price points are assigned across a range from 0 to 5 cents per kwh, then each rating point is worth only 0.1 cents per kwh. In the first case, the maximum environmental score is 5 cents per kwh, while in the second the maximum is 2.5 cents per kWh.

Thus, even if qualitative environmental impacts are given scores, they can be translated into monetary terms. While scoring systems that weight and/or rank both price and non-price environmental factors without monetization may appear to avoid the problem of valuing externalities, in fact, these systems merely obscure monetary values. The valuation of all of project impacts, both price and non-price, is essential to the selection process. The values are either implicitly hidden in a point system or explicitly stated in a monetized system.

One could argue that costs and environmental impacts should be separately represented as different attributes of alternative resources or resource plans. Than, a plan could be selected whose overall attributes are the most desirable. But this simply shifts, but does not eliminate, the problem of monetization. If the higher cost of two plans is selected, it means that the difference in the environmental impacts of the two plans has a value in monetary terms that is at least as great as the difference in direct costs. While such an approach has merit, it could introduce *post hoc*, and potentially arbitrary, valuation decisions and result in an unpredictable playing field for planners.

In addition, monetization makes it clear that we are willing to spend ratepayer dollars in order to achieve desirable environmental objectives. With a non-monetized system, it may not be clear that favoring environmentally attractive options in the resource evaluation process will result (eventually) in higher utility revenue requirements. By stating the values used for environmental impacts in terms of dollars, it is made evident that these decisions will result in electric price impacts, with the magnitude of the possible price increases made implicit. The imputed environmental costs represent an upper limit on the amount that society is willing to pay for resources that avoid environmental impacts. Avoided costs and purchased power prices are commonly stated in cents per kilowatt-hour. It is likely more intelligible to state that "the maximum penalty to a project for toxic air emissions is 0.1 cents per kilowatt-hour" than to say that "4 percent of the environmental points will be based upon air toxics." Thus, monetized environmental impacts could be readily incorporated into both integrated planning analyses as avoided cost calculations, as additional variables and/or fixed costs.

Consistency among evaluations is the second main reason why we argue for monetizing environmental externalities whenever possible. Monetization provides for consistency among various kinds of resource use decisions that the utility must make. When good data for damage or abatement costs are available, they will be assigned accurately by the system. Also, judgements must be made explicit. For example, if a utility uses a value of $1,500 per ton for SO_x in its planning decisions, then it could use the same value in its operating (i.e., system dispatching) decisions. If the values are not explicitly stated, then various practices of the utility could be inconsistent, and thus could undermine the objective of providing service at least societal cost.

Monetization also provides for equivalence among utilities. Using explicit prices for externalities would help to ensure that environmental impacts are valued equally by different utilities. If a point system is used, the value of an environmental point can depend upon the scoring procedures adopted for price. If utilities use different approaches on price, then the goal of consistency for environmental externalities would be undermined.

Monetization provides an opportunity to benchmark the values used for environmental impacts with the values adopted in other sectors of society. Decisions made by other industries or by the federal government may rely, either implicitly or explicitly, on a particular valuation of an environmental impact. Using explicit cost figures for the electric utility decision process allows comparison with those costs assumed in other decisions, helping to provide guidance and consistency.

Systems that do not use monetization do not allow for consistent treatment of externalities over a range of economic contexts. At the same time, monetization, like other systems, does make possible the use of judgemental policy weights in the application of costs. Thus, using monetization need not be viewed as incompatible with the subjective/judgmental aspects of other environmental evaluation methods. However, it is important that monetary values and policy weights be kept distinct, to the extent possible. This permits the basis of their overall impact (value times weight) to be better understood. It also permits separate development of the contribution of value and weight to new overall impacts, based upon further research and/or new policies, respectively. Prematurely combining these functions into a single number can obscure their basis and impede orderly updating.

Once acceptable values for environmental externalities have been established (e.g., based on damage or control costs) policy considerations may require that other factors be taken into account, either to attenuate or increase those values. For example, the New York State Public Service Commission chose to weight its carbon dioxide value by 20 percent to reflect the uncertainties and still nascent state of national and international policy developments on the global warming issue. Weighing about factors greater than 1.0 could also be chosen, e.g., to reflect concerns about irreversibility. By distinguishing between the valuation and its policy weight rather than combining them and obscuring their difference, greater clarity can be preserved and a basis laid for future improvements or changes.

Given the benefits associated with monetization of environmental impacts, we believe that full monetization based on both direct and indirect methods ought to be used in utility system planning and decision-making. While weighing/ranking systems may be useful as an instrument for developing implicit cost estimates for various environmental loadings, there is no reason to use such systems in the ultimate valuation system. As we have noted above, once a weighing/ranking system includes price, it is possible to translate the system into a fully monetized system.

Damage Costs vs. Control Costs for Valuation

The art and science of attributing costs to the environmental impacts of energy production and use are still in their developing stages. At this time, two general approaches have been taken to the problem, one emphasizing the costs associated with environmental damage and/or its remediation, the other emphasizing the costs of prevention.

There are other valuation methods which can be used. For example, one could develop relative weights for environmental issues by polling of experts or consumers and then multiply these weights by an overall environmental cost to determine the value placed on particular issues. New England Electric System (1990)[9] has proposed this type of polling system for valuing environmental externalities in New England. For the purpose of this section, we will consider only damage costs and control costs, as these are the two most commonly proposed techniques for valuation.

While some significant efforts have been made using the damage costing approach, we are not convinced that they currently provide a suitable basis for valuing air emissions. We would like to note, however, that studies along the lines of Hall, et al. (1984)[4] could be of great relevance. That study estimated the economic benefits of improvements to human health that would result from improvements to air quality in the South Coast Air Basin of California (essentially the Los Angeles area). In our opinion, the study is an impressive effort at illustrating the economic value of improvements to air quality. Unfortunately, from the perspective of energy planning, the study was not designed to allow for the calculation of health benefits (e.g., per ton of NO_x). For that reason, it is difficult to use the study results to generate air emissions values for use in energy planning. The study results are also only suitable for setting a lower bound on environmental/health benefits, because the study did not include all human health impacts, nor did it value such effects as visibility, materials damage, crop damage, etc. Finally, since the study focussed only on the Los Angeles area, it would be difficult to transfer the results to the air quality situation in other parts of the country.

In general, damage costing approaches face two sets of problems. First, estimating the direct *physical impacts* of environmental insults (air emissions, water emissions, etc.) can be a large and exceedingly complex task, fraught with uncertainty. These impacts depend on the *emissions loadings* themselves, the *transport* of pollutants through various media, the *exposure* of receptor populations, and the *dose/response* relationship of those populations. Moreover, there can be a series of subsequent non-physical (e.g., socio-economic) as well as physical impacts that ensue in both the near and long term. The local, regional, and global consequences of air emissions, for example, depend upon atmospheric, biological, chemical, geophysical, ecological and physiological relationships across time, space and socio-economic and cultural conditions. In modelling such relationships, elements of scientific uncertainty, availability of accurate data, the ecological response to perturbation, including non-linearities and feedback effects, among other factors, must be taken into account. The net result is that the prediction of the direct impacts of environmental perturbations may be very difficult.

Second, the *valuation* of environmental damages can be both difficult and controversial. For economic goods that are traded in a market, estimation of environmental damages presents significant analytic challenges. Valuation of non-market goods, moreover, provides even greater difficulties. Even if physical impacts can be estimated in terms of, for example, premature human deaths, increases in respiratory problems or loss of species, significant uncertainties and questions arise when we attempt to place a dollar value on such impacts. Significant uncertainties remain although a variety of methods of contingent valuation, hedonic pricing, willingness-to-pay and willingness to accept.[12] Moreover, it may be impossible, in principle, to directly establish costs for some of the consequences of environmental degradation, such as loss of habitat, loss of species or socio-cultural disruption.

Given the significant difficulties that arise in attempts to directly value environmental externalities, many analysts have chosen to use control or prevention costs as surrogates for

California Energy Commission recently adopted costs for certain air emissions that reflect the *average* cost of certain emissions control strategies that may be required under proposed air quality regulations for the South Coast Air Quality Management District (California Energy Commission 1990[1]). Similarly, the New York State recently developed costs based on *average* control costs to use in developing utility bidding systems (NY SEO et al. 1989a[10], b; [11] Putta 1989[14]).

While such cost estimates are relatively easy to implement in practice, the rationale for using average costs is not compelling. In general, there is no reason to believe that the costs of preventing a specific externality bear any simple relation to the actual damage costs for which they are used as proxies. Particularly noxious pollutants may be inexpensive to avoid, while relatively benign pollutants might be very expensive.

Because of this problem with simple cost of control valuation, a more nuanced control cost approach has been proposed, based on the notion of regulators' "revealed preferences" (Schilberg et al. 1989)[15] or "shadow pricing" (Chernick and Caverhill 1989)[2]. In this approach, existing and proposed environmental regulations are analyzed in order to estimate the value that society implicitly places on specific environmental impacts. For example, acid rain legislation may be analyzed in order to estimate the costs society is willing to impose on itself to reduce emissions of SO_x. In analyzing the regulations, we can identify the highest (or marginal) cost reduction strategy required by the regulations. This can be taken as an estimate of the value that regulators (and, by implication, society) have placed on air emissions. As the very least, it can be argued that this value represents the "revealed preferences" of regulators, and that, to be consistent, it ought to be applied when decisions affecting these environmental impacts are made.

The relationship between the two concepts of shadow prices and revealed preferences has a technical expression in certain analytical procedures that find optimal solutions to problems under constraints. For example, if a least cost energy plan subject to a CO_2 constraint is constructed using linear programming techniques, the shadow price is the implied value or benefit of the last step in reaching the target (i.e., the marginal cost of pollution reduction). Choosing a target, however, itself requires or implies some perception or estimation of damages, if not damage costs, either as an external norm or as the benefit to which reduction costs are compared. In general one would compare the shadow prices for a series of targets and select that point at which the marginal cost and benefit are equal. Since damages may not always be expressible in cost terms directly, the choice of that point reflects a willingness to pay (i.e., a willingness to incur additional costs to avoid the risk). By exploring such relationships one could reveal the implied preferences of regulators who have set such targets. In turn, such explicit exploration could serve to inform future policy discourse so that new preferences become revealed.

If one is willing to accept the notion of regulator rationality, moreover, then an argument can be made that the marginal control costs actually represent the cost to society of air emissions, because regulators, in their rationality, would have set their regulation so that marginal costs approximate marginal benefits. In practice, of course, the revealed preferences cost may be either higher or lower than actual damage costs, either through lack of information and insufficient analysis or because a risk-averse or risk-taking margin is adopted.

The revealed preferences approach thus does have a number of difficulties. In particular, it is based on the notion that regulators and legislators base their decisions on some implicit and reasonably accurate conception of marginal costs and benefits, rather than other political and social concerns, including, for example, the total costs or the impact on certain groups. In recent comments before the Massachusetts Department of Public Utilities, state air quality regulators themselves questioned the assumptions about the process of regulation that underlie the revealed preference approach to monetization.[8]

Moreover, we know that society's revealed preferences can change over time, as information, analysis, values and policy change. Thus, a limitation of this approach is that past or current revealed preferences of regulators may bear little relation to the actual impacts and their value to society, that may be identified by further research and public debate. As examples, one could take the value placed by United States society on acid precipitation twenty years ago or on greenhouse warming today.

Despite these problems, we believe that the revealed preferences approach is still a useful way to estimate the values that society currently places on air emissions, in lieu of further research on actual damages/impacts and their costs. While that research is being pursued, as it should be, the use of emissions valuation figures based on regulators' revealed preferences can be very useful. A protocol that can provide a large additional benefit to conservation and low-polluting supply options can be put into place and can be part of resource decisions that are soon to be made.

Application to Utility Planning and Operation

Consistent with our belief that the economic decision-making would be improved by the internalization of external costs, we recommend that environmental costs be used within energy planning and operating decisions. In this section, we briefly describe some of the planning issues and contexts in which environmental costs may be useful.

Integrated Resource Planning

Environmental costs can be imputed to resources within traditional resource selection procedures. Thus, if a utility is considering resource additions or retirements and demand-side management (DSM) in its long-range planning, regulation can require that it include environmental costs within its planning evaluation and decision-making. We have recommended this approach in recent studies in Rhode Island[23] and California[22] where this approach was subsequently adopted. We have also recommended this approach before the Nevada Public Service Commission which is considering promulgating a rule on these matters.[24]

More generally, monetized environmental costs could be imputed to their sources in system-wide least cost evaluations. The inputs to that procedure would be the facility-specific emissions rates times their value/costs, along with the costs traditionally ascribed to energy facility options, such as capital, fuel, operating and maintenance, taxes and financial. Environmental loadings and their costs could then be treated as additions to variable costs where appropriate (as with emissions rates) or to fixed costs (as with aesthetic and land use impacts). In general, one should track both direct cost and environmental attributes of alternative resources and resource plans separately, as well as together with a common *numeraire*. This permits examination of the multi-attribute character of the alternatives. It also permits a more comprehensive damage and damage cost analysis, if it is desired and the necessary information and analytical tools are available. In that case the loadings, transport, exposure, dose-response and valuation steps in the process would be pursued. Alternatively, as we have suggested here, loadings could be ascribed costs directly, using such approaches as revealed preferences to chose the appropriate control costs as values.

With integrated planning, the system interactive effects of supply resources and demand side resources would be evaluated. System interactive effects can be captured through dispatch analyses and related overall PVRR calculations. These effects include the accurate representation of baseload, cycling and peaking operation and, thereby, fuel costs and emissions, as reflected in capacity factors, and seasonal factors in plant availability, fuel type and usage, and emissions rates. Head to head comparison of the levelized busbar costs of facilities, including their emissions costs, is useful, especially in screening and for comparisons in a bidding system. But, this entails a level of approximation that is not necessary in fully integrated planning evaluations. A fully integrated system planning approach could also include the direct and emissions costs impacts on non-electric sources (e.g., in fossil-fired end uses that replace electric generation or in steam boilers that are displaced by cogenerating facilities) that can be affected by system resource planning decisions.

Finally, consistent with the goals of environmental preservation, a full incorporation of externalities within integrated resource planning should include environmental impacts of all resources that a utility employs to meet its energy requirements, including purchases from afar. We do not wish to set up a system that exports environmental risk. Power purchases should be ascribed their environmental impacts.

Full Cost Dispatching

Ultimately, it would be appropriate to include environmental costs not only within utility planning decisions, but also within utility operating decisions. In particular, it would be appropriate to add variable environmental costs to the variable economic costs that are used to determine the dispatch of electric generating facilities. The result would be least *full cost dispatching* that would reduce overall pollution (or the aggregate costs of such pollution) at some increase in direct costs. Such an approach would best be taken at the power pool level, where conventional economic dispatch is performed. In this context, of course, the allocation among utilities of the additional costs resulting from the altered dispatch will be of particular concern.

Full cost dispatching is a natural extension of conventional dispatch protocols, as it adds additional variable costs to fuel and operating and maintenance costs. Dispatch affects planning decisions as well as operating decisions, especially when evaluating new resources. With conventional dispatch one option that may be desirable in an overall plan is to install new facilities in the dispatch. If the overall economics (capital and variable cost) were improved, new facilities could be used to "back down" existing generation as well as to help meet load growth. These considerations could be extended to encompass overall costs, including the environmental cost of electricity generation. Units could be backed down for

In general, utilities that wish to comply with an amended Clean Air Act by meeting their acid gas limitations, will attempt to do so at the lowest cost feasible. This would be consistent with the goals of least cost planning, to which utilities and their ratepayers aspire. One option is the full range of options available to meet these goals is acid gas reduction through modified dispatch. Thus, utilities may find that full cost dispatching will to some degree become a part of their operating protocols.

Bidding Systems

Electric utilities are increasingly seeking new generating capacity and, in some cases, demand-side management resources through competitive bidding systems. Until now, evaluation of bids has primarily focussed on the bid prices of the projects, with some adjustments made for likely viability, ease of dispatching and similar characteristics. As we have argued in Massachusetts (Shimshak et al. 1990a,[16] b,[17] c[18]) it is relatively straightforward to include costs for environmental externalities within such utility bidding systems. Bid projects should be required to specify their environmental characteristics (e.g., pounds of NO_x per MWH), upon which they would then be imputed environmental costs. These costs can then be used in a protocol for comparing and ranking the projects, again subject to concerns about viability, etc. For example, they may be added to the direct monetary costs (bid prices) of a proposal, in order to determine that proposal's overall *societal cost*. Note that use of the bidding system encourages proposed projects to pursue all pollution control that is less expensive than the imputed environmental costs. To ensure that projects do not provide misleading environmental data about themselves, contract terms can be specified that allow for adjustment to contracts if projects perform less well than originally claimed.

It is important to note in this context that, even with relatively high environmental adders, competing fossil fuel units will still produce substantial emissions. It is therefore very important that the bidding process be set up to encourage and ensure that a vigorous market of competing proposals is developed. This would help to ensure that the lowest overall cost options would be elicited and that low-polluting supply resources would be drawn into the process. It is important, moreover, that energy conservation and, more generally, demand-side management be permitted to compete within an all resources process, in order that their relative pollution merits are adequately reflected in resource acquisition decisions.

Collecting External Costs

As discussed above, we believe, that the external costs associated with environmental impacts should be imputed to their sources when evaluating utility planning (including bidding) and operations. In the electricity sector, in particular, this could involve adding the external costs of electricity production to direct costs in a systematic way.

One question that arises in this context is whether external costs should be merely imputed in the planning process, without actual collection from ratepayers ("phantom" costs) or whether they should actually be collected ("real" costs). Consistent with the traditional notion of economic efficiency, we believe that it may be appropriate to collect external costs from energy consumers. Consumers would then be presented with "correct" price signals, rather than the artificially low prices that result from only imputing external costs to planning decisions.

By collecting actual externalities costs, the energy-related decisions of both electricity consumers (through increased conservation and fuel switching) and utilities (through changed priorities for dispatch and power purchases) could become more efficient from a total cost standpoint, to the degree that these decisions are price elastic. Moreover, modified (i.e., decreased) consumption can, in turn, affect planning decisions. Inclusion of externality costs and the goals of such valuation in the planning and behavior of utilities and their customers.

The revenues collected from such a system could be placed in a fund that would be used to remediate the very problem that was the source of those revenues: the emissions of pollutants. Such remediation could include investments in pollution control, in environmental remediation, in conservation and renewables, in further research on environmental damages, damage costs and control techniques, and in other related options. Of course, the design and administration of such a system would require careful thought, in order to ensure that it meets its objectives.

A similar approach that is receiving attention with regard to CO_2 is one ascribing pollution taxes to energy purchase prices. This has the merit of being economy-wide, crossing all sectors in a *prima facie* consistent manner.

An Alternative Approach: Environmental Targets

There are instances where direct valuation of certain impacts is not possible or desirable, because information (on damages, revealed preferences, etc.) is lacking or because there are policy constraints. It may sometimes be desirable to compare resource plans by estimating and considering their economic and environmental attributes without reducing them to a single factor. Thus, the cost of avoiding certain environmental or public health risks would be derived by comparing those costs of alternative resource plans with different environmental attributes.

Furthermore, energy and environmental targets could be specified and a least cost analysis can be performed to meet both goals. This is, in effect, what will likely occur to some degree in meeting the acid gas restrictions in the pending amendments to the Clean Air Act. It may occur soon if an international protocol on greenhouse gas limits is established. Indeed, analysts in some countries have been exploring the cost curve for reducing CO_2 to various levels. These "shadow prices" of carbon dioxide targets are the direct counterparts of monetary values for CO_2 emissions.

In concluding, we should note that in one study (Tellus Institute 1990a),[21] we adopted a different approach to these issues. In that study, we asked the following question: How much would it cost, over a thirty year planning horizon, to meet a predetermined air emission limit (e.g., no net increase in emissions)? We then evaluated competing resource plans in regard to the costs that would be required to meet the emissions standard. In this "standards" approach, the focus is not on valuing air emissions externalities, as it has been in the work described above, but rather on the cost of avoiding such externalities. This distinction is important when we attempt to project the impact of imputing a value, for example, \$3.25 per pound of NO_x to energy resources. In general, use of such value does not mean that we will actually pay \$3.25 to eliminate each pound of NO_x. Rather it means that we will pay up to \$3.25 per ton to reduce NO_x emissions. Some reductions will come at lower cost. Thus, to project the actual cost to society of using the \$3.25 per pound estimate, we would need to project the distribution of NO_x reduction potential available for less than or equal to \$3.25 per pound. The weighted average of these costs would be the correct cost to use for *projecting* the actual costs of meeting the emissions standard.

This *environmental standards* approach sets environmental targets and poses the question for resource planning: what is the least cost system (consistent with reliability and uncertainty considerations) to achieve two objectives, satisfying end-use energy requirements and meeting environmental targets. This least cost energy/environment planning approach places environmental objectives as an up-front consideration in energy planning, with the same analytical status as energy goals. There have been a number of recent instances of this approach, including our least cost planning study in Vermont (Tellus 1990a)[21] and recent studies of greenhouse gases reduction strategies (Kram and Okken 1989;[6] Krause, Bach, and Koomey 1989[7]).

Of course, in the energy/environment planning approach environmental targets (or constraints, in an optimization procedure) need to be specified. These must come, in large measure, from outside that analytical process. In setting those targets (e.g., for CO_2, or SO_x emissions) some notion of the damage or risk of damage must be involved. Thus, some assessment of environmental and related damages and the public policy debate on the relative importance of those risks is necessary. The more general form of the energy/environment planning approach, then, would be to explore a range of environmental targets for a given pollutant and to compare the costs of meeting energy requirements while satisfying the environmental constraint. A curve of energy costs could then be constructed and compared with a "curve" of potential damages. The damage curve can be both quantitative and qualitative (especially where there is great uncertainty) and also might refer to monetized costs where these can be appropriately and reliably estimated.

This analysis would thus compare costs to benefits in a way that could illuminate the issues and the decisions that would be required. The question of where the target should be set becomes: how much more are we willing to pay for our energy in order to avoid increases in certain sets of risks? This can be done for specific facilities as well as for entire system plans. For example, if a small hydro facility falls within a least cost plan, one could remove it and find the new least cost plan. The higher cost of electricity can then be compared to the environmental damages or risks, with the policy question posed as one of willingness to pay.

Ultimately, then, this approach is linked with at least the idealized form of the revealed preferences approach, with certain key differences. Rather than using only existing revealed preferences of regulators, it would encourage explicit further exploration of all aspects of the cost/impact relationship, that are at best only implicit and imperfectly implicated in existing revealed preferences. Such exploration could reveal the shadow prices and the benefits of different levels of environmental protection, and inform the policy and planning processes.

Acknowledgements

This paper draws upon work funded by a number of sources. Our original work was sponsored by the Vermont Department of Public Service (Tellus Institute 1990a). Subsequent contracts with the Massachusetts Division of Energy Resources (Shimshak et al. 1990a,b,c), the California Public Utilities Commission (Tellus Institute 1990b), and the Rhode Island Division of Public Utilities and Carriers, the Rhode Island Division of Stadewide Planning, the Rhode Island Governor's Office of Energy Assistance (Tellus Institute 1990c), and the Nevada Office of Consumer Advocate have allowed us to expand our efforts substantially. The views expressed in this document are those of the authors, however, and should not be attributed to these agencies.

References

1. California Energy Commission (CEC), 1990. "Committee Order for Final Policy Analysis", Docket No. 88-ER-8 (the 1990 Electricity Report), March 27, 1990.

2. Chernick, Paul and Emily Caverhill, 1989. *The Valuation of Externalities from Energy Production, Delivery, and Use, Fall 1989 Update*, a Report to the Boston Gas Company, December 22, 1989.

3. Chernick, Paul and Emily Caverhill, 1990. "Report to the Boston Gas Company on Including Externalities in 89-239", April 12, 1990, Boston, Massachusetts.

4. Hall, Jane et al., 1989. *Economic Assessment of the Health Benefits in the South Coast Air Basin, from Improvements in Air Quality*, Final Report to the South Coast Air Quality Management District, Contract No. 5685, June 1989.

5. Hohmeyer, Olav. 1988. *Social Costs of Energy Consumption: External Effects of Electricity Generation in the Federal Republic of Germany, Springer-Verlag, New York.*

6. Kram, T. and P.A. Okken. 1989. "Two 'Low CO_2' Energy Scenarios for the Netherlands," paper presented at the IEA/OECD Expert seminar on energy technologies for reducing emissions of greenhouse gases (April 12-14, 1989), Paris.

7. Krause, Florentin, Wilfrid Bach, and John Koomey. 1989. *Energy Policy in the Greenhouse, Volume One, From Warming Fate to Waring Limit: Benchmarks for a Global Climate Convention*, International Project for Sustainable Energy Paths, El Cerrito, California, September, 1989.

8. Massachusetts Department of Environmental Quality Engineering (MA DEQE). 1990. "Supplemental Comments of Division of Air Quality on Environmental Externalities for DPU #89-239," May 9, 1990.

9. New England Electric System (NEES). 1990. *Integrating Environmental Externalities into Resource Planning at New England Electric*, study prepared by Temple, Barker, and Sloane.

10. NY SEO, NY DPS, and NY DEC. 1989a. "Air Impacts/Electricity" in *(Draft) New York State Energy Plan, Staff Report Volume IV: Electricity Issues*. Albany, New York.

11. NY SEO, NY DPS, and NY DEC. 1989b. "Environmental Externalities" in *(Draft) New York State Energy Plan, Staff Report Volume IV: Electricity Issues*. Albany, New York.

12. Organisation for Economic Co-operation and Development (OECD). 1989. *Environmental Policy Benefits: Monetary Valuation*, Paris, France.

13. Pace University Center for Environmental Legal Studies et al. 1990. *Draft Final Report: Environmental Externality Costs From Electrid Utility Operations*, a Report to the New York State Energy Research and Development Authority and the U.S. Department of Energy, February 28, 1990.

14. Putta, Sury N. 1989. "Competition in Electric Generation - Environmental Externalities," paper presented to the Power Planning Committee of the New England Governers Conference (October 1989), Boston, MA.

15. Schilberg, Gayatri, Jeff Nahigan, and William B. Marcus. 1989. "Review of CEC Staff's Revised 'Valuing Emissions Reductions for Electricity Report 90' Staff Issue Report #3R," prepared for the Independent Energy Producers Association, December 5, 1989.

16. Shimshak, Rachel, Bruce Biewald, Harvey Salgo, Donald Marron, and Steve Bernow.

Massachusetts Department of Public Utilities, Case No. DPU 86-36-G and DPU 89-239, February 23, 1990.

17. Shimshak, Rachel, Bruce Biewald, Harvey Salgo, Donald Marron, and Steve Bernow. 1990b. "Comments of the Division of Energy Resources on Environmental Externalities," presented before the Massachusetts Department of Public Utilities, Case No. DPU 86-36-G and DPU 89-239, March 2, 1990.

18. Shimshak, Rachel, Bruce Biewald, Harvey Salgo, Donald Marron, and Steve Bernow. 1990c. "Final Comments of the Division of Energy Resources," presented before the Massachusetts Department of Public Utilities, Case No. DPU 86-36-G and DPU 89-239, May 4, 1990.

19. South Coast Air Quality Management District (SCAQMD). 1989. *Air Quality Management Plan, 1989 Revision: Final Appendix IV-A, Tier I, Tier II, and Contingency Control Measures*, March 1989.

20. South Coast Air Quality Management District (SCAQMD) and Southern California Association of Governments (SCAG). 1989. *Air Quality Management Plan: South Coast Air Basin, Executive Summary*, March 1989.

21. Tellus Institute. 1990a. *The Role of Hydro-Québec Power in a Least-Cost Resource Plan for Vermont.* A report to the Vermont Department of Public Service. January 19, 1990.

22. Tellus Institute. 1990b. *Conservation and Capacity Optimization Alternatives to the PGT/PG&E Gas Pipeline Project*. Prepared for the California Public Utilities Commission under contract to Jones & Stokes Associates, Inc. May 1990.

23. Tellus Institute. 1990c. *Evaluation of Repowering the Manchester Street Station*. A report to the Rhode Island Division of Public Utilities and Carriers, Rhode Island Division of Statewide Planning, and Rhode Island Governor's Office of Housing, Energy and Intergovernmental Relations. May 1990.

24. Tellus Institute. 1990d. *Incorporating Environmental and Economic Goals into Nevada's Energy Planning Process*. A report to the Nevada Office of the Consumer Advocate. July 1990.

25. Therkelsen, Robert L. 1989. "Valuing Emissions Reductions for Electricity Report 90," California Energy Commission Staff Paper #3R, November 21, 1989.

26. Therkelsen, Robert L. 1990. "Consideration of Multiple Attributes in Electricity Planning, Resource Selection and Siting," California-Energy Commission, Docket No. 88-ER-90 Hearings, June 29, 1990.

4.2 Basic Problems in Valuating Environmental Damages: The Case of Forest Damage

Prof. Dr. Hans-Jürgen Ewers
Director, Institut für Verkehrswissenschaften
Universität Münster
Am Stadtgraben 9
D-4400 Münster

1. The Problem of Uncertainty

The problem of "technology and environment" is often compared with the story of the sorcerer's apprentice. Our society is the sorcerer, the progress of technology is the conjuring trick and it seems as if we cannot stop this trick. This technical development is accompanied by environmental damage, for example forest damage or global warming.

There are two solutions to this problem. The first one is to completely forbid magic performances or to allow only such performances which can prove that they are absolutely harmless. The second one is to improve the performances by giving incentives to reduce the risks of the conjuring tricks.

Both solutions are alternative strategies in the political discussion about technologies with environmental impacts. The first one would stop nearly all innovative activities, because there will always be some problems of uncertainty with new technology. We will never know with absolute certainty, what environmental consequences technologies will have on complex ecological systems such as forests of the global climate. Nobody can guarantee, that his activity does not have any harmful environmental impact, especially when it is new. I think we would even have problems proving the harmlessness of the substances we emit when we breath. So if we do not want to stop all innovative activity, we must develop strategies to deal with this uncertainty.

2. Environmental Damages as Complex Phenomena

The problem of uncertainty is caused by the fact that environmental damages are complex phenomena. The case of forest damage demonstrates the impossibility of valuating the damage in common cost-benefit-analyses (see Ewers et al. 1986; Tampe-Oloff 1985). There are three reasons for this impossibility (see Ewers 1985).

(1) The polluter can not be identified adequately. Forest damages ar caused by widespread groups of emitted (industry, foreign countries, households, traffic, power stations). Complex chemical processes and transmission mechanisms in the air avoid the identification of the specific part of emissions caused by each group.

(2) There is no direct connection between cause and effect of the damages. Although scientists agree that air pollution is the main cause of forest damage, they must admit that they are not absolutely certain about it. Forest damage is a complex phenomenon without one specific chemical substance which can be identified as the main cause. The thesis of "environmental stress" (a combination of air pollution, climatic influences and "normal" damages, such as insect- or mushroom infestation) seems to be plausible. In this complex structure, the main factor differs in each case, especially when synergisms between different substances exist. These synergisms multiply the isolated effects of each harmful substance.

(3) The damage itself cannot be adequately identified. Particularly when accumulation of related effects plays a role, the presently visible damages might have been caused decades ago. In other words: What we call forest damages seems to be the last step of a long process: loss of vitality, stability and changes in ecological systems caused by "environmental stress".

These characteristic features of complex environmental problems were the main results of an empirical study initiated by the German Umweltbundesamt six years ago. The object of our research was the demonstration of methods for environmental damages. As an example we analysed the problem of forest damage.

The essential result of the study has been that the difficulty of valuating environmental damages is not a methodological problem of cost-benefit-analyses. The economic methods of valuating environmental damages are not exact but do give adequate cost estimates. The problem lies in the quantification of physical effects, because we cannot determine with sufficient certainty the chain of causation. It cannot be expected that this situation of uncertainty about the ecological consequences of human actions will basically change, even if we invest much more time and energy in environmental research. We therefore need to develop political strategies which take this uncertainty into consideration (see Ewers 1984, Flassbeck/Maier Rigaud 1982).

3. Management of Environmental Risks as an Alternative Strategy

Taking uncertainty into consideration, we cannot expect that new technologies can be invented without risks. We therefore have to develop a Management of environmental risks in order to keep them on a reasonable level.

(1) Alternative Allocation of Risks
The theory of environmental economics describes different principles of risk allocation (see Siebert 1987). The "polluter pays-principle" states that the polluter is liable for his activities and consequently bears the environmental risks. The "public burden-principle" says that the state is liable for environmental pollution and therefore has to pay for it. Although we hear much about the "polluter pays-principle", in reality the "public burden-principle" still dominates. Not only does the state take the risk when the polluter cannot be identified or when no legal rules or limits exist. But also governmental limit is a guarantee by the state that is liable for damages caused within this limit. This guarantee offers no incentives to private enterprises to use ecologically harmless technologies.

(2) Alternative Methods for Technology Impact Analysis
Environmental problems are normally urgent problems. When damages are recognized, there is often no time to wait for long scientific research about cause and effect. Considering the fact of uncertainty means to accept heuristic methods which can give quick answers in such situations (see Kleindorfer/Kunreuther 1987, 1982; Center for Energy and Environmental Studies 1984; Morgan/Amaral/Henrion/ Morris 1982, Electric Power Research Institute 1982). Such methods are:

- Use of probability calculus or subjective probabilities as substitutes for unknown relations between cause and effects
- Use of scenarios
- Use of the Delphi-method

These methods don't try the impossible, namely discovering the unknown causation, but try to guide the decision-making process despite uncertainties. The cost/benefit analysis of forest damages initiated by the Umweltbundesamt has demonstrated the use of such heuristic methods. The forest damages in the next decades have been quantified under several different assumptions. We used a combination of expert interviews (Delphi-method) and the scenario technique. These mainly qualitative methods were supported by a simulation model for the development of the population of one special tree: the spruce. The results for this specific tree can be used to forecast the development of other three populations. Therefore we took the statements of experts into consideration. In this way we were able to quantify the future tree populations on the basis of a simulation model and of extrapolations which were justified by experts. All damages have been quantified for a simulation period of 77 years (from 1984 until 2060).

Politicians must accept uncertainty and therefore make decisions on the basis of heuristic methods. Otherwise the chance to act successfully will disappear. The case of global warming is an example where such heuristic estimations exist, but where political leaders - including the President of the United States . do not accept the uncertainty of such methods and wait for better scientific results.

(3) Reducing Environmental Risks by Reducing Emissions

If one cannot distinguish systematically between harmful and harmless emissions, there is only one rational strategy. That is to reduce emissions as quick by as possible and therefore to integrate incentives for the reduction in the emissions as well as the legal limits for pollutants.

Political measures to achieve this aim are (see Bundesminister des Innern, Ewers 1984):

- automatic and regular reduction in emission and immission limits
- spreading information about dangerous environmental impacts of technologies (information and environmental engagement correlate positively, as empirical studies show (see Schulz 1985; Winter 1987)
- increased use of economic incentives (polluter charges, trade of emission certificates, etc.) to reduce pollution.

It is important that the measures offer incentives to do more for the protection of the environment than it is required by law. This should include old factories as well as new ones. The aim is to support new kinds of technologies that have fewer harmful environmental effects.

These new technologies should be part of an integrated solution of environmental problems and should replace of the end-of-pipe solutions which are currently used much too often. Integrated technologies with fewer environmental impacts can solve the problem of a trade off between economic growth and environmental aims. A basis of such efficient solutions are economic instruments in environmental politics. Only they can offer the needed incentives to invent intelligent technologies without harmful emissions.

4. The Forest Damage Scenarios

The example of the forest damage study initiated by the German Umweltbundesamt has demonstrated that heuristic methods are necessary in the analysis of environmental impacts of technologies. The study was based on three scenarios:

- One trend-scenario assumed a development of emissions and immissions in the next decades which extrapolates environmental activities possible under the political conditions at the beginning of 1984. For example we assumed reduced energy consumption in the near future and compulsory use of catalytic convertor. So we quantified the estimates of the harmful substances which are mainly responsible for forest damages: sulphur dioxide and nitrogen oxide. We calculate a possible reduction in sulphur dioxide of 40 percent by 1900 and a reduction in nitrogen oxide of 30 percent. We now know that these estimates were too optimistic. By the year 2060 the trend-scenario calculates a level of only 25 percent of current levels of sulphur dioxide and 35 percent of nitrogen oxide.

- One status-quo-scenario assumed the 1980 level of emissions for the whole simulation period (until 2060). The aim of this scenario is to document the forest damages that will occur if nothing is done for the environmental protection. Only then can we calculate the costs we would have to pay if nothing was done to fight pollution. These costs are the opportunity costs of environmental protection.

- One reference-scenario assumed environmental conditions as they were in the thirties and the beginning of the forties: A time without significant pollution. Even if the word forest damage was still unknown in the vocabulary of the late forties, fifties and sixties, there were already early symptoms of the damage in these decades. The reference scenario describes a development that would have happened without harmful emissions.

The difference between the reference-scenario and the two other scenarios reveal the damages which are caused by emissions.

The results of the scenarios were alarming. Here are a few examples:

- Even on the basis of the optimistic trend-scenario, there will be a reduction of forest area of nearly fifteen percent. The corresponding figure of the status quo-scenario is 30 percent.

- The most impressing result is the development of the population of the different trees. The volume of the spruce will be reduced to 50 percent by the end of the period in the trend-scenario. In the status-quo scenario, only 20 percent of the volume of the spruce will be left by 2060.

- In the status-quo scenario, by the year 2010 there will not be any spruce left which is older than 80 years.

- Experts state that the other trees, for example the pine, will have the same development with a time lag of five or ten years. The reduction of the other trees will have the same dimension as the reduction in the spruce.

Graphic 1: Reduction of wood (spruce) caused by forest damages (1984 to 2060)

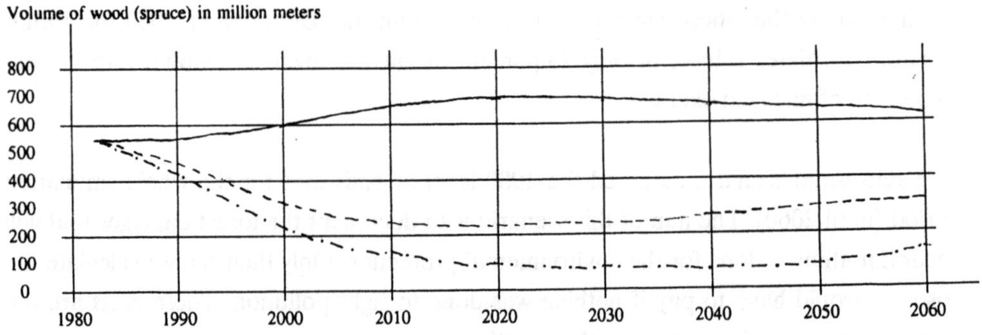

1 >reference-scenario<: no significant air pollution (situation before the Second World War).

2 >trend-scenario<: clean-air-aims of the german government are realised.

3 >status-quo-scenario<: air pollution remains on the state of the beginning of the eighties.

source: Ewers (1986, S. 129)

The valuation of costs and benefits in these different scenarios was divided into three categories:

- forestry
- leisure time and recovery activities
- water and land

(1) Forestry

The reduction in wood production seemed to be the main cost factor in the forestry. Consequences in other sectors caused by this reduction had to be neglected. We analysed two different kinds of costs in the forestry: the reduction in the output and the costs of compensatory activities (for example compensatory fertilizing). As a result of this estimate, there are costs amounting from 1,7 to 4,8 billion DM (trend-scenario) and from 2,3 to 6,3 billion DM (status-quo scenario) per year. The minimal and maximal limits depend on the discount rate on which the estimation is based (zero percent as a minimum, three percent as a maximum). The estimate has neglected some risks, for example changing prices of wood. So we can say that the valuation has been conservative and does not quantify all possible costs which have to be borne by forestry. The study gives only the minimal value of forest damages.

(2) Leisure Time and Recreation Activities

In the area of leisure and recreation we considered several kinds of cost categories: Costs which are payed by the users of forests (user costs), by the leisure industry and by people who do not actually visit forests but are willing to pay for the possibility of using them as a place of recovery (non-user costs). The quality of forests being used as a recreation area will be reduced as a consequence of forest damages. The preference of people to visit forests will decline because the beauty of the forests will have disappeared. We estimated fewer visits to the forest and fewer holidays in forest areas as a consequence of those damages.

On the basis of these assumptions we calculated a yearly damage in the field of leisure time and recreation activities of 2,3 to 6,3 billion DM (trend-scenario) and of 4,2 to 11,3 billion DM (status-quo-scenario). As I have already mentioned, the minimal and maximal limits depend on the discount rate the estimate is based on.

Table 1: Tourism (overnight stayings) in German forest regions (1983)

german forest regions	tourism (overnight stayings per year)
Bavarian Alps	15 021 552
Bavarian Forest/Oberpfälzer Forest	5 235 530
Fichtelgebirge/Frankenwald	1 526 616
Rhön/Spessart	6 388 907
Black Forest	22 013 600
Pfälzer Wald	881 899
Eifel/Hunsrück	9 665 512
Westerwald/Taunus/Odenwald	7 757 098
Sauerland	11 762 146
Harz/Weserbergland	5 841 727
Lüneburger Heide	1 406 054
overnight stayings (total)	87 500 647

source: Ewers et al. (1986, S. 4-73)

Graphic 2: Reduction of forest visits caused by forest damages (1984 to 2060)

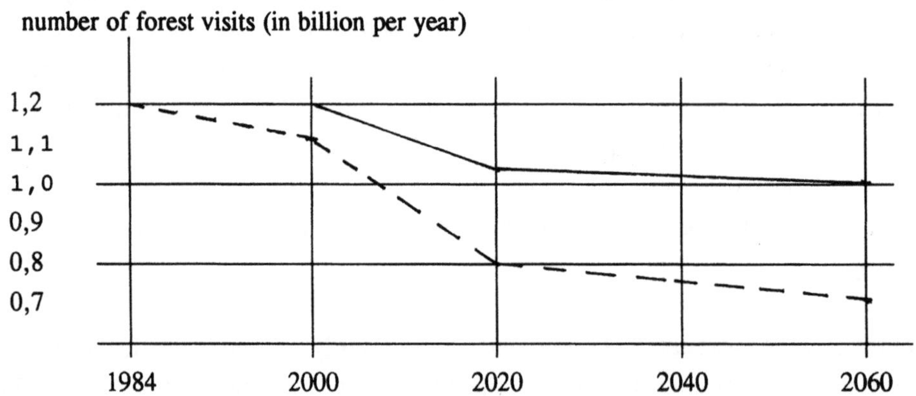

[1] >trend-scenario<: clean-air-aims of the German government are realised.

[2] >status-quo-scenario<: air pollution remains on the state of the beginning of the eighties.

source: Ewers (1986, S. 4-25)

(3) Water and Land

Based on our calculations, the value of the damages to the quality of water and land did not occur in the same dimensions as the damages in the two other cost categories. One reason for this is the fact that it turned out to be extremely difficult to analyse the effects of complex chemical and physical processes. Especially, we had problems in evaluating the risk of a changing water quality caused by forest damages. Beside methodological problems, another reason is that most regions in Germany do not have to fear risks of a diminished water quality caused by forest damages.

Table 2: Costs of investments in protection against the dangers of high water, avalanches, **erosion and torrents caused by forest damages**

costs of investments for	region	trend-scenario[1] (in billion DM)	status-quo-scenario[2] (in billion DM)
	Hesse	0,185	0,493
	Baden-Württemberg	0,522	1,243
	Bavaria	0,949	2,552
protection against high water	other federal states	0,619	1,651
	Federal Republic of Germany (total costs)	2,275	5,939
protection against avalanches and erosion	Bavarian Alps	6,500	7000
total costs		8,775	12,939

[1] clean-air-aims of the german government are realised.

[2] air pollution remains on the state of the beginning of the eighties.

source: Ewers et al. (1986, S. 5-80 and 6-28)

The complete costs in the area of water and land quality were evaluated at a level from 281 to 374 million DM (trend-scenario) and a level from 433 to 616 million DM (status-quo-scenario). In this case the minimal and maximal limits are also dependent on the discount rate of the estimate.

Summing up all categories of forest damages, the total costs per year are:
- 4,3 to 11,4 billion DM (trend-scenario)
- 7 to 18,3 billion DM (status-quo-scenario)
- On the basis of an adequate real discount rate of two percent, the total costs per year amount to 5,5 billion DM (trend-scenario) and 8,8 billion DM (status-quo-scenario).

To avoid misunderstandings in judging these results: The trend scenario assumes a strenghtening of our efforts in environmental protection. If we do not reinforce our activities in environmental protection, the costs will correspond to those estimated in the status-quo-scenario.

The example of forest damage shows the necessity of quick and efficient political activities against air pollution. Passivity cannot be justified with the excuse that we have to wait for clear proof of cause and effect. We can in fact do both:

- Scientific research to improve our knowledge about these complex ecological systems
- and creation of political measures based on the current knowledge while at the same time research is being conducted.

Table 3: Total costs of forest damages (monetized damages in billion DM per year)

cost category	trend-scenario[1] (in billion DM)	status-quo-scenario[2] (in billion DM)
forestry	2,3	2,9
leisure, recreation	2,9	5,4
water and land	0,3	0,5
total monetized damages	5,5	8,8

1 clean-air-aims of the German government are realised.

2 air pollution remains on the state of the beginning of the eighties.

Valuation of the estimation : Very conservative, especially in the cost category >water and land<. For example, costs of protection against the dangers of global warming are neglected.

5. Conclusion

In the early diagnoses of harmful environmental impacts there will always be some amount of uncertainty. Scientists and politicians have to accept this uncertainty. Otherwise they would be unable to solve urgent ecological problems. As I have explained, forest damage and global warming are examples for this. It is inevitable that environmental politics contain certain risks, but they can be managed and reduced to a bearable limit with economic solution such as the creation of politics which offer incentives to detect, reduce and avoid risks.

This means that our political systems have:

- to accept "soft" analysis methods for the estimation of environmental impact of technologies and to create new institutions to detect environmental risks. "Soft" heuristic methods must be accepted by politicians.
- They must also implement automatic and regular reduction in emission and immission limits.

References

Bundesminister des Innern (Hrsg.) (1983). Umweltbrief vom 28. Oktober 1983: Abschlußbericht der Projektgruppe "Aktionsprogramm Ökologie", Argumente und Forderungen für eine ökologisch ausgerichtete Umweltvorsorgepolitik, Bonn.

Center for Energy and Environmental Studies (1984), A Conceptual Framework for Integrate Assessments of the Acid Deposition Problem, Carnegie-Mellon-University, Pittsburgh

Electric Power Research Institute (1982), Acid Deposition: Decision Framework, Volume 1, Palo Alto

Ewers, H.-J. (1984), Lehren aus dem Waldsterben. In: Allgemeine Forstzeitschrift 20 (19. Mai 1984), S. 506-509

Ewers, H.-J. (1986), Zur Monetarisierung der Waldschäden in der Bundesrepublik Deutschland. In: Kosten der Umweltverschmutzung, Tagungsband zum Symposium im Bundesministerium des Innern am 12. und 13. September 1985, Berichte 7/86 des Umweltbundesamtes, Berlin, S. 121-144

Ewers, H.-J., et. al. (1986), Zur monetären Bewertung von Umweltschäden, Berichte 4/86 des Umweltbundesamtes, Berlin, S. 121-144

Flassbeck, H., Maier-Rigang, G. (1982), Umwelt und Wirtschaft, Tübingen

Kleindorfer, P.R., Kunreuther, H.C. (1987), Insuring and Managing Hazardous Risks: From Seveso to Bhopal and Bayond, Berlin-Heidelberg-New York

Maas, C. (1986), Determinanten der Entwicklung und Nutzung umweltfreundlicher Neuerungen in Industriebetrieben, Diskussionspapier 108, Wirtschaftswissenschaftliche Dokumentation, Technische Universität Berlin

Morgan, M.G., Amaral, D.A.L., Henrion, M., Morris, S.C. (1982), Technological Uncertainty in Policy Analysis. Final report on a case study directed at exploring long-range transport and possible health impacts of sulfur air pollution from coal-fired power plants. Carnegie-Mellon-University, Pittsburgh

Schult, W. (1985), Der monetäre Wert besserer Luft, Frankfurt-Bern-New York

Siebert, H. (1987), Umweltschäden als Problem der Unsicherheitsbewältigung: Prävention und Risikoallokation. Reprint no. 4 des Sonderforschungsberichts "Internationalisierung der Wirtschaft", Universität Konstanz

Tampe-Oloff, M. (1985), Zur Komplexität als Hindernis problemorientierter Reaktionen auf das Waldsterben, Diss., Freiburg

Wicke, L. (1986), Die ökologischen Milliarden, München

Winter, G. (1987), Das umweltbewußte Unternehmen, München

4.3 The Valuation of Environmental Externalities in Utility Regulation

Paul Chernick and Emily Caverhill
Resource Insight, INC. (formerly PLC, Inc.)
18 Tremont Street, Suite 1000
Boston, Massachusetts 02108
USA

Introduction

Monetizing environmental externalities provides an approximation of the societal value of reducing impacts on human health and the environment from electirical enrgy supply, including utility and non-utility fossil-fired generation, demand-side management and off-system power purchases. A dollar estimate of the full societal cost of the supply option is established by placing a value on the air, water and terrestrial effects of each supply option and adding that to the option's capital, operating and maintenance costs. Repeating this exercise allows different electric power supply options to be compared on a full-societal cost basis. In monetization, these externalities are expressed as a cost/unit of the externality, such as $/lb of emissions of $/gal of water consumed, and the full societal cost is usually stated as a cost/kWh generated.

In this paper, a rationale for monetizing externalities is provided, and the strenghts and weaknesses of four techniques for monetizing externalities are discussed with examples of the application of each method. The authors' preferred technique for incorporating externalities into utility planning in the near term, implied valuation through the estimation of the marginal cost of abatement, is discussed in detail.

Valuation Methods

In general, four basic approaches have been used to estimate environmental externalities:

1. Estimating the relative physical, chemical, or toxicological potency of various pollutants;
2. Polling of experts or other people;
3. Direct estimation of the environmental effects of a pollutant, and the valuation of each of those effects; and
4. Determination of the implied societal value of reduction of the pollutant, from the maximum cost society has comitted (or appears about to commit) to pay for reductions of this pollutant.

In the first two approaches, relative potency and polling, the relative values of specific externalities are estimated. These relative values can then be translated into cost (monetized) by estimating the value of one of the externalities, and relating the other related externalities to it.

Relative potency works quite well for estimating the relative importance, or value, of the major energy-related greenhouse gases (CO_2, methane, N_2O, CO and NO_x). The relative values of reducing emissions of the greenhouse gases can be related to the value of reducing CO_2 emissions, by correcting for the shorter lives and lower molecular weights of the other greenhouse gases, and properly discounting the effects of the gases over time. Lashof and Ahuja (1990) estimate that the instantaneous global warming potentials (GWP) of the greenhouse gases relative to CO_2 are: 10 for CH_4, 180 for N_2O, and 2.2 for CO. (That is, CH_4 is ten times as potent as CO_2 on a weight basis.) Chernick and Caverhill (1990a) discounts the value of future emissions of the greenhouse gases and uses Lashof and Ahuja's results to estimate the relative *value* of reducing emissions of the greenhouse gases as: 76 for CH; 320 for N_2O; and 6 for CO. These results are based on a discount rate of 6 % real.

Relative potency might also be used for estimating the relative importance of other externalities. Data in Hendrey (1986) implies that the acid rain damages caused by NO_x are 83 % those of SO_2, per pound of emissions (Chernick and Caverhill, 1989). Hohmeyer (1988) cites work by Grupp (1986), which finds that the toxicity of SO_2, particulates, and VOCs are equal per pound of emissions, and that NO_x is 25 % more toxic per pound. Grupp's estimates appear to refer only to acute human toxicity, although Hohmeyer applies them for all purposes. While relative potency provides probably our best estimates of the relative value of reducing greenhouse gas emissions, it may be inferior to the more direct monetizing technoques, described below, for the other major air pollutants.

The second approach, polling, can be employed in valuation if the polling instrumetn is designed carefully, and the poll respondents are experts in the field in question and are informed of the intended use of the poll results. For example, it may be useful to poll epidemiologists on the coefficients of the dose-response functions relating human illness and death to pollution levels. However, the results of a polling effort can be useless for valuing externalities, or worse, misleading. A good polling effort must: (1) poll a representative sample of experts in the field of interest, and publish the names and credentials of the respondents for public review; (2) provide a clear statement of the question to be answered in the poll, and the intended use of the answer; and (3) make the polling instrument and the respondents' answers available for public review. Polling is unlikely to provide better information than the other methods of estimating externalities presented here. Certainly, its usefulness is critically dependent on the quality of the polling instrument and credentials of the respondents.

Direct assessment and valuation, the third approach, attempts to identify all quantifiable effects of an externality to society. In general, each effect must be quantified, and a unit cost must be established. This procedure is generally treated as if it were a technical exercise, but the definition of relevant externalities, selection of important effects, choice of quantification measures and techniques, and determination of unit cost are often highly judgmental and subjective.

For environmental externalities, such as air pollution and water pollution, the relevant effects might include impacts on: human life and health; human comfort and pleasure including effects of noise levels, visibility, and odors; domesticated plants and animals, such as commercial forests, livestock, crops, and lawns; wild plants and animals in a variety of habitats; and non-living materials, such as those exposed on buildings (especially stone and paint), at archeological and historical sites, on monuments, on vehicles, and in natural landforms.

Determination of at least some of the social costs of some environmental externalities is feasible, although this process is complex. In principle, direct estimation requires the determination of how the pollutant moves through the atmosphere, how it changes chemically (especially for precursors to ozone and to particulates), how it is deposited on various surfaces (lungs, soil, water, paint, etc.), how much it affects each system (including visibility, human health, crops, materials, wildlife, etc.), and the unit value of the effects on each system. In fact, the transport calculations are very complicated, and the atmospheric chemistry and dose-response relationships may be highly uncertain.

Even when these difficult technical issues have been resolved, togh policy questions remain in the valuation of effects. Costing of market-valued goods such as agricultural production and paint can be relatively straightforward. It is more controversial to assign values to human morbidity and mortality (which are themselves complicated by many subcategories including age, health, duration of illness, and the extent of suffering and dread), as well as to visibility, historic monuments and forests, wildlife and ecosystems.

Direct estimation of environmental costs is a daunting and time-consuming task, given the multi-disciplinary background needed by those responsible for assigning values to human life and health, visibility and other non-market costs. Yet if the assessors exercise caution in understanding data sources, correcting errors and reflecting regional differences, all to the extent feasible, direct estimation serves as a highy clear and workable method to value what seem to be intangible external costs. For a more detailed discussion of direct estimation of externalities see Ottinger, et al. (1990), Hohmeyer (1988) or Chernick and Caverhill (1989).

The fourth approach, implied valuation, relies on the costs of required or anticipated regulations or control measures to estimate a societal value of reducing emissions of particular externalities. For example, the new U.S. Clean Air Bill, once adopted, gives us specific information about what U.S. residents are willing to pay, on a national basis, to reduce SO_2, NO_x, VOCs and toxic air emissions. Similarly, other international, federal or state regulations provide estimates of the value of reducing externalities at the margin. This method is also referred to as shadow pricing, revealed preference and marginal cost of abatement. There is increasing use of the cost of abatement to determine the value of reducing externalities, specifically in California, New York and most recently Massachusetts. The strenghts and limitations of this method are discussed in more detail in the following sections.

In the third and fourth approaches, direct estimation and implied valuation, the value of reducing each externality is monetized. Typically, values are expressed in such terms as $/pound of pollutant emitted, or $/unit of the externality. The combined value of all of the externalities of a power source (expressed in $/kWh, for example) is then the sum across externalities of the following product:

Amount of each pollutant emitted x Unit value of that pollutant

In the next section we discuss the implied valuation technique in more detail through a discussion of its application and associated uncertainties and analytical difficulties. For simplicity, we will discuss generic approaches in the context of externalities for which $/lb valuation is appropriate.

Implied Valuation and Cost of Abatement

As used in this method, the cost of pollution control is sometimes described as if it were a proxy for the cost of emissions. In fact, the costs of control provide direct information on the societal value of emission reductions, under two theoretical approaches.

First, the cost of required control measures serves as an estimate of the price that society is willing to pay to reduce the pollutant. If legislators and regulators require measures that cost as much as $2/lb to reduce sulfur emissions, it seems reasonable to assume that reducing emissions from those sources must be worth at least $2/lb, and that reductions from other sources (due to conservation or fuel choice) must also be worth $2/lb at the margin. This is the rationale behind the "implied valuation or revealed preference" approach to the use of control costs for valuing externalities.

Second, the costs of required controls may directly establish the social benefits of reducing emissions, to the extent that they define the direct pollution-control costs that can be avoided by an exogenous reduction in emissions. If the objective of environmental regulation is to maintain a given level of ambient air quality, the construction of a less polluting plant or the reduction in output requirements due to energy conservation will allow regulators to back out the most expensive control measures that would otherwise have been required.

One important aspect of the implied valuation approach is that only the *marginal* cost of control matters. From the "implied preference" perspective, the fact that many required controls are inexpensive, or even that some inexpensive controls have not yet been required, is irrelevant to the determination of the marginal price society is willing to pay. From the avoided-control-cost perspective, the appropriate estimate of the social cost of control is the highest-cost item that will be reqired, since conservation (or low-pollutant generation) will allow that most expensive measure to be avoided.

One general criticism of this approach is that the highest cost of control is a reasonable estimate of the value of reducing emissions only under a mandate requiring an additional reduction. The costs of controls required in the past to correct historical environmental problems are not directly relevant to determining the value of reductions in pollution today, especially if the environment is now much cleaner. In our earlier work, and most recently in our analysis of the implications of the New York State Energy Plan regarding the costs of required controls, we have used current and future costs to maintain or improve air quality.

Further, for the foreseeable future there *is* a clear mandate for reductions in ambient levels of SO_2 and NO_x, and in CO_2 emissions in some jurisdictions. For instance, the goals and objectives of the New York State Energy Plan including a 50 % reduction in SO_2 emissions from 1980 levels, a 25 % reduction in NO_x emissions from 1988 levels for stationary sources and possibly a 25 % or greater reduction in NO_x emissions from mobile sources. For the greenhouse gas CO_2, the recommended target is a 20 % reduction in emissions by 2008. These targets respond to the deficiency of current *emissions* regulations to meet *ambient air quality* standards. In other word, an individual source may be in compliance with its emission permits, but its remaining emissions, no matter how small, contribute to degraded ambient air quality.

Wherever future measures (including application of current regulations to future facilities) will be necessary to achieve or maintain acceptable pollution levels, the incremental cost of the required control for that pollutant provides a reasonable value for reducing emissions of that pollutant at that margin. Note that we are not simply urging the use of the highest cost we can imagine, or that has been proposed; we refer to the highest cost among those that are likely to be imposed in the absence of reduced energy generation. Further, we are interested in the incremental cost of control. We use the term "incremental" because pollution control options tend to occur in discrete blocks, corresponding to particular rules, although the term"marginal" would also be appropriate. By definition, the incremental cost is the highest cost of required controls.

Given this definiton, the marginal control cost represents the highest cost society has explicitly demonstrated a willingness to pay for recent or future reductions of an externality. Obviously, society is happy to pay less, if it can get the reductions for a lower cost. However, the amounts which are effectively available at the lower price must be exhausted, or else, the higher-cost measures would not have been required. However, the marginal cost of control is probably lower than the value of reducing an externality at the margin. The next-more-expensive option may be so much more expensive that it is deemed

Examples of abatement options which are at least partially constrained by technical and policy constraints include the use of CO_2 scrubbers or the use of nuclear power to replace fossil fuels. The estimated marginal cost of abatement is below the highest cost of control currently required under regulations, significantly below the cost of control option rejected because it is currently too expensive, and approximates the "true" societal value.

Our estimates of the marginal control costs are actually understated by even more than implied by this argument. We have deliberately selected values somewhat lower than the very highest marginal cost we have identified, so they are probably lower than the actual value of reducing externalities at the margin.

It seems unlikely that the marginal value of reducing externalities would drop sharply in the range of reductions individual states are considering. Indeed, the cancer-promoting characteristics of NO_x, the carcinogenic properties of NO_x reaction products, and many other externalities, probably have no threshold for their effects. Our use of the implied valuation method does *not*, as some critiques have assumed, estimate the value of eliminating an externality. Even if the more ambitious of the utility DSM programs were replicated state-or country-wide, no major externalities would be eliminated. We are estimating the value of marginal improvements in air quality by relating emissions reductions to air quality improvements. Therefore, the value of reducing the *next* unit of an externality is close to (and perhaps indistinguishable from) the value of reducing the *preceding* unit.

Even if the environmental damage of emissions dropped sharply as successive emissions reductions were achieved, the marginal cost of abatement *still* represents a minimum estimate of the marginal value of reducing emissions, to the extent that society will be able to back out or avoid expenditures on the most-expensive control measures on new plants or future retrofits. If a conservation program is adopted, the demand for supply side emissions reductions will be reduced by a conservation program, again avoiding the most expensive control measures. Until all of the SCR, which is and will be required on new plants in most states to control NO_x, is no longer necessary because of dramatic improvements in local and national air quality, the marginal cost of SCR will be a reasonable estimate of the value of reducing NO_x emissions.

In addition, the use of expected costs of control is also a way of reflecting the additional costs of uncertain but likely future regulations, such as SO_2 scrubber retrofits and allowance purchase requirements under the Clean Air Bill or the adoption of a carbon tax on fossil fuels.

Implied Value of Unregulated Externalities

The previous discussion is presented as if all externalities under consideration were regulated. For externalities that are not yet regulated, such as the greenhouse gases, costs of control are also a valid way of estimating the value of reducing emissions. First, there is considerable discussion of adopting a federal carbon tax, and several states are moving to reduce CO_2 emissions by a target percentage. While these measures may not be immediately adopted, primarily due to concerns about the macro-economic effects of taxes and the lack of an international structure for requiring reductions, some measures to control global warming are likely to be adopted in the utility planning horizon.

Second, the weight of evidence indicates that we must stabilize the CO_2 concentration in the atmosphere in order to slow the current accelerated rate of global warming. The stabilization effort is going to come either from reducing CO_2 emissions or increasing CO_2 sinks. Krause, Bach and Koomey (1989) estimate that reductions (or abatement) of CO_2 emissions on the order of 80 % are required to stabilize global climate.

Our options for stabilizing global climate include: reducing CO_2 emissions through improved energy efficiency; decreased use of fossil fuels by eliminating wasteful uses and increasing use of non-fossil energy sources; halting the destruction of forests; and increasing carbon sinks, including tree-planting and CO_2 scrubbing, to the extent feasible. While we are beginning to get an idea of what certain mitigation strategies will cost, we do not yet have a complete supply curve for CO_2 abatement strategies showing how much abatement is available at what cost.

However, we can attempt to describe the supply curve for CO_2 abatement. At negative and very low cost, there is some potential for reduction of CO_2 emissions through halting the destruction of forests. Some of the chaeaper or negative cost options of both of these strategies are currently being pursued, however there is potential for further reductions from these measures. Higher cost energy efficiency options are also available. We know also, based on our analysis in (Chernick and Caverhill, 1989), that reasonable tree planting costs fall in the range of 2-10 cents/lb carbon requested. The premium for non-fossil energy technologies ranges from a slightly positive (for some hydro, and some geothermal, wind and solar-thermal energy in California) to moderate, including CO_2 scrubbing at a cost of greater than roughly 10 cents/lb carbon, and expensive for nuclear, renewables located in less optimal sites, and other technologies.

Given the current uncertainty of the availability of cheap efficiency and renewable options, the limited potential for new hydro sites, the cost and general opposition to nuclear, and the moderate cost of some tree planting sites, we expect that some tree planting will be done as part of a CO_2 mitigation strategy designed to reduce ambient CO_2 levels by any meaningful percentage. For this reason we use the costs of abatement, in this case reasonable costs of tree planting, to estimate the marginal value of reducing CO_2 emissions.

Summary

In summary, externalities can be monetized for near-term use in utility regulation using at least one method, implied valuation, or marginal cost of abatement. Monetizing environmental externalities provides an approximation of the social value of reducing impacts on human health and the environment from electrical energy supply. The externalities estimates generated through the use of marginal cost of abatement ar conservative (low) estimates of the actual marginal values of reducing the externalities.

References

1. Chernick, P., and E. Caverhill. 1989. The valuation of Externalities from Energy Production, Delivery and Use. PLC, Incorporate, Boston.

2. --------. 1990a. Report to the Boston Gas Company on Including Externalities in 89-239.

3. Grupp, H. 1986. "Die sozialen Kosten des Verkehrs". Verkehr und Technik, Teil 1 und 2, in.

4. Hendrey, G.R. 1988. "Acid Deposition: A National Problem" Acid Deposition: Environmental, Economic and Policy Issues. Plenum Press, New York.

5. Hohmeyer, O. 1988. Social Costs of Energy Consumption: The External Effects of Electricity Generation in the Federal Republic of Germany. Springer-Verlag, Berlin.

6. Lashof, D.A., and D.R. Ahuja. 1990. "Relative Global Warming Potentials of Greenhouse Gas Emissions." Nature. 344(6266):529-531.

7. Ottinger, R. et al. 1990. Environmental Externality Costs from Electric Utility Operations.

8. Krause, F. W. Bach and J. Koomey. 1989. Energy Policy in the Greenhouse, Volume 1.

4.4 Integrating Energy and The Environment in The Marketplace

Daniel J. Dudek
Environmental Defense Fund
New York, New York
USA

1. Introduction[1]

Economists have long believed that the internalization of environmental externalities is a positive and effective policy prescription for managing these problems. Although internalization can be accomplished through a wide array of policy instruments,[2] economists have preferred the use of economic incentives. Most frequently these prescriptions have taken the form of charges or fees for emissions, in essence a special corrective tax.

While convenient in theory, effluent fees have not met with success in the U.S. The fee concept has not been able to overcome the daunting practical problems of political acceptability and implementation. In many cases the problem is typified by the ethical concern that the effluent fee policy would allow companies to "pay to pollute". In addition, the fee imposes an information burdon upon regulatory agencies similar to the requirements of the traditional command approach. The fee setting authority must know the shape and location of the aggregate marginal treatment cost curve and either the marginal benefit function or a socially accepted emissions reduction target. Alternatively, it must be able to experiment with different fee levels to trace out polluters' responses. For industries, like electric power, characterized by high fixed costs of pollution control, uncertainty in the fee level compounds the difficulty of the investment problem.

An alternative, based upon transferable permits dominated in physical quantities of pollutants has flourished in the U.S. however. Transferable permits have been applied to

---

1 The author gratefully acknowledges the support of the Joyce Foundation.

2 In fact, any policy which requires the emitter to take account of emissions would produce internalization. Command-and-control regulations requiring the use of specific control technologies is one extreme form of internalization.

everything from land development rights to atmospheric discharges (Tripp and Dudek, 1989). Marketable permit programs require the regulatory authority or the political process to explicitly define the goals of the pollution control policy in terms of allowable emissions. Once total levels are defined, emissions are parcelled out to individual emitters. In contrast to fee or charge systems, the market in these transferable permits performs the valuation function, not the regulator. The regulatory authority records transfers of title and monitors compliance.

The tremendous political appeal of these systems is based upon the policy's focus on environmental objectives. At their root is a solid regulatory program which prescribes emissions reductions to each source. Sources are free to determine their individual control strategy, including purchasing extra reductions from more distant sources, but they are not free to avoid their control obligations.

The electric power sector has been the source of significant and persistent environmental externalities. This paper will focus on those externalities associated with the discharge of transboundary atmospheric pollutants. In particular, the design and operation of a marketable permit system for SO_2 emissions in the U.S. developed by the Environmental Defense Fund will be discussed. This policy approach will then be extended to CO_2, perhaps the greatest environmental challenge confronting the electric power industry.

2. Short History of Emissions Trading

Emissions trading is actually a generic term for a collection of flexible ar pollution policies created to overcome the conflicts produced by the failure to meet air quality goals in urban areas in the U.S. in the late 70's. The EPA's final emissions trading policy includes provisions for bubbling, offsetting, and netting. Tietenberg (1985) provides an excellent overview of the development of these policy reforms for the management of air emissions. Over time, these basic concepts have been applied to a wider number of pollutants and circumstances. Emissions trading is ideally suited for the management and transboundary pollutants such as acid rain.

The concept of emissions trading is based on the premise that pollution reductions can be accelerated if polluters are allowed to optimize pollution control investments. The evidence from the experience with trading clearly shows significant cost savings for emitters however. Both Hahn and Hester (1986) and Liroff (1986) in their reviews of the performance of netting and bubbling respectively observed no significant deterioration in air quality. However, these analyses focused on narrow local air quality problems and on the directly regulated individual pollutants. For large markets, such as acid rain, early reductions to create banked credits and the stimulus to innovation in emissions reduction are expected to produce significant environmental benefits. Flexible control policies also create the opportunity to reduce cost-effectively more than one pollutant at a time. For example, Dudek, et al. (1990) identified significant CO_2 reduction benefits from an acid rain emissions trading program. Cost-effective SO_2 control investments could eliminate between 5.4-6.0 % of expected utility CO_2 emissions over the decade 1995-2005.

The emissions trading system operates quite simply. Polluters are assigned a baseline for each emissions source which establishes the initial conditions from which reductions can be assessed. Baselines are established by regulation by the appropriate air quality management authority. The polluter must meet the minimum standards in force for each pollutant at each source. However, these regulatory obligations can be met by over-controlling at another emissions point and transferring the over-control. The flexibility provided by the ability to transfer over-control is the heart of emissions trading. It encourages firms to search for cost-effective reductions. When coupled with the prospect of trade outside the firm the incentive of revenues for emissions reductions is added. Institutional innovations such as emissions reduction banking and recordation functions reduce the transactions costs of extra-firm transactions.

It is important to point out that a tradeable permit system is a particular policy tool built upon a regulatory program that establishes overall limits on emissions of CO_2 or, as a surrogate, limits on combustion of fossil fuel or deforestation. The trading system is designed to introduce flexibility in economic decision-making by allowing regulated sources to increase emissions above regulatory limits through the purchase of emission credits or to acquire income from the sale of such credits when they emit regulated pollutants below prescribed limits while encouraging investment in pollution reductions beyond minimum requirements. This is the essence of "harnessing the power of the market".

3. Acid Rain Emissions Trading

The U.S. Congress stands on the verge of enacting a sweeping reauthorization of the Clean Air Act. The most innovative aspect of this new legislation is the acid rain provisions. These provisions would require a 10 million ton reduction in SO_2 emissions from 1980 levels (see Figure 1), 8.9 million of which would be required from electric utilities. The reductions would be accomplished in 2 roughly equal 5 year steps.[3]

While there are differences between the House and Senate versions, both would use emissions trading as the main policy vehicle. Each would create virtually a national market for SO_2 emissions allowances. The essential ingredients of the policy are represented in the schematic in Figure 2. The first step is to determine a baseline for each source. In this case, historic fuel use in btu's was used. The baseline is multiplied by an allowable emissions rate per Btu to yield an annual SO_2 tonnage allowance which is assigned to each utility plant by permit. In effect, this can be thought of as the initial deposit into a checking account which the plant draws upon as it discharges pollutants. Discharges are monitored in real time through continuous emissions monitoring systems (CEMS) which are required for all sources within the system.

The regulatory agency, in this case the Environmental Protection Agency, assures compliance by comparing each plant's actual emissions as tracked by CEMS with the annually permitted emissions. Plant emissions beyond annual allowances that are not compensated by reductions at other plants are subject to a \$2000/ton fine[4]. In addition, the offending source must produce additional reductions in order to compensate the environmental account. The combined financial implication of these penalties is 2-3 times the highest marginal cost of control. These penalties also depart from previous practice in that they are designed like income tax bills which are due and payable upon receipt. In the past, polluters were able to defer fines pending adjudication before an administrative law judge. Frequently this was a protracted process with time and inflation eroding the economic force of any penalty. In addition, penalties were based upon a judicial interpretation of any profits made as a result of the violation.

The relatively draconian penalties in the system are feasible due to the incredible fexibiity granted to SO_2 sources. For example, the ability to over-control at one plant and do nothing at another means that utilities will be able to optimize pollution control investments on a system-wide basis. In effect, SO_2 emissions characteristics by plant will be another criteria under which plants are dispatched. The first form that trading can be expected to take then is intra-utility reallocation of allowances. Research indicates

---

3 Since the precise form of the legislation is not known currently, the reduction paths shown in Figure 1 are only approximate.

4 The level of the fine is automatically adjusted over time according to changes in the consumers' price index in order to maintain its economic force.

Figure 1: SO_2 emissions in the U.S.

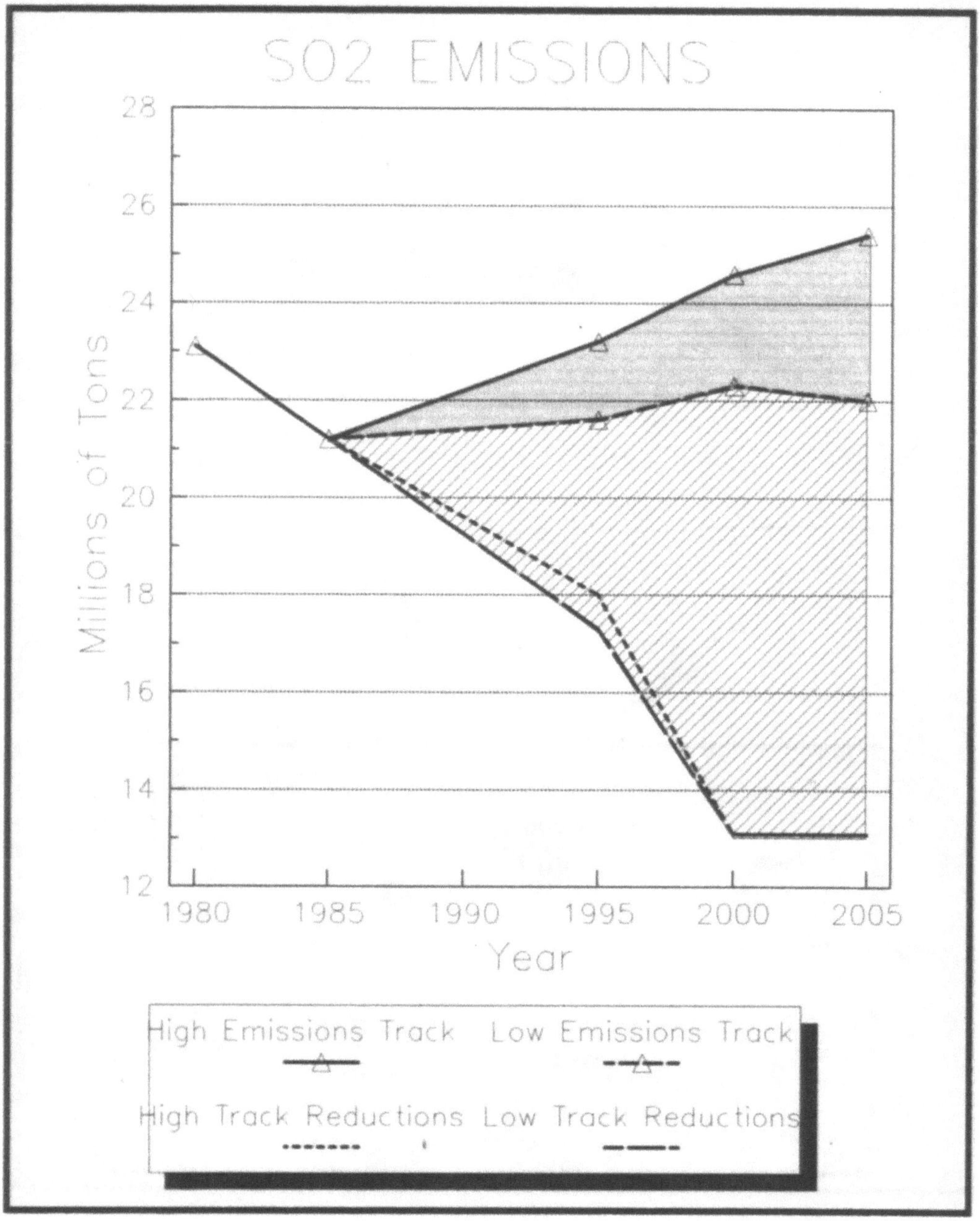

Figure 2: Elements of emissions trading

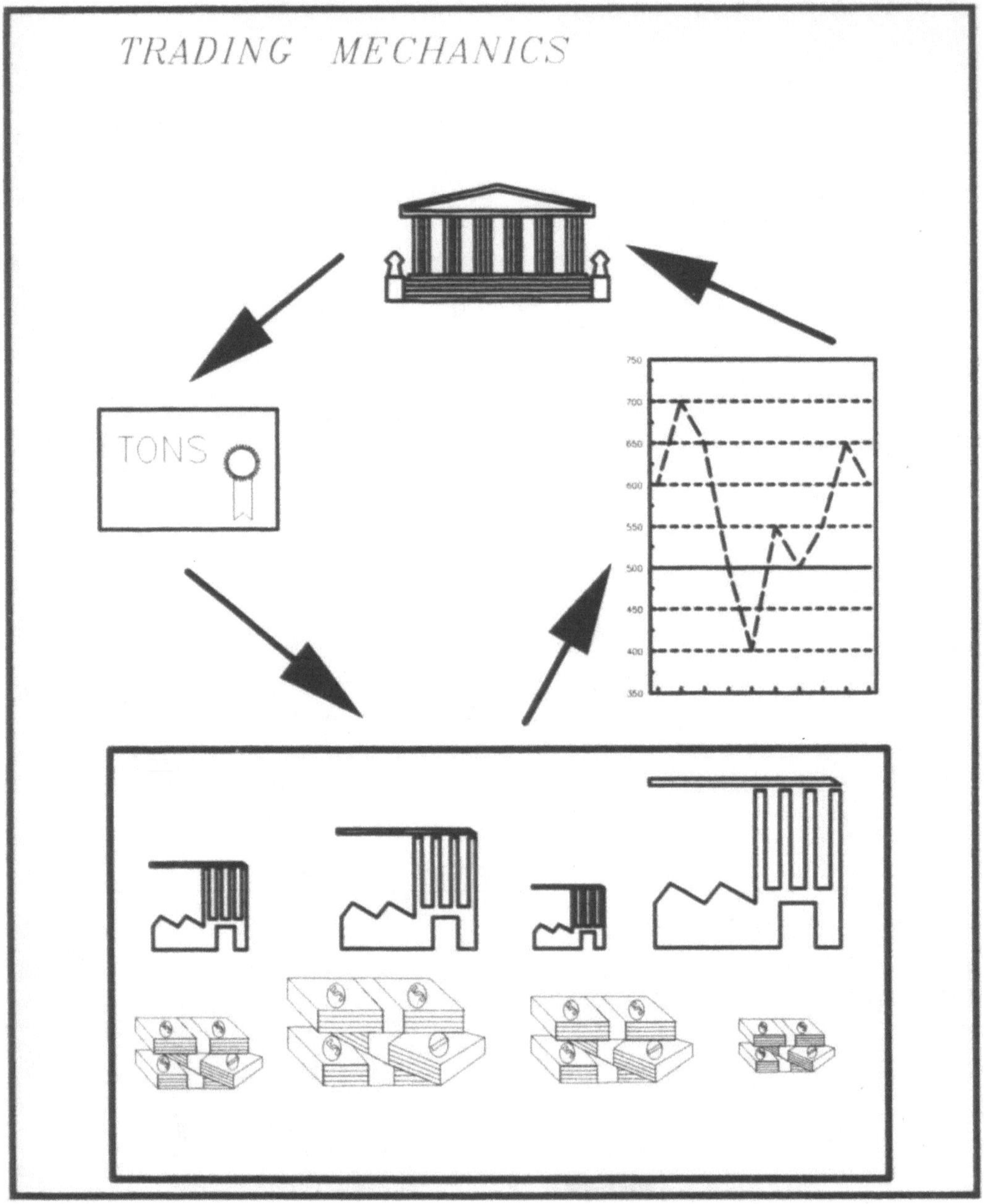

that even if no interregional trading occurs, the cost savings from trades will reduce overall program compliance costs by at least 50 %.[5]

One of the other unique features of the program's design is a nationwide cap on total SO_2 emissions. The cap is designed to ensure that once the required aggregate emissions reductions are achieved that they are maintained. This is accomplished through the requirement that all new plants must offset completely their proposed emissions. In effect, new plants must buy into the system. Concern has been expressed that existing sources would hoard allowances and prevent the competition that new entrants would bring by denying them access to allowances. As a stopgap against such possibilities, there are precedent setting auction provisions. Between 2 - 5 % of the annual allowance entitlements will be withheld from existing sources and auctioned. Auction revenues are cycled back to the sources from which the allowances were tapped. The annual auctions are open to anyone.[6]

Clearly, the combined force of these provisions is to generate strong incentives for the development and use of low or non-polluting generating technologies and fuels. In particular, the offset provisions are likely to pressure new sources to construct that exactly optimize between the value of SO_2 emission credits available in the market and the marginal costs of control technology. Since new plants will likely need to demonstrate access to offsetting emissions reduction for several years of operation in order to receive debt financing if not permitting from the air pollution control authority, this will be a powerful internalizing force for new generating capacity.

The emissions reductions potential of this program reaches beyond the precursors of acid rain however. Figure 3 depicts projected increases in utility CO_2 emissions over the next decade. Both a high and low emissions scenario were developed from EPA data (EPA, 1989). The scenarios differ not in overall economic activity but in assumptions concerning energy fundamentals such as the relative price of oil and natural gas and the rate of energy

---

5 Because the acid rain emissions trading program will create a market, precise estimates of compliance costs are difficult. The market responses of individuals to the opportunities provided by the program will undoubtedly drive costs lower over time.

6 One of the proposed auctions would create in effect a futures market. Utilities making SO_2 control investments could offer for sale in a Department of Treasury sponsored auction future emissions reductions. The Treasury would pass the results to EPA which would debit and credit the allowance accounts of the buyer and seller. This institutional arrangement would reduce the risk of non-performance on two party contracts.

demand growth. In addition, energy efficient technologies such as cogeneration are much more aggressively exploited in the low base case.

Figure 3: Utility CO_2 emissions

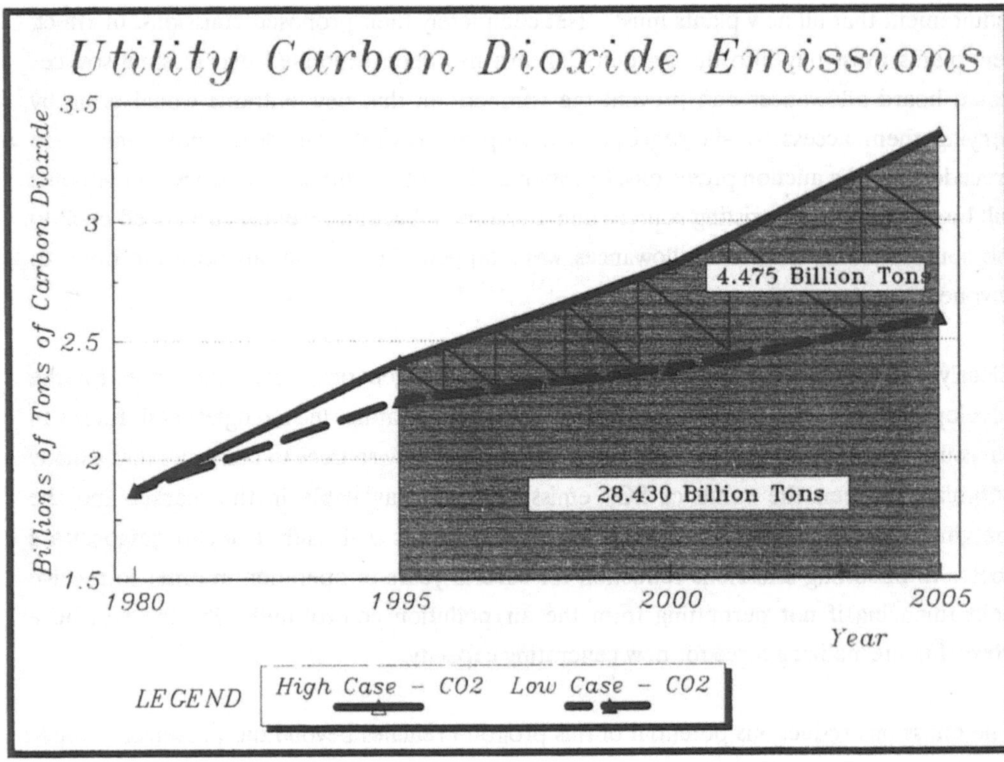

Since the CO_2 profiles of these two scenarios are so radically different, the question naturally arises as to what policies can help society achieve the low base case? One way would be for emitters to be required to take account of all of their emissions and optimize their operations over all costs. Unfortunately, no policy is so systematically rational. Pollution control policies tend to be focused on a pollutant-by-pollutant basis without concern for any synergy or antagonism. Nonetheless, policies do differ in their effects on several pollutants despite the ostensible focus on a single pollutant. For example, in the case of acid rain controls, requiring the use of flue gas desulphurization devices will likely add to CO_2 emissions. However, if utilities were left to determine their own control strategy, cost-effective SO_2 controls that also reduced CO_2 could be undertaken. Table 1 summarizes the expected results for CO_2 if utilities are free to choose control strategies as under emissions trading (Dudek, et al., 1990).

Table 1: The CO_2 Balance Sheet

SOURCE of REDUCTIONS	CUMULATIVE AMOUNT (mm tons)	SOURCE of INCREASES	CUMULATIVE AMOUNT (mm tons)
Resid to Natural Gas	411.440-459.600	Added Scrubbing	30.480-124.700
Cogeneration	423.920	Sub-Bituminous Coal	10.010-19.200
Demand Response	128.850		
Conservation	728.620		
TOTAL	1,692.78-1,740.99		40.580-143.900

Cumulative CO_2 Reductions 1995-2005 --
1,548.88 - 1,700.41 Million Tons

More flexible policies such as emissions trading can produce reductions in several pollutants simultaneously because sources are not prevented from optimizing over several pollutants as is the case with the normal command-and-control rather than exclusive. However, if we are not left to rely solely upon the powers of strategic positioning to elicit such investments, opportunities to grant regulatory credit for simultaneous reduction must be created. Unfortunately, this approach also cuts across the normal regulatory grain of expropriating any serendipitous reductions for the environmental account.

4. Alternative CO_2 Trading Systems Designs

Before describing a trading program at either a national or international level, the regulatory framework must be described. Initially, as a simplifying assumption, this paper considers only one greenhouse gas - carbon dioxide. If, through an international convention/protocol or even a series of bilateral conventions, nations want to achieve controls on emissions of CO_2 one approach, based on the Montreal Protocol experience, would be for those countries to establish goals and then allocate the total loadings authorized by the goals among the cooperating nations. The goals could be developed on the basis of historic data as in the case of the U.S. acid rain trading program or they could be arbitrary based upon factors such as GNP and population. These goals and allocation system must take into account environmental objectives, economic efficiency needs and international equity concerns.

For an international agreement to achieve limits on global CO_2 emissions, it must establish overall emission and/or fossil fuel use reduction goals, reduction schedules or formulae, allocation formulae, monitoring, reporting and enforcement provisions and mechanisms for financing and technology transfer. Such an agreement must do all of these things whether or not it provides a mechanism for trading of CO_2 emission credits among sources in different nation states. Of course, CO_2 trading could be implemented on a separate national basis irrespective of whether any international convention/protocol authorized international trading. One of the most likely and useful examples for allocating national environmental performance responsibilities is given by the Montreal Protocol. By stating national requirements in terms of consumption requirements, all nations would have a role within the framework of the agreement irrespective of whether they produced significant fossil fuels or not. In addition, the market power that the main fossil fuel producers might otherwise wield would be diluted.

There are a large number of alternative designs for a greenhouse gas trading program. Only two specific alternatives are described in this sort discussion paper. Two polar extremes are discussed. One option would allocate permits to fossil fuel producers while the other would distribute permits to end users. In both cases the limiting, but simplifying assumption of CO_2 trading is introduced. This discussion pertains both to national as well as international trading systems.

The first case involves allocating permits to fossil fuel producers. The second would assign permits to fossil fuel end users. The designs differ largely in effect, extent of trading, and administrative ease.

4.1 Allocation to Fossil Fuel Producers

Allocating permits to fossil fuel producers as indicated in the diagram below, would restrict the bulk of trading activities to energy producers and sink managers. In the case of greenhouse gas trading, the capture and destruction of banked CFCs and halons would also be a valuable source of reductions as would the recovery of coal-bed methane. In both designs the control point for compliance is the fossil fuel producer who would be required to turn in a sufficient quantity of permits to cover annual production.

End users would perceive the program in economic terms much as a tax on carbon, i.e. they would face higher energy prices. In fact, users would be offered energy at two prices - one price with the permit bundled in and another with the permit to be supplied by the fuel purchaser. Buyers would evaluate the costs of generating new permits versus the price differentials charged by the energy producers.

As indicated above, the permit program, although behaving somewhat as a tax policy in effect, would nonetheless differ in at least two critical aspects. First, the "tax" would be determined by the market in permits, i.e. by the decisions of private actors rather than through legislative or bureaucratic means. Second, sinks are encouraged.

Figure 4: CO_2 permits to producers

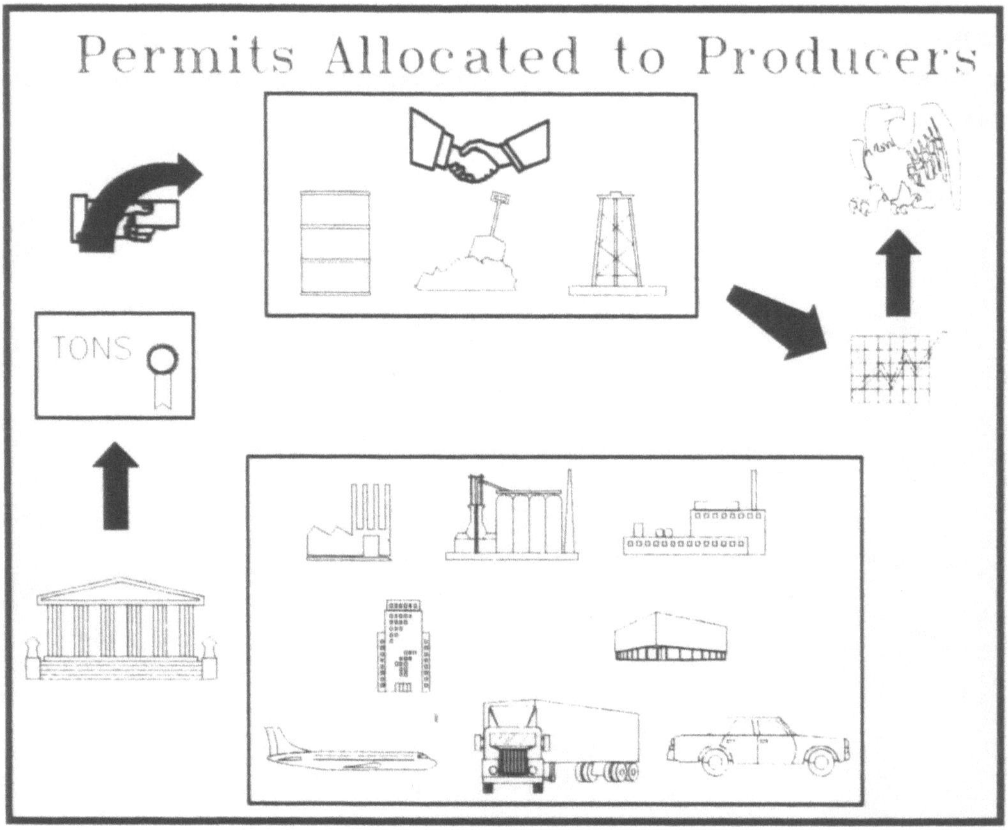

4.2 Allocation to End Users

An alternative design for a CO_2 trading program would allocate permits to end users of fossil fuels as depicted in the figure below. The big liability of this approach is the creation of baselines for all of the source involved. Even if a line were drawn at major commercial and industrial consumers, literally thousands of firms' records would have to be evaluated. In effect, the baselines would be consumption entitlements denominated in CO_2 equivalents. The big asset to this approach is the direct stimulus to energy efficiency. Any energy use reduction below the individual baseline is by definition a credit that can be transferred. This direct revenue mechanism would be a dramatic boost to fossil fuel conservation efforts.

This design would still encourage sinks and reductions in other greenhouse gases for the purposes of trading. Fossil fuel producers could still offer two tiered prices, but the main permit supplies would come from end users who would pass permits covering their purchases to producers. Compliance would still be measured at the point of production of fossil fuels.

Figure 5: CO_2 permits allocated to users

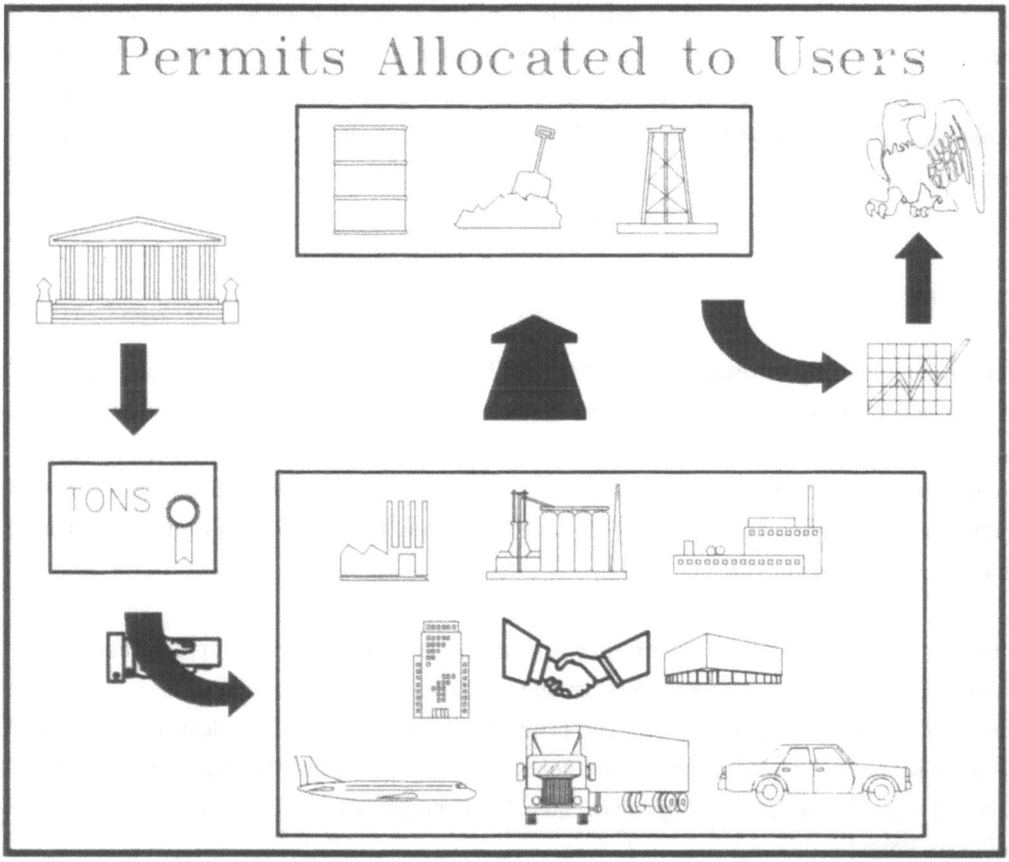

5. Application to CO_2

The primary advantages of using marketable permits were distilled recently in the process of the IPCC discussions concerning response policies. Those identified included:

flexibility
direct control of emission levels
efficiency in pollution control
mechanism for trading among different greenhouse gases
incentives for development of non-or low-polluting technologies.
(Working paper by Bertram et al. For IPCC/WG3-II/Doc 3)

In addition, permits would decentralized pollution control decision-making commensurate with the numbers of sources involved. An international CO_2 (or greenhouse gas) trading program also can accommodate a variety of separate national implementation strategies and policies. No nation would be compelled to achieve its control obligations through trading. Lastly, by linking sources and sinks in one comprehensive strategy, an explicit mechanism of self-interest is established between along the often divisive North and South axis.

5.1 Sink Stimulus

CO_2 trading is likely to produce the same proliferation in supply options as generated in the deregulation of the financial services industry. The huge variety of precisely tailored investment vehicles available in the current market are likely to be mirrored in the CO_2 (or greenhouse gas) permit markets. After all, a CO_2 trading program would create a new form of environmental currency. One direct effect would be the stimulus to develop sinks as well as source reductions.

Since tropical forests are a major sink for carbon and their burning or destruction at current rates constitutes some 15 % of global CO_2 loadings, the question is how an international greenhouse agreement can promote the retardation of the destruction or the re-establishment of any forests and whether a trading program can facilitate attainment of those goals. Melding these objectives may be complicated by the development of a separate Forest Protocol.[7]

The destruction and burning of tropical forests in countries such as Brazil and Indonesia contribute a significant fraction of their CO_2 budgets. Thus, if such countries allocated their CO_2 emissions targets proportionately to the forest sector and fossil fuel sectors, Indonesia would have to take steps to prevent increases in the destruction of its forests: Brazil would have to implement a program to decrease the rate of loss of its forests by 2 % per year.

If tropical forests acres can be converted into CO_2 and energy efficiency equivalents, countries like Brazil and Indonesia might find it economically attractive, as well as environmentally sound, to retard the destruction of their forests even faster or to implement restoration programs where, in the process, they can earn CO_2 credits which their own industry could use or which they could sell to foreign firms or governments.

---

7 The Houston Summit meeting of the G7 countries in July of 1990 ended with a commitment to develop such a protocol.

Those countries in turn could use the revenues from the sale of such credits to finance programs to retard forest loss. A trading program authorized under an international convention/protocol could therefore provide such countries with substantial economic incentives to protect their forests over and above the regulatory mandates of such a convention/protocol.

While a convention/protocol should encourage the creation and protection of carbon sinks, such as forests, defining quantitatively the magnitude, size and rates of loss of those sinks is an enormous undertaking. Satellite monitoring is a critical tool But this effort must be made whether or not international trading of credits involving sinks is authorized so long as CO_2 emissions from destruction or creation of such sinks is incorporated into a convention/protocol.

One of the primary advantages of greenhouse gas trading is the fact that it is quantity denominated. Everyone can readily determine what is required and what has been accomplished. The program is denominate in environmental terms. There are no conflicts over revenue and its distribution internationally. It is completely consistent with any underlying national implementation policy. It is voluntary, no one is compelled to trade. Depending upon the initial distribution of CO_2 permits (reduction responsibilities), it also contains an explicit mechanism for North-South flows of funds. Tropical forests in less developed nations would be linked to fossil fuel combustion in developed nations.

6. Implementation Issues

International trading in greenhouse gas permits can be established in a limited number of coordinated phases. The first phase would build the infrastructure necessary to facilitate such a global market. This phase would be devoted to the creation of international monitoring systems, tracking and recording systems, and market clearing-house. The second phase could be a limited permit system more narrowly focused as an offset program targeting new sources. The final phase would be the development of full scale trading perhaps only for CO_2 initially with other gases included as the capacity of the trading system is expanded. The phased sequence above could be easily pursued by individual nations irrespective of any international greenhouse gas control agreement. In the event of an agreement with significant reduction schedules, the second and the third phases would be collapsed into one. In any event, the administrative apparatus required to facilitate international greenhouse gas trading is no more complex than that required for any other effective control system and would impose no time delay in implementation.

For an international convention/protocol that imposes limits on emissions of CO_2 from burning of fossil fuels and destruction of forests to function effectively, the participating nation states must be willing to and be able to obtain and report reliable information about fuel use and forest loss. The convention/protocol must also provide for a credible enforcement mechanism and the establishment or strenghtening of an international agency to receive and assimilate data and to help assure compliance.

If the convention/protocol authorizes bilateral or multinational trading of CO_2 emission credits, that international agency or some international body would have an additional role to play. More specifically, at the national level, any country with a domestic trading program would have to have in place an administrative mechanism for certifying that individual firms were entitles to credits or had to acquire credits to comply with standards. If those credits wee to be tradeable to a source in another country, some international agency should have the authority to record those credits and to assure that the national credit certification program was viable and accurate. Since a country might have an incentive to inflate its eligibility for credits, this international agency would need additional resources to assure reliable national reporting of information and valid credit issuance programs.

A trading system can also help to assure the integrity of any system of emissions limits. As industrial enterprises - or even governments - attempt to comply with CO_2 limits, technical or economic factors will inevitably prevent some from complying at some times. Without a trading system, a variance system (formal or simply by non-enforcement) will inevitably develop. A formal system will in turn require technical or economic criteria for evaluating variance requests and bureaucracies to apply those criteria. Inevitably also, those bureaucracies may not discipline needed economic enterprises for the sake of what may appear "temporary" violations of CO_2 limits. The problem could lead to distrust between nations and diminishing willingness to make needed sacrifices. By allowing dischargers the flexibility of choice in reduction decisions and the ability to bank extra reductions, there is no credible justification for violation of emissions limits.

A trading system should eliminate the need for variances by allowing industries and nations that cannot comply with limits to purchase credits. No criteria would be necessary to evaluate variance requests, and the sole task for bureaucracies would be the registration and verification of the transfer. Enforcement officials would not need to negotiate complex compliance orders, but simply to order the purchase of credits. Enforcement could also be more credible because the purchase of credits should often be less draconian than shut-down of enterprises. In addition, the failure of some enterprises to comply with initially set mechanism will replace bureaucratic judgement, those enterprises that do not comply with initial limits should do so because of real economic needs not political influence.

6.1 The Offset Program

One of the first direct implementation steps for CO_2 trading is an offset requirement for new CO_2 sources. This is a program which the major industrialized nations should undertake unilaterally as a first rational step toward the significant reductions required. It is a minimally step which can be established now without any regulatory lag or lead-in. Major new sources would include new fossil fuelled electric generating stations, new industrial boilers, and cogeneration facilities selling power into regional electricity grids. Such a program would:

* begin reduction of accumulation of greenhouse gases in the atmosphere
* establish administrative mechanisms to facilitate trading
* stimulate private market development of CO_2 reducing technologies.

The heart of the program is the ability to create and transfer CO_2 credits. Any CO_2 reduction meeting the certification would constitute a legitimate CO_2 credit available for offset. The program would create explicit procedures for the certification of several identified offset credit sources. In addition, these procedures would allow any source to receive credit for CO_2 reductions for any such reductions that can be demonstrated to have been explicitly created for this program and which constitute CO_2 reductions that would have normally been emissions.

Examples of sources of CO_2 offsets include energy efficiency investments, vehicle fuel efficiency improvements beyond the Corporate Average Fuel Economy (CAFE) baseline, appliance efficiency improvements beyond the mandated standards, fuel switching to less CO_2 intensive fuels, and the planting of trees (see Dudek and LeBlanc (1990) for offset costs). Putting the burden of proof of new reductions upon the potential offset supplier allows the program to be open-ended and to encourage the widest diversity of offset sources to be developed, demonstrated, and engaged. These sources should include other greenhouse gases, e.g. methane, and N_2O. The greenhouse trading ratios to represent radiative differences between gases could be established in consultation with WMO or any other internationally recognized atmospheric scientific authority.

Certification for some sources will require the development of specialized rules and methods. For example, in the case of automobiles, reductions beyond the CAFE standard would be translated into a quantity of CO_2 through a statutory assumption concerning the actuarial life of a vehicle. In the case of trees, the uptake rates of individual species by location and soil type would need to be specified. Improved uptake rates could be prospectively established through demonstration projects. For example, new more intensive management procedures could improve uptake rates and should not be precluded from being developed.

In order to ensure that a sufficient supply of offset credits is available, an offset bank could be created and maintained at the appropriate environmental authority. Sources generating reductions can be certified for credit as soon as the reductions are produced. These reductions can then be maintained as an available supply. This bank can function like an information exchange identifying the owners of certified reductions. Information maintained would be contact, quantity and minimum bid price.

For some offset supplies such as trees, it will be critical to assure that the CO_2 sequestering value of the offset is maintained. For example, for trees, it is critical that the stored carbon credited as an offset is maintained as stored carbon. For example, if a forest used for an offset is subsequently harvested, only those amounts of wood transferred into a relatively permanent use like housing stock should be credited with maintaining the stored carbon.

Each year, each source subject to the offset provisions would be required to deposit with the environmental authority sufficient CO_2 offsets to fully compensate for its CO_2 emissions in that year. Each year all sources regulated by the program would submit a report of CO_2 emissions. Each source would be required to adopt CO_2 emissions monitoring or a monitoring method of equivalent accuracy. Each year, the environmental authority would publish a report of such emissions.

Sources failing to comply would be subject to a penalty per ton of CO_2 not covered by an offset. The penalty should be adjusted by the producer price index in order to assure its continued economic force over time. In addition, the emitter would be required to submit a plan for the production of offsets to compensate for the uncompensated emissions.

7. Conclusions

In the debate about greenhouse gas policy much energy has been expended in the prices example, if the penalty for failing to hold sufficient CO_2 permits is a fine levied per unit of excess CO_2 discharged as in the acid rain case, then that fine becomes the upper limit of the price for CO_2 permits. On the other hand, if not only a fine is levied but a requirement to produce a compensating CO_2 reduction, then there is no price ceiling and the systems do not necessarily diverge. As indicated in the discussion on alternative designs, when permits are issued to fossil fuel producers, the tradeable permit program behaves exactly like the carbon tax except for one crucial difference. Under the permit system, the level of the tax is set by the market and varies continually depending upon market conditions and not upon political or bureaucratic whim.

International greenhouse gas permit trading has several critical advantages unmatched by alternative policy options. First, marketable permits would provide the greatest spur to innovation in both low polluting technologies, emission control options, and pollutant sinks. This substantial economic incentive for forest protection is an important and often overlooked feature of any greenhouse policy. Providing the option of producing additional permits over time through forest management changes lessens the impact of greenhouse gas control on economic performance. Permits also provide an explicit mechanism for reallocating scarce resources, namely CO_2 permits, over time and space. Certainly fixed international obligations could be periodically renegotiated, but it is difficult to imagine that such a process could even begin to approach the efficiency of a genuine global market. Lastly, a system of marketable permits provides all parties with a relatively simple means of determining performance. The sum of permits transferred to control authorities at any one point in time measures total emissions.

In the international context, the primary gains from trading include:

1) powerful incentives for extra reductions
2) stimulus to technical change and innovation
3) creation of a mechanism for international transfer of reductions reducing the burden of any individual emissions limit
4) consistency with any national control program.

International trading in greenhouse gases would build directly upon the successful model of the Montreal Protocol. The administrative apparatus required to implement trading is no different than that required to track CFC production for example. The administrative requirements to track and monitor any greenhouse gas protocol will be very similar. The definition of control obligations within the Montreal Protocol include the tracking of exports and imports as well as conversion factors between different CFCs to account for differences in ozone depletion potential. The Montreal Protocol has provisions which allow or the economic rationalization of production which by any other name is trading. nternational greenhouse gas trading would function in a like manner.

3. References

Albanese, Anthony S. and Meyer Steinberg, "Environmental Control Technology for atmospheric Carbon Dioxide", Brookhaven National Laboratory, report DOE/EV-0079, May 1980, 51 pp.

Dudek, Daniel J., "Chlorofluorocarbon Policy: Choices and Consequences", Environmental Defense Fund, New York, April 1987, 26 pp.

Dudek, Daniel J. and Alice M. LeBlanc, "Offsetting New CO_2 Emissions: A Rational First Greenhouse Policy Step", *Contemporary Policy Issues*, Vol. VIII, July 1990, pp. 29-42.

Dudek, Daniel J. and John Palmisano, "Emissions Trading: Why is this Thoroughbred Hobbled?", *Columbia Journal of Environmental Law*, Volume 13, Number 2, pp. 217-256.

Dudek, Daniel J., Alice M. LeBlanc, and Peter Miller, "CO_2 and SO_2: Consistent Policy Making in a Greenhouse", Environmental Defense Fund, January 1990, 31 pp.

Elman, Barry, "Economic, Environmental, and Coal Market Impacts of SO_2 Emissions Trading Under Alternative Acid Rain Control Proposals", U.S. Environmental Protection Agency, Washington, D.C., March 1989.

Flavin, Christopher, "Slowing Global Warming: A Worldwide Strategy", Worldwatch Paper 91, October 1989, Table 7, p. 69.

Goldemberg, Jose, "Brazil's Small Share of the Greenhouse", *New York Times*, July 29, 1990.

Grubb, Michael, "The Greenhouse Effect: Negotiating Targets", Royal Institute of International Affairs, London, 1989.

Hahn, Robert and Gordon Hester, "Where Did All the Markets Go? - An Analysis of EPA's Emissions Trading Program", *Yale Journal on Regulation*, Volume 52, 1986, pp. 48-49.

Hahn, Robert and Gordon Hester, "Marketable Permits: Lessons for Theory and Practice", *Ecology Law Quarterly*, volume 16, no. 2, 1989, pp. 361-406.

Hahn, Robert and Robert N. Stavins, "Incentive-Based Environmental Regulation: A new Era for an Old Idea?", Energy and Environmental Policy Center Discussion Paper, Harvard University, June 26, 1990, 40 pp.

Liroff, Richard, *Reforming Air Pollution Regulation: The Toil and Trouble of EPA's Bubble*, Conservation Foundation, Washington D.C., 1986, pp. xvi-xvii.

Mellilo, J.M., and J.R. Gosz, "Interactions of Biogeochemical Cycles in Forest Ecosystems", in *The Major Biogeochemical Cycles and Their Interactions*, edited by Bolin, Bert and Robert B. Cook, SCOPE 21, John Wiley, 1983, pp. 177-222.

Monastersky, R., "Nations to Ban Ozone-Harming Compounds", *Science News*, vol. 138, July 7, 1990, p. 6.

Nordhaus, William D., *The Efficient Use of Energy Resources*, Cowles Foundation Monograph 26, Yale University, New Haven, 1979, 161 pp.

Nordhaus, William D., "How Fast Should We Graze the Global Commons?", *American Economic Review*, vol. 72, no. 2, May 1982, pp. 242-6.

Nordhaus, William D., "Count Before You Leap", *The Economist*, July 7, 1990, p. 22.

Shapiro, Michael, and Ellen Warhit, "Marketable Permits: The Case of Chlorofluorocarbons", *Natural Resources Journal*, vol. 23, July 1983, pp. 577-91.

Tietenberg, T.H., *Emissions Trading: An Exercise in Reforming Pollution Policy*, Resources for the Future, Washington D.C., 1985, 222 pp.

Tripp, J.B.T. and D.J. Dudek, "Institutional Guide-lines for Designing Successful Transferable Rights Programs", *Yale Journal on Regulation*, Vol. 6, pp. 381-403, June 1989.

Tripp, J.B.T. and D.J. Dudek, "Comments of the Environmental Defense Fund on the IPCC Working Group III Economic Measures Paper", Environmental Defense Fund, New York, NY, January 1990.

U.S. Environmental Protection Agency, "Acid Rain Emission Base Case Forecasts", May 1989.

Victor, David G., "Tradeable Permits and Greenhouse Gas Reductions: Some Issues for U.S. Negotiators", Department of Political Science, Massachusetts Institute of Technology, February 22, 1990, 61 pp.

Whittaker, R.H. and G.E. Likens, "Carbon in the Biota", in Woodwell, G.M. and Pecan, E.V. (eds) *Carbon and the Biosphere*, AEC Symposium Series No. 30, Springfield, VA, National Technical Information Service, 1973, pp. 281-302.

Woodwell, George M., "The Carbon Dioxide Question", *Scientific American*, Vol. 238, No. 1, January 1978, pp. 34-43.

World Resources Institute, *World Resources 1990-91*, Oxford University Press, New York, 1990, p. 144.

4.5 Valuation of Environmental Cost by Heat Emissions from Pollution Control

Reiner Kümmel and Uwe Schüssler
Physikalisches Institut der Universität Würzburg
Am Hubland
D-8700 Würzburg

1. Introduction

Recently, several attempts have been made to estimate the external costs of the West German energy system, i.e. the harmful effects of energy conversion processes in the country not charged to the producers but to third parties or the general public. The studies of Herz et al. [1], Wicke [2], Hohmeyer [3] and Friedrich et al. [4] indicated annual pollution damages of many billions Deutsche Mark. They also showed how sensitively external cost estimates depend on controversial valuation criteria [3,4]. Apart from the difficulties of valuating in monetary terms very different risks like radiation from nuclear waste, on the one hand, and global warming because of carbon dioxide (CO_2) emissions from fossil fuel burning, on the other hand, there is the big problem of how to determine the full intertemporal external costs of a given energy system, i.e. the opportunity cost to future energy users resulting from limitations on energy consumption [3,4]. The important and theoretically novel problem here is that limitations will not only result from depletion of energy resources but also from the finite capacity of the environment to absorb the pollution associated with energy conversion. The seriousness of this problem is illustrated by the greenhouse effect and the limits it may impose on the burning of carbon which presently satisfies about 60 % of the world's energy needs (whereas approximately 30 % of the world's energy consumption is covered by the oxidation of the hydrogen in the fossil fuels): In order to stabilize the climate the "fossil carbon budget for the period from 1985 to 2100 (should)... be about btC (billion tons of carbon...), equivalent to about 55 years of fossil releases at current rates" [5]. Thus, the environmental limit on the use of carbon in fossil fuels seems to be much closer than the limit from the depletion of all coal resources.

Given these problems of risk and opportunity cost assessment in monetary terms it has been proposed to develop a thermodynamic pollution indicator for environmental accounting [6-8]. For each energy conversion process this indicator balances the environmental burden, which cannot be avoided because of fundamental laws of physics, on a unified heat scale and shows the approach to the ultimate limit to energy use drawn by the "heat barrier" (see below). In this paper we describe the construction of the pollution indicator, considering the example of electricity generation, and indicate how it may help to develop monetary incentives for energy conservation, fuel substitution and pollution control.

2. Heat Equivalents of Noxious Substances

Pollution is the emission of noxious substances and heat by human activities. A substance is considered noxious, if it is fed into the biosphere in quantities which significantly alter its natural concentration to which the living species have adapted themselves in the course of evolution and human history. Pollution control is the process which by suitable technical installations keeps the injection of noxious substances into the biosphere below limiting values. (These limiting values must be determined by risk assessment and social consensus, which is the harder part of the problem we rather don't touch here). Each pollution control process requires additional energy and produces additional (waste) heat emissions. Thus, with the help of the "Heat Equivalents of Noxious Substances" (HEONS) - defined as the waste heat generated by pollution control in a production process of given utility output divided by the primary energy input into a not pollution controlled production process of the same utility output [7,8] - one can aggregate the heterogeneous pollutants in terms of thermal pollution. Since thermal pollution should remain below $3x10^{14}$ Watts of anthropogeneous heat emissions in order to avoid severe climatic disturbances ("heat barrier"), the HEONS show the varying degrees at which pollution abatement with respect to different noxious substances accelerates the approach to the heat barrier. Thus, even in the most optimistic case of nearly perfect pollution control, an energy system in which the heat emissions from the production process itself and from pollution control add up to a big number has less expansion capacity in the future than one in which this sum is small. In this sense the total of the heat emissions is a unified measure of the opportunity costs of energy systems. Furthermore, the HEONS or the corresponding energy requirements of pollution control may play a similar role in assigning levies or taxes to energy carriers and conversion processes as do the oxygen equivalents of noxious substances in industrial sewages according to the German sewage law: The noxiousness of the water pollutants which can be neutralized by oxidization is measured in terms of the required amount of oxigen, which in turn determines the levy on the sewage.

The heat emissions from a not pollution controlled production process itself can usually simply be set equal to the heating value (enthalpy) of the total primary energy input. This can be determined by the methods of energy analysis [9]. In order to compute the HEONS for a given pollution control process a methodology similar to that of energy analysis has to be developed: One has to fix the physical and chemical process boundaries, decide which indirect energies and heat emissions shall be taken into account and establish a set of rules from which to calculate the energy and waste heat balancew [8]:

Physical and chemical boundaries of pollution control processes

Substances may be proper to the biosphere or noxious. By definition proper substances are present in their natural concentrations as a permanent passive background. If they are involved in pollution control processes their consideration in energy and materials balances begins and ends with the form they assume under standard environmental conditions. For instance, water vapor emitted into the environment during a process of pollution control is taken into account in the energy balance as waste heat given by the energy of condensation to water at the temperature of the environment. Similarly, the water vapor stream fed into a process is considered as an energy flow resulting from the heating of water (at initially environmental temperature) and its evaporation. Noxious substances, on the other hand, have to be traced in principle all along their way from entering the system, by being mined or produced in a chemical or nuclear reaction, until they leave the system by being deposited properly or transformed into proper substances. In practical calculations this tracing and accounting is stopped, if one arrives at emissions of quantities or concentrations which are below the limiting values determined and accepted by society.

Pollution control terminates with waste disposal. The following disposal options are admitted in the analysis:

a) Open disposal sites for inert and ecologically harmless substances in a suitable, non volatile form which is stable for long times under environmental conditions (e.g. calcium sulphate from desalination installations).
b) Closed disposal sites for inert but ecologically possibly harmful substances conditioned into non volatile forms stable for long times under disposal site conditions (e.g. heavy metal dust from power plant filters).
c) Deep space for aggressive, volatile, ecologically dangerous substances if society does not tolerate storage deep in the earth (e.g. plutonium and other radioactive waste).

Waste heat analysis in first order

This analysis determines the heat emitted in the pollution process itself and all subsidiary processes which provide the necessary materials and energies and remove and dispose of all noxious substances which may be generated during pollution control. However, the heat emissions due to the production of the pollution control installations, e.g. desulfuration plants or disposal sites, are neglected. (Of course, they can be included in a higher order analysis, if one invests more work.) This so defined waste heat analysis in first order corresponds to the second order energy analysis classification scheme proposed in 1979 by Boustead and Hancock it would correspond to a mix of first and second order, [9].

Results for energy and waste heat balances

In order to calculate the HEONS one needs the energy balances of all pollution control processes in first order waste heat analysis. The resulting heat emissions are assigned to the different noxious substances in a unique way according to the following rules.

Rule 1: Mechanical and electrical energy is directly transformed into waste heat, if it is used to surmount friction. It is also counted as waste heat, if it is transformed into potential energy which does not subsequently perform useful work but rather dissipates more or less slowly. The total waste heat related to the use of mechanical and electrical energy (= exergy), also includes all the waste heat generated during the conversion of primary energy into the considered exergy at known technical conversion efficiencies [10].

Rule 2: Chemical processes with endothermal reactions emit that fraction of process heat as waste heat which is not transformed into chemical binding energy. In exothermal chemical processes the reaction enthalpy determines the waste heat. An exception are chemical processes which occur right during primary energy conversion itself, e.g. in a furnace, because all heat involved here is part of useful heat.

Rule 3: The decrease of energy conversion efficiencies by pollution control shall be compensated, if possible, by increased primary energy input Q'. The resulting additional waste heat ΔQ of the production process enters the HEONS. It is determined by the initial primary energy or process heat Q_0, the conversion efficiencies η_0 without pollution control, and the reduced conversion efficiency η' with pollution control, under the condition of constant utility, i.e. $Q'\eta' = Q_0\eta_0$:

$$\Delta Q = Q'(1-\eta') - Q_0(1-\eta_0) = Q_0(\eta_0 - \eta')/\eta'. \quad (1)$$

If this compensation af a reduced conversion efficiency is not possible, one has to include in the heat equivalents the waste heat of an additional, pollution controlled production process which makes up for the diminished productivity.

Rule 4: The energy input into the transportation of materials within the biosphere is completely counted as waste heat, because it is all dissipated by friction some times or other.

3. Heat equivalents of chemical pollutants and radioactive waste in electricity generation

The outlined first order waste heat analysis has been applied to the calculation of the HEONS for pollution control in hard coal power plants and boiling water reactors [7,8]. As an example we discuss cryogenic CO_2 removal from conventional coal power plants in some detail, whereas for the other pollutants and radioactive waste we just give the results.

3.1 NO_x, SO_2 and CO_2 abatement

We consider a "state of technology" power plant fired by low sulfur hard coal. The flue gases of such a plant typically contain per normal cubic meter (m_n^3) 2000 mg sulfur dioxide (SO_2) and 1000 mg nitrogenous oxide (NO_x, computed on the basis of NO_2); carbon dioxide (CO_2) occupies about 15 % by volume of the total dry flue gas. The limiting values for power plants of more than 300 MW_{th} (megawatts thermal) presently valid in Germany are 200 mg NO_x and 400 mg SO_2 per m_n^3 after at least 85 % desulfurization. For smaller plants the limiting values are less stringent. For the "greenhouse gas" CO_2 there are not yet any limiting values. However, the German Physical Society and the German Meteorological Society [11] recently recommended that the CO_2 emissions should be globally reduced to one third of the present 20 Gt/year in order to avoid dramatic climatic changes. Thus, we assume that pollution control with respect to CO_2 has to be done in such a way that the present amount of electrical energy is available to the consumer at one third of the present CO_2 emissions.

NO_x is removed from the flue gas by a catalytic reaction which involves TiO_2 as a catalyst and ammonia (NH_3) as the reducing agent. The waste heat analysis shows that for 80 % NO_x removal the heat equivalent is 23.71 Megajoules (MJ) per Megawatthour thermal (MW_{th}), or **0.66 %** of the primary energy input into a plant of same electricity output without NO_x abatement. In a good approximation the waste heat is equal to the required primary energy.

Desulphurization by wet limestone scrubbing has a market share of over 90 % in West Germany. For 95 % SO_2 removal the waste heat analysis yields the heat equivalent of SO_2 to be 158.4 MJ per MW_{th} or **4.4 %**. In a good approximation the calculated waste heat is equal to the required primary energ.

Separation of CO_2 from the flue gases of conventional fossil fuel power plants has been studied theoretically on the basis of chemical binding and stripping with an alkanolamin based solvent [12,13] and refrigeration under pressure [13,14]. In integrated coal gasification plants physical absorption of CO_2 by a selexol solvent is possible after CO shifting of the flue gas [15,16]. The energy requirement for separation and liquefaction of 90 % of the CO_2 lowers the power plant efficiency from 38 % to 29 %, if one uses alkanolamin scrubbing [12,13] and from 44 % to 38 % in integrated coal gasification plants [15]. If, on the other hand, the chosen disposal option, e.g. injection of solid CO_2 from ships into the deep ocean, requires a freezing of CO_2, power plant efficiency drops to about 19 % in the first and 32 % in the second case (16). A discussion of the various CO_2 disposal options like storage in the deep sea or in empty oil and gas fields and carbon recycling by solar carbon technologies and biomass conversion into fuels is given in Refs. [17-19].

CO_2 removal by refrigeration under pressure has the advantage of not requiring large quantities of chemical or physical solvents and of delivering solid SO_2 without additional processes and energy inputs. In principle the required installations can be added to any conventional power plant with SO_2 and NO_x removal. There may be corrosion problems because of the remaining traces of SO_2, NO_x and other sour gas components, but the use of the special corrosion resistant materials may help. Whether these measures will in fact lead to overall expenditures comparable to those of chemical scrubbing as conjectured in [16] should be explored in a pilot plant. Such a plant is shown schematically in Fig. 1: The flue gas, sufficiently purified from all pollutants except CO_2, passes through cooling sections A and B where nearly all remaining water condenses. The following compressor C, driven by an engine M and an expansion turbine F, generates the pressure in the gas which is required by the chosen freeze-out temperature, the phase diagram of CO_2 and the pressure losses in the system [14]. In the cooling section D the flue gas is being divided into two streams which are cooled down to the inlet temperature of the separation unit E in a heat exchanger WT and a combination of water and heat pump cooling. The separation unit E further cools down the gas by heat pumping, until CO_2 freezes out, and removes the sublimation heat. The CO_2 lean, cold, high pressure gas is used as a refrigerant in the heat exchanger WT and then powers the expansion turbine F before it leaves the plant.

Figure 1: Process scheme for carbon dioxide removal by refrigeration under pressure

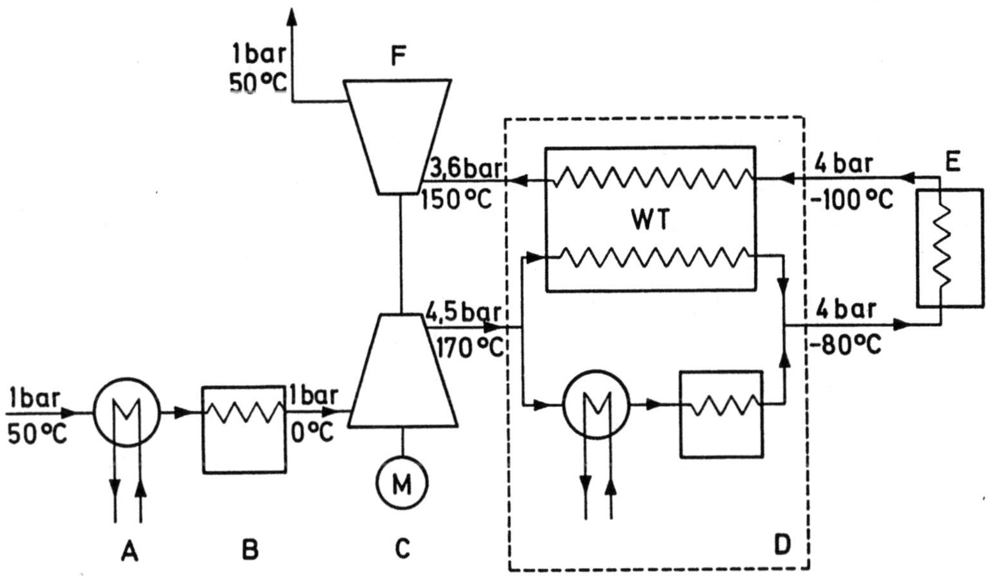

A typical coal power plant of 2000 MW_{th} has a flue gas flow of $2.4x10^6$ m_n^3 per hour, where 1 m_n^3 is defined as 1 m^3 dry gas at 1 atm. and 0 °C. The volume shares of N_2, CO_2, and O_2 are 69 %, 13 %, and 5 %, respectively, so that the initial concentrations are for nitrogen: $N_{N2} = 35.2$ mol/m_n^3, carbon dioxide: $N_{CO2} = 6.7$ mol/m_n^3, and oxigen: $N_{O2} = 2.7$ mol/m_n^3. If p is the total pressure of the flue gas which is kept constant in cooling section E, and $p_s(T)$ is the sublimation pressure of solid CO_2 at temperature T, the degree x of CO_2 removal (defined as the difference between the initial and the final CO_2 concentrations divided by the initial CO_2 concentration) turns out to be [14,18]

$$x = 1 - [(N_{N2} + N_{O2})/N_{CO2}] \times p_s(T)/[p - p_s(T)]. \qquad (2)$$

It is convenient to fix the freeze-out temperature at T=-100 °C. Then the desired degree of removal determines the total gas pressure p in unit E and the pressures and temperatures in the other process steps. The numbers shown in Fig. 1 refer to a removal degree of x=80 %. The indicated assumed temperature differences of 20 °C between hot and cold streams and the pressure losses of about 10 % between inlet and outlet of unit D (and in inlet and outlet of turbine and compressor) could be reduced to small values in an ideal plant of very large size. The electrical energy required for compression and cooling is calculated assuming an isentropic efficiency of 85 % for turbine and compressor and 60 % of the Carnot efficiency for the cooling heat pumps. It is supplied by the power plant. Therefore, consuming a part ΔP of the electrical power P_0 originally produced in the power plant at an efficiency η_0, the freezing out of CO_2 reduces the net power plant efficiency to $\eta' = \eta_0 x (1 - \Delta P/P_0)$. In order to curb the overall CO_2 emissions from electricity generation in coal fired power plants by 66 % at unchanged electricity output, one has to increase the gain of electrical energy per kg of emitted CO_2 from initially 1.20 kWh_E (at $\eta_0 = 38$ % without CO_2 removal) to 3.60 kWh_E. Assuming that CO_2 disposal requires 0.02 kWh_E per disposed kg and that 5 % of the CO_2 is lost by leakages we find that 82 % of the CO_2 is lost by leakages we find that 82 % of the CO_2 in the flue gases must be removed and that this lowers the power plant efficiency from $\eta_0 = 38$ % to $\eta_0 = 28$ %. According to Eq. (1) in Rule 3 this requires an enhanced input of primary energy and additional heat emissions of $P_0(\eta_0 - \eta')/\eta'$. these are 0.36 MW_{th} = 1.29 GJ per MW_{th} primary energy input. The electrical energy of 7.2 kWh_E = 25.8 MJ consumed by transporting 358 kg CO_2 per MW_{th} is generated with the diminished efficiency of 28 %. According to Rule 1 this causes the additional waste heat of 92.1 MJ. Thus, the heat equivalent of CO_2 at 66 % reduced emissions is 1.38 GJ=0.383 MWh per MWh_{th} or <u>38.3 %</u>. The heat emissions are in good approximation equal to the primary energy necessary for CO_2 abatement.

3.2 Nuclear waste disposal

A major unsolved problem of electricity generation from nuclear power is the lack of socially accepted disposal sites and mechanisms. The problem is not one of quantity - a boiling water reactor of 1300 MW_E produces only about 56 t of radioactive waste during 7000 hours of annual operation whereas a coal power plant of the same electricity output emits 8.5×10^6 tons of CO_2 in the same time - but rather of risk assessment. The HEONS concept cannot project uncertain risks into one single number, (and either can any other concept of pollution indicators). However, it is possible to calculate the HEONS for different pollution abatement scenarios appropriate for different risk categories (where the acceptance of a risk of a certain category is again a matter of social consensus). For nuclear waste this means calculation of the HEONS for, e.g., the often proposed (and often opposed) disposal in the earth crust, on the one hand, and for shooting the radioactive material into interstellar space where it can do no harm whatsoever, on the other hand.

We did not bother to determine the HEONS for the deposition of nuclear waste in empty rock salt caverns. It is obvious that they will be much smaller than in case of deep space disposal for which we have done the calculations [7,8] assuming rockets and electromagnetic launchers as the means of accelerating the well packed, burnt-out fuel rods to escape velocity. Considering hydrogen rockets each of which would carry 350 kg of

Mass drivers, operating like the linear motors which power the magnetically levitated trains running in Germany and Japan, have been proposed by H. Kolm of MIT [20] for nuclear waste disposal and built in prototypes by H. Kolm, G.K. O'Neill and coworkers at Princeton for space manufacturing purposes [21]. If 40 fuel rods of a total mass of 175 kg constitute the payload of one 1000 kg steel projectile with silicon carbide shielding, the heat equivalents of nuclear waste are **0.075 %**, and the energy consumed in deep space disposal is 0.097 % of the thermal energy produced by the nuclear reactor. It exceeds the HEONS, because part of it is lost as kinetic energy in space. If one wants to minimize the risks, due to a possible crash of the projectile because of a failure during acceleration along the 7.8 km long track, and puts only one fuel rod instead of 40 into one 1000 kg projectile, the heat equivalent increases to 3 % and the required specific energy to 3.9 %. Thus, in this scenario the HEONS of nuclear waste are an order of magnitude smaller than that of CO_2. However, there are some points to be considered before a comparative judgment of fossil and nuclear fuels can be made on the basis of the HEONS. Among them are the safety of the launching process, the disturbance of the atmosphere and the safety of the nuclear power plant itself. The latter may be treated by calculating the HEONS associated with extensive security measures like underground construction (although this may be far too expensive). Such calculations are certainly possible. One just has to invest enough time and manpower.

4. Incentives for energy conservation and pollution control

Pollution control is expensive. For instance, carbon dioxide removal and disposal may be double the price of electricity [13]. An alternative to pollution control, and the better one, is pollution avoidance by energy conservation. However, Second Law Analysis and vector optimization show that at present energy prices energy conservation measures like heat recovery by heat cascading, heat pumps and cogeneration with the possibilities of reducing primary energy consumption in the industrial sectors of West Germany, The USA, the Netherlands and Japan by 30 %-50 %, would enhance the total cost of the energy system by about 30 % [22]. On the other hand, if energy prices would double or triple, energy conservation would actually save cost. Thus, sufficiently high energy prices combined with stimuli of environmental protection are necessary in order that market forces favor energy conservation and pollution control.

We propose the following scheme to move the market in the right direction:

1. A public institution, e.g. the parliament, determines surcharges on energy carriers and their conversion processes. These surcharges may be in the form of taxes, levies or assurance bonds [23].

2. Part of the surcharge is based on the thermal pollution potential (TPP) of the energy carrier.
2.1 For fossil and nuclear energy sources the thermal pollution potential is equal to the enthalpy content, i.e. the heating value, of the energy carrier. This is shown in the national energy balances.
2.2 For biomass the TPP is determined by the enthalpy content and the primary energy consumed in the farming of the biomass.
2.3 For solar energy the TPP is determined by the reduction of the earth's albedo due to the increased absorption of solar radiation in solar power plants and the (non solar) energy consumed in the construction and operation of these plants.

The TPP (or enthalpy) charge stimulates energy conservation and the resulting reduction of emissions.

3. Another part of the surcharge, the pollution control (or HEONS) charges, is based on the HEONS for a given energy conversion process with a given energy carrier. It stimulates the use of the environmentally least damaging energy carriers and conversion processes.

If an energy user proofs that he has established pollution control in an energy conversion process such that his emissions of noxious substances above the limiting values are reduced by x %, y %...., then his HEONS should be reduced by $\alpha x\%$, $\beta y\%$,... $\alpha, \beta < 1$. This stimulates the installation of pollution control processes.

4. The energy surcharge is designed as a zero-sum-game for the state. It may be used to subsidize energy conservation and pollution control technologies an/or it may be partly or totally refunded to the consumer according to schemes like the ones discussed by Hannon et al. [24] and Herendeen et al.[25].

5. Conclusions

The energy requirements and heat emissions of pollution control which keeps the for any industrial process according to the rules of (first order) waste heat analysis. Model calculations for coal and nuclear power plants yield for NO_x, SO_2, CO_2 and burnt-out nuclear fuel rods heat equivalents of 0.7 %, 4.4 %, 38 % and 0.08 % - 3 % of the thermal power of the not pollution controlled power plants, respectively. More thorough calculations for improved pollution control processes may change these numbers somewhat. Risks associated with the pollution control processes themselves can partly be included in the HEONS. A calculation of the HEONS for the construction and operation of solar energy installations would allow a comparison of the (most benign) environmental impacts of fossil, nuclear and renewable energy sources. On the basis of this one may design incentives for the build-up of an energy system compatible with long-term sustainable development.

Acknowledgments

This work was supported by the Deutsche Forschungsgemeinschaft and is part of the research project "Ökologische Rahmenbedingungen der Sozialen Marktwirtschaft" of the "Studiengruppe Entwicklungsprobleme der Industriegesellschaft" (STEIG) e.V..

References

1. H. Herz, E. Jochem, M. Schön, Umweltaspekte des Energiesystems, Paper presented at the Symposium "Preisbildung bei natürlichen Ressourcen" der Deutschen Forschungsgemeinschaft, Bonn, February 1, 1985
2. L. Wicke, Die ökologischen Milliarden, München 1986
3. O. Hohmeyer, Social Costs of Energy Consumption, Springer, Berlin, 1989
   s.a.: O. Hohmeyer, Eurosolar Journal 1, 6 (1990)
4. R. Friedrich, U. Kallenbach, E. Thöne, A. Voß, HG.-H. Rogner, D. Karl, Externe Kosten der Stromerzeugung, VDEW Verlag, Frankfurt, 1989
   s.a.: R. Friedrich, A. Voß, Eurosolar Journal 1, 11 (1990)
5. F. Krause (Principal Investigator), W. Bach, J. Koomey, Energy Policy in the Greenhouse, Vol. 1, IPSEP, El Cerrito, CA, and European Environmental Bureau, 1989
6. R. Kümmel, Ecological Economics 1, 161 (1989)

7. U. Schüssler, R. Kümmel, ENERGIE, Jahrg. **42**, 40 (1990)
8. R. Kümmel, U. Schüssler, Ecological Economics, in the press
9. E. Nieuwlaar, Developments in Energy Analysis, Ph.D. Thesis, Rijksuniversiteit Utrecht, 1988; unpublished
10. H.-M. Groscurth, R. Kümmel, and W. van Gool, Thermodynamic limits to energy optimization, Energy **14**, 241 (1989)
11. German Physical Society and German Meteorological Society, The threat of man made global changes in climate, Bad Honnef, 1989. German text in: Physikalische Blätter **43**, 347 (1987)
12. M. Steinberg, H.C. Cheng, and F. Horn, A systems Study for the Removal, Recovery and Disposal of Carbon Dioxide from Fossil Fuel Power Plants in the US, BNL-35666 Informal Report 1984, Brookhaven Nat. Lab., Upton/N.Y., 1984
13. J. Fricke, U. Schüssler, R. Kümmel, Physik in unserer Zeit **20**, 168 (1989)
14. U. Schüssler, R. Kümmel, "Carbon Dioxide Removal from Fossil Fuel Power Plants by Refrigeration under Pressure", in: Proc. 24th Intersociety Energy Conversion Engineering Conference (W.D. Jackson Ed.), IEEE, New York, 1989, pp. 1789-1794
15. C.A.Hendriks, K. Blok, W.C. Turkenburg, "The Recovery of Carbon Dioxide from Power Plants", in: Climate and Energy (P.A. Okken, R.J. Swart, and S. Zwerver Eds.), Kluwer, Dordrecht, 1989; pp. 125-142
16. Report on the expert meeting on "Entsorgungsmöglichkeiten von CO_2 aus Kraftwerksprozessen", in the Bundesministerium für Forschung und Technologie, Bonn, March 14, 1990.
17. P.A. Okken, R.J. Swart, and S. Zwerver (Eds.), Climate and Energy, Kluwer, Dordrecht, 1989.
18. R. Kümmel, Removal and Disposal of Carbon Dioxide, in: Balances in the Atmosphere and the Energy Problem (E.W.A. Lingeman Ed), European Physical Society, Geneva 1990, pp. 115-135.
19. U. Schüssler, Physik in unserer Zeit, 21. Jahrg., 155 (1990)
20. H. Kolm, L5 News, Vol. 5, No. 9, 10 (1980)
21. G.K. O'Neill, Summary Session Opening Remarks. In: B. Faughnan and G. Maryniak (Editors), Space Manufacturing 5, Proc. 7th Princeton/AIAA/ SSI Conference, American Institute of Aeronautics and Astronautics, New York, 1985, pp. 9-12
22. H.-M. Groscurth, R. Kümmel, Energy **14**, 685 (1989)
23. R. Costanza, Ecological Economics **2**, 57 (1990)
24. B. Hannon, R.A. Herendeen, P. Penner, Energy Syst. Policy **5, 141 (1981)**
25. R.A. Herendeen, F. Fazel, Resour. Energy **6**, 277 (1984)

4.6 Contingent Valuation Study of The Environmental Costs of Electricity Generating Technologies[1]

Shepard C. Buchanan
Bonneville Power Administration
Portland, OR 97213

Introduction

The environmental costs of electricity generation technologies are the total value of risks imposed on society by the environmental effects of those technologies. The economic value of environmental risks is the amount that society would be willing to pay to avoid those risks or the amount society would require as compensation to accept those risks voluntarily.[2]

The usual approach to determining society's willingness to pay to avoid environmental risks of electricity generating technologies is to establish the significant environmental effects, measure them, and then apply economic valuation techniques to each of the effects.[3] For example, one might determine that sulfur emissions create health risks, reduce visibility, and damage materials. Then, each of those risks would be quantified in units that can be valued in monetary terms. Next, economic values for those risks would be determined and summed to yield an estimate of the economic value of the risks associated with sulfur emissions. The process would be repeated for the other emissions and other environmental intrusions with all costs totalled to obtain a complete estimate of the environmental cost of a particular electric power resource.

There are many advantages to this approach but I will mention just three. First, it provides for an objective analysis of each component of the electric power resource's environmental

---

1 Presented to German-American Workshop on External Environmental Costs of Electric Power, Ladenburg, Federal Republic of Germany, October 23, 1990.

2 Pace University, *Environmental Costs of Electricity*, Oceana Publications, Dobbs Ferry, New York, 1990.

3 See, for example, ECO Northwest, *Generic Coal Study*, for Bonneville Power Administration, DE-AM79-86BP62884, Portland, Oregon, 1987.

costs. Such an analysis provides meaningful and replicable information for weighing the costs and benefits of alternative resources. It displays the data, logic, and assumptions used, permitting other parties to debate the analysis while avoiding the unresolvable philosophical disagreements which are typical of subjective processes.

Second, by estimating the environmental costs of each environmental effect associated with an electric power plant, a utility can focus mitigation efforts efficiently. Controls can be implemented where gains in economic efficiency would be the greatest. Or, pollution offsets could be purchased elsewhere as envisioned by pending versions of the Clean Air Act.

Third, identification of significant elements of environmental costs can guide future research. Future research would be of particular value for those effects that appear to be costly to society as well as for effects that are highly uncertain and need further refinement. Uncertainty could be reduced whether it lies in the physical quantification of the effects or in the economic valuation of the risks.

There are, however, some disadvantages to the usual segmented approach to estimating environmental costs. Two general difficulties are worth nothing. First, it is sometimes difficult to obtain reasonably certain estimates of the actual environmental risks posed by a generating technology or a power plant. I don't want to over-emphasize this phenomenon because uncertainty is a fact of life for resource planners. Who, for example, can claim to foretell future rates of load growth or future fuel prices? Despite the impossibility of forecasting accurate values for these and other variables, we nevertheless have little choice but to try to do so, coping with uncertainty as well as we can. The same is no less true for environmental cost estimates. (I might also note that, in my opinion, environmental costs are likely to be underestimated because one can be reasonably sure that some environmental effects will be either overlooked or just plain unknown.)

Second, the use of the expected values based on "expert" or scientific assessments of risk can be problematic under some circumstances. Expected values are typically used in economic analyses because they are unbiased estimators. Unfortunately, expected values do not represent the variability of possible outcomes. In the extreme, the use of expected values masks the potential for low probability, high cost outcomes, or catastrophes.[4] Some argue that if probabilities are low enough, such costs should be coded as zero and

4 Pace, *supra*.

dismissed.[5] Others argue that risks of large scale catastrophes, no matter how improbable, are risks not worth taking and therefore must be treated as having infinite costs. The resulting dilemma of whether to use near-zero or near-infinite costs makes for fascinating philosophical discussions but frustrates analysts who must finally choose some value.[6]

Recognition of these difficulties led Bonneville Power Administration (BPA) to seek a different method for estimating the environmental costs of certain generating technologies. The remainder of this paper describes the method and the results of its initial application.

The Contingent Valuation Study

Perhaps the most important goal of an environmental cost study is to obtain an estimate of a resource's total environmental cost which, when added to its internal costs, results in an estimate of the resource's total cost. The utility can then select the resource(s) with the lowest total cost. As I mentioned at the outset, the environmental cost of a resource is society's willingness to pay to avoid the resource's environmental risks, or the compensation society would demand to incur those risks voluntarily. The contingent valuation method can estimate either of those values using a question phrased something like, "How much would you pay to avoid (some resource)?"

In a study sponsored by BPA, Hinman et al. of Washington State University conducted a survey of public attitudes about various risks. The study included contingent valuation questions about various risks. The study included contingent valuation questions about ratepayer's willingness to pay to avoid specific generating technologies.[7]

---

5 Rescher, Nicholas, Risk: *A Philosophical Introduction to the Theory of Risk Evaluation and Management*, University Press of America, Lanham, Md., 1983.

6 While the analyst could perform a sensitivity analysis using both a zero cost and some extremely high cost, the result would not be particularly illuminating nor would it be of any real help to a decisionmaker.

7 Hinman, George, Eugene Rosa, and Randall Kleinhesselink of Washington State University, *Public Attitudes Toward Risks Associated with various Activities, Substances, and Technologies*, for Bonneville Power Administration, DE-AI79-89BP98129, Portland, Oregon, July, 1990.

The survey was conducted by mail in January, 1990. Questionnaires were mailed to a random sample of 1600 households in the BPA service territory which includes all of Washington, Oregon, and Idaho, and a portion of Montana. The response rate for the survey was 52 percent.

Before discussing the contingent valuation portion of the survey, let me briefly summarize the other findings. The survey asked about levels of concern for ten environmental issues, levels of environmental activism, general opinions about risk, comparisons of 30 items involving various risks, and the contingent valuation questions. Among the ten environmental issues, air pollution from coal-fired plants, water pollution, storage of nuclear wastes, transportation of radioactive materials, and ozone depletion in the upper atmosphere were of most concern to people. The greenhouse effect, new nuclear plants, fish losses, radon in homes, and additional dams were of less concern.

The respondents, especially women and those most concerned about the ten environmental issues, collectively think the world is riskier than it was ten years ago. They also believe, however, that it is possible for individuals to take actions to reduce and control the risks they face.

When comparing concern about energy related items involving risks, respondents indicated the most dread for risks associated with transporting nuclear materials, radioactive waste disposal, and nuclear accidents, although nuclear power plants, themselves, were not especially dreaded. Similar expressions of dread were given for AIDS, crime, and alcohol-related accidents. Low levels of concern were expressed for hydropower and solar electric power. Additional questions regarding people's attitudes about these and other risks were also asked. Approximately half of the respondents (49 percent) indicated they would pay an additional charge on their electric bills to avoid the addition of one or more of the three types of generating facilities in the region. Of those who would not pay an additional charge, about one-fourth were not concerned about the environmental consequences of the three technologies while the rest were unwilling to pay because of other reasons.

The average amount respondents would be willing to pay to avoid the three technologies, including those who gave responses of zero, was $3.11 per month or $37.32 per year. The total number of residential customers in the region in 1988 - the last year for which such information is available - was 3.58 million. Therefore, the total amount the region's residential customers would be willing to pay to avoid the three technologies, assuming the sample is representative, is $133.6 million per year. This figure could be interpreted as the amount the region[8] would be willing to pay each year to meet future load growth with energy conservation measures and non-hydro renewable resources.

The respondents were also asked to indicate the amount they would pay to avoid each of the technologies. The average annual willingness to pay to avoid new hydro was $6.24; for fossil fuel it was $14.16; and, for nuclear it was $14.76. Expressed as regional annual willingness to pay values, the totals were $22.3 million to avoid hydro, $50.7 million to avoid fossil fuel, and $52.8 million to avoid nuclear. (Theses do not add to the overall total of $133.6 million because some respondents failed to preserve the equality in their individual answers on the questionnaires.)

Regression analyses were conducted to test whether the demographic backgrounds of the respondents were predictive of the additional charge on the utility bill respondents were willing to pay.[9] The regression analyses were conducted for a group (A) consisting of *all* respondents and for a group (B) consisting of only the respondents who indicated a positive willingness to pay.

The regression equation was:

$$(1)\quad Y = A_0 + A_1{}^{*}\ AGE + A_2{}^{*}\ SEX + A_3{}^{*}\ INCOME + A_4{}^{*}\ EDUCATION$$

Where

Y = Willingness to pay to avoid one or more generating technologies
AGE = Respondent's age in years
SEX = Respondent's sex (1 if female; 0 if male)
INCOME = Family income in thousands of dollars per year
EDUCATION = Highest grade completed

The results for each of the two groups were:
Group A: Only those respondents willing to pay a positive amount

---

8 Actually, the region could be willing to pay a higher amount after accounting for potential willingness to pay by business and industry. Firms, as well as households, benefit from reductions in environmental risks.

9 Regression analyses were also conducted for the willingness to pay value for each of the three technologies using the same explanatory variables. To avoid overwhelming the reader with numbers, those results are not presented but are similar to those shown here.

(2) $$Y = 8.33 - .07^{*}\ AGE - 1.00^{*}\ SEX + .09^{*}\ INCOME + .82^{*}\ EDUCATION$$
$$(-3.16)^{*}\ (-1.34)\ (0.28)\ (2.74)^{*}$$

$R^2 = .07$ $F = 6.3^{*}$

(t values shown in parentheses; * indicates significant at 99 % confidence level or better.)

Group B: All respondents

(3) $$Y = 3.32 - .05^{*}\ AGE - 0.19^{*}\ SEX + .12^{*}\ INCOME + .89^{*}\ EDUCATION$$
$$(-3.82)^{*}\ (-0.42)\ (0.67)\ (4.82)^{*}$$

$R^2 = .08$ $F = 14.0^{*}$

(t values shown in parentheses; * indicates significant at 99 % confidence level or better.)

The results indicate that younger people and those with more education are more likely to pay higher amounts to avoid the three technologies. The coefficients on income and sex are not significant, implying that neither variable can be shown to be related to the willingness to pay value. These results held true for both groups. The demographic variables, while collectively significant as predictors of willingness to pay as evidenced by the significant F values, do not explain much of the variation in willingness to pay. Together, the four variables explain only eight percent, or less, of the variation in willingness to pay.[10]

Discussion

The results of the study show that it is feasible to use the contingent valuation method to derive estimates of society's willingness to pay to avoid the risks of various energy resources. The application of environmental cost estimates we typically envision is for a specific resource consideration; for example, a utility needs to add 200 MW of capacity and has a menu of resources from which to choose. Whether the utility engages in a bidding

---

10 While other variables could have been regressed against the willingness to pay values, and in the study some were, the study was not designed to test hypotheses of the relationship between potential explanatory variables and respondent's willingness to pay to

process, a negotiating process, or simply builds its own resources, least-cost planning principles would lead the utility to select the resource or combination of resources capable of meeting need while having the lowest total social costs. In such a process, the environmental cost of each plant or project would be added to its costs to produce an estimate of its total social cost.

The nature of the BPA study was somewhat different. It did not examine the cost of different projects or plants. Instead, the required respondents to evaluate their willingness to pay to avoid entire technologies. Respondents were not provided any information about the technologies, where such plants might be located, or what the effects of the plants might be; they had to rely on their own knowledge to make judgments regarding their willingness to pay additional amounts on their utility bills to avoid the technologies.

Consequently. the elicited values can not be used to determine resource environmental costs on a cents-per-kilowatt-hour-basis. (Such was not the intent of the survey.) The estimates could be used, however, in the planning context to evaluate the cost-effectiveness of different resource strategies. For example, a resource strategy based on the lowest internal costs could be formulated first. (This would be the strategy that would emerge from a least-cost planning process that did not include environmental costs or other externality costs.) Then, an alternative resource strategy could be formulated relying only on resources other than the resource technologies which people indicated a willingness to pay to avoid. In the case of the BPA study, such a strategy would consist of energy conservation measures and non-hydro renewables such as geothermal, wind, and cogeneration using waste fuels, for example. If such a strategy could be implemented for less than the $136 million annual willingness to pay value, it would be a cost-effective investment. Thus, although the BPA study does not result in levelized cost estimates by resource, it does provide information that could be of use in developing resource strategies.

Despite the lack of levelized cost estimates, the BPA study could be modified easily to examine the specific electric power plants or projects. The principal advantage of the approach is that one does not need to develop separate cost estimates for all of the different environmental risks that may be associated with a power resource. People are asked to determine the amount they would pay to avoid the entire plant or project, each individual is free to consider any combination of attributes they wish. Moreover, the analyst does not need to decide on expected values or how to treat catastrophic risks. The amounts that people are willing to pay to avoid such risks are included in their expressed payment amounts. Those who deem catastrophic risks to be unacceptable may provide high bids, while those who treat them as zero cost risks may give bids of zero.

It would, of course, be desirable for people to make informed choices about electric power resources to meet their future needs. Information regarding the characteristics of different power resources, including the environmental and other risks associated with those resources, could be provided to potential respondents. A balance needs to be struck to avoid overloading respondents with information, providing information that may bias responses, and not providing sufficient information. A discussion of bias, which is a significant issue for contingent valuation studies, is beyond the scope of this paper, but the reader is referred to Mitchell and Carson[11] for a thorough review of not only bias, but all issues related to contingent valuation methodologies.

Conclusion

Environmental costs are measured by the amount that society would be willing to pay to avoid environmental risks or the amount society would require as compensation to incur the risks voluntarily. The usual approach to measuring society's willingness to pay is to quantify and price each of the environmental risks associated with an electric power resource. This approach has significant advantages.

Two disadvantages, however, are the uncertainty of physical estimates and the potential for catastrophic risks. Both present difficulties to the analyst attempting to quantify environmental costs.

To overcome those difficulties, BPA sponsored a contingent valuation study to query the public directly as to its willingness to pay to avoid different generation technologies. The study found that the region's households would be willing to pay $133.6 million each year in additions to their utility bills to avoid nuclear, fossil fuel, and hydroelectric power plants.

This result may have implications for resource strategies in that the estimated amount would be an additional system benefit of strategies that meet future load growth without recourse to any of the three resources.

---

11 Mitchell, Robert and Richard Carson, *Using Surveys to Value Public Goods: The*

Finally, the methodology could be adapted to estimating costs of specific resource plants or projects. Such studies would result in environmental cost estimates on a levelized cost basis. Bonneville Power Administration and the author seek comment on this approach and its appropriateness for use in the resource planning process.

4.7 A Comparison of Emissions and Externality Costs of Waste-to-Energy and Coal Fired Electric Power Plants

David R. Wooley
Professor for Environmental & Energy Law
Center for Environmental Legal Studies
Pace University School of Law
78 North Broadway
White Plains, New York
USA

1. Introduction and Summary

A controversial energy issue in the United States concerns the use of household and commercial waste as a fuel for producing electric power. Interest in waste-to-energy power plants began in the late 70's, but accelerated as federal and state governments began to close down old style solid waste landfills to protect groundwater.[1] Initially, waste-to-energy plants were seen as a panacea for solid waste and energy problems. One argument frequently advanced by combustion technology vendors is that incinerators are a cleaner source of electric power than conventional fossil fuel fired power plants. Today Waste-to-Energy plants are now increasingly opposed by citizen groups, local governments and energy advocates as serious new air pollution threads and as a misdirection of public or consumer funds that would be better spent on energy conservation and waste recycling facilities.

Electrical power production from incinerators is growing.[2] The draft New York State Energy Plan predicts that electric generating capacity at waste-to-energy plants in New York will grow from about 200 megawatts in 1989, to about 341 megawatts by 2008.[3] Nationally, the number of waste incineration plants, most of which will generate electricity, is expected to grow from 111 plants in 1987 to 300 plants by the year 2000.[4]

The regulatory controls over incinerator emissions are changing rapidly. Prior to 1987, during the period of greatest growth, incinerators were virtually unregulated. The only federal restriction was an obsolete limit on particulate emissions, and few states made any from New York State Attorney General and the Natural Resources Defense Council (NRDC), EPA used it's authority under the 1977 amendments to the federal Clean Air Act to impose some restrictions on new incinerators. This was accomplished by an EPA guideline, which for the first time required scrubbers and modern particulate matter controls on new incinerators. In December of this year EPA will promulgate comprehensive national regulations for new and existing incinerators.

During 1988-1990, some states began to implement new capacity bidding schemes which assigned a value for environmental externalities of power resources, a change that tended to advantage conservation services over combustion technologies in competitions to supply growing electric power demand.

Valuation of incineration externalities and the treatment of those values in bidding programs may affect whether Waste-to-Energy plants can compete successfully with other supply options for new capacity. The Public Utility Regulatory Policies Act (PURPA)[5] provides a limited guarantee that incinerator operators, along with other independent power producers, may sell electricity to local utilities at the utilities' avoided cost (the price utilities would bear to generate an equivalent amount of electricity themselves).[6] A recent order of the New York State Public Service Commission determined, however, that waste-to-energy plants are subject to new capacity bidding procedures.[7] Failure to prevail in bidding may greatly reduce the revenues a plant may earn from utilities under PURPA. A reduction in electric sales revenue would require a corresponding increase in garbage tipping fees, potentially affecting the economic viability of some plants.[8]

This evolving regulatory climate is changing the economics of incinerators and causing solid waste officials to take a closer look at the wisdom of producing energy from garbage.

Environmental Externalities of Incineration

Waste-to-energy plants emit a wide variety of conventional and toxic pollutants. As shown in the tables below, toxic emissions include dioxins, furans, cadmium, lead, chromium, mercury, arsenic, and nickel, as well as the conventional pollutants, sulfur oxides SO_x, hydrogen chloride (HCL), nitrogen oxides (NO_x), hydrogen fluoride, particulate matter, carbon monoxide and carbon dioxide.

State-of-the-art waste-to-energy plants emit SO_x, NO_x, HCL, lead, cadmium, copper, arsenic and mercury at higher rates per kWh generated than do well controlled coal-fired plants. Waste-to-energy plants and new coal-fired plants emit about the same amount of chromium per kWh generated. Of the metal emissions, only nickel is emitted in smaller amounts per kWh in waste-to-energy plants.

Waste-to-energy plant emissions also cause material damages. Acid gas emissions, particularly HCL, are corrosive to a wide range of metal, paint, stone, and other materials used in consumer products and structures. These damages vary widely depending upon stringency of acid gas control and proximity to valuable structures.[9]

Incinerators have both positive and negative effects on land use and ground and surface water quality. Waste-to-energy plants produce ash which is 30 % of the original volume of municipal solid waste. This volume reduction reduces land devoted to fill and possibly lowers landfill methane emissions. Incinerator ash residue, however, must be disposed of in landfills and is high in soluble toxics. Indeed, incinerator ash is higher in toxic metals than ash from coal combustion. Incinerator ash disposal practices vary widely and determine whether the overall threat of ground and surface water contamination is increased or reduced as a result of incineration of solid waste. At present, it is not internalized in the costs of current disposal practices. We have not, however, discovered a valuation of these impacts.

Like any steam-driven electric generation plant, a garbage incinerator requires a significant amount of cooling water. Although no figures on this usage were obtained, it is likely that waste-to-energy plants would utilize roughly the same amount of water per unit of power generated as a coal-fired power plant. The impacts on aquatic environment in terms of thermal pollution and fish impaction, would depend upon the size of the receiving water body and the cumulative stresses from other steam generating plants.

2. Allocation of Environmental Externality Costs of Waste-to-Energy Plants

For this reason, the environmental externalities of garbage incinerators may not be solely attributable to the plant's electricity production. There are several ways to allocate environmental costs between a plant's solid waste function and its electric power production. One way is to estimate the cost of environmental externalities of a Waste-to-energy plant, subtract from that figure the cost that would otherwise be incurred in landfilling solid waste, and assign the entire difference to the electricity production from the plant in levelized mills per kilowatt hour. This was the approach taken in 1986 in a final report prepared by the Bonneville Power Administration.[10]

A variation on this method would be to calculate the environmental externalities of land filling refuse and to use that figure as an avoided environmental cost, representing the solid waste share of total incinerator externality costs. This figure could be subtracted from total incinerator externality costs. The remainder would represent the environmental externality of electrical production from Waste-to-energy plants. Such an inquiry would need to look at groundwater impacts, land use, odor, methane emissions, and other landfill effects.

Other important factors include whether incinerators reduce landfill methane emissions and whether CO_2 emissions from incinerators are greater than CO_2 emissions from landfills. If so, the value of the avoided methane (a potent global warming agent) might be used as an offset to incinerator externalities. We did not find any studies which estimate the effect of incinerators on landfill methane emissions, nor a social value for methane emissions and other avoided landfill effects. We did not discover any studies which describe the relative CO_2 emissions of incinerators and landfills, though presumably the former are much greater.

Another approach assigns none of the environmental externality costs of waste-to-energy plants to electricity production. This approach assumes that the sole purpose of the plant is to avoid placing garbage in a landfill and that the electrical production is incidental. It is unusual, however, for solid waste disposal to be the only purpose of consideration in building an incinerator. Most communities have the option of either building a landfill or constructing a waste-to-energy plant to handle municipal refuse which cannot be disposed of through recycling, and the costs are often comparable. Nationally, the average tipping fee at waste-to-energy plants is approximately $39 per ton, while fees to dispose of waste in a state-of-the-art landfill are approximately $35 per ton.[11] Waste-to-energy plants in the United States are financed through revenues received from municipalities or haulers for disposing of garbage ("tipping fees"), and by sale of electric power or steam. Sale of electric energy is usually a major part of the revenues received by waste-to-energy plants. Each ton of municipal refuse, when burned in a plant equipped with steam generators, can produce approximately 500 kWh of electrical energy.[12]

In most cases, therefore, the motivation and financial ability to build a waste-to-energy plant (as opposed to other solid waste management options such as landfills or recycling facilities) is based upon a mixture of solid waste disposal and electric power functions.

The importance of electric sales to the decision to build an incinerator is also demonstrated by the size of typical electric sales. In 1986 the average electric power rating of waste-to-energy facilities in the U.S. was 25.537 megawatts.[13] This is roughly the amount of power produced by a 1000 ton per day mass burn facility. Assuming that the 1000 ton per day plant operates at 76 % capacity, it would sell over 138 million kWh a year[14] and achieve annual electric sales revenue of $8,322,000 (at $.06 per kilowatt hour).[15] By comparison, annual revenues from tipping fees would be $10,818,600 (at $39 per ton).[16] This is confirmed by the draft New York State Energy Plan which states that electric sales revenue for waste-to-energy plants are only "slightly lower" than revenues from garbage tipping fees.[17] In some areas , where garbage tipping fees are artificially low, electric sales can be as high as 70 % of plant revenues.[18]

Table 1 provides data for selected waste-to-energy plants showing the typical range of electric power generating capacity for different sized plants.[19]

Table 1: Electric power from municipal waste

Plant Location	Tons/Day	Rated Electric Power Capacity (MW)
Westchester, NY	2,250	60
Baltimore, MD	2,250	60
Tampa, FL	1,000	22.5
Bitterford, SC	607 (RDF)	22.5
Marion Co, OR	550	13.3
Commerce, CA	300	11.44
Oswego, NY	200	4.0

Electric sales revenues help keep tipping fees down, so that incinerators may compete with landfills as a cost effective solid waste option. The electricity sales, therefore, often provide the additional revenues needed to make incinerators an economically viable waste disposal choice.

In some areas, however, landfills may be prohibited (either by local geologic /groundwater conditions, or by lack of land). On Long Island, for example, the $100 per ton cost of trucking waste to remote out-of-state landfills is probably the primary justification for constructing a waste-to-energy plant.

Variable local landfill tipping fees, and electric sales rates make it difficult to reach a

One solution is to allocate externality costs on the basis of the ratio of electric sales to total plant revenue. If electric sales are expected to produce half of the revenues, then one might reasonably assume that if half of the motivation for building the plant is to receive electric sales revenue. In areas where landfill shortages have escalated tipping fees, a proportionately smaller percentage of the social costs could be assigned to the electricity production from the plant.

3. Incinerator Ash

Ash from waste-to-energy plants consists of fly ash trapped by air pollution control equipment, along with combustion products and non-combustible residues left behind in the combustion chamber ("bottom ash"). Both fly ash and bottom ash contain hazardous substances. Exposure to these toxics in incinerator ash may occur through contamination of groundwater, exposure of workers to ash in waste-to-energy plants or at landfills, an exposure of the public in general windblown ash dust (particularly during transportation of ash from incinerator to landfill). When ash is disposed of in municipal solid waste landfills, the mildly acidic conditions in the fill promote movement of cadmium and lead into groundwater. These metals may be leached at rates which are sufficiently high to classify the ash as hazardous waste.[20]

The concentration of toxic metals in incinerator ash is much greater than in unburned garbage.[21] Heavy metal concentrations and leachability characteristics of incinerator ash are generally much higher than for ash generated by coal-fired power plants. This is particularly true for incinerator fly ash.[22] In addition, some toxic organic chemicals, including dioxins and furans, are formed in the combustion process and become concentrated in ash residues.

An unresolved question is whether groundwater is more at risk of contamination by leachate from unburned solid waste or from incinerator ash. This, of course, depends largely on the type of landfill used for each. A full accounting of the externalities of electricity generation from an individual incinerator would require an assessment of whether the incinerator is having the effect of diverting solid waste away from preexisting, poorly designed landfills and directing incinerator ash to state-of-the-art landfills (with proper liners and groundwater monitoring safeguards). In such a case the overall effects on groundwater quality may be positive. In many cases, however, incinerator ash is disposed of in the same adequately controlled landfill space that would have held the less toxic unburned solid waste.[23]

There is a great deal of confusion regarding legal requirements for disposal of incinerator ash.[24] Clarifying legislation is being considered in proposed amendments to both the Clean Air Act and the Resource Conservation and Recovery Act (RCRA). EPA considers ash from municipal incinerators to be exempt from hazardous waste disposal requirements unless it fails a standard EP toxicity test. But it has not enforced the requirement nor issued guidelines for disposal of ash that passes the test but contains some leachable toxic metals. This controversy is well summarized in OTA's recent report.[25]

The continuing confusion over waste-to-energy ash disposal requirements and the hodgepodge of current disposal practices make it impossible to generalize about the externalities of incinerator ash. No one has attempted to value the health or environmental risks associated with incinerator ash. Moreover, no single value would apply in all circumstances, since some disposal techniques are much less risky than others. For now, therefore, no externality value can be assigned.

4. Air Pollution Emissions: Estimates of Environmental Externality Costs

Two studies were found relevant to assigning social costs to electricity produced by waste-to-energy plants. The 1986 Bonneville Power Authority[26] study estimated human health, visibility and materials damages costs of a generic waste-to-energy plant. The generic plant was assumed to be a 10 megawatt generating capacity expected to burn 500 tons per day of refuse. Emissions characteristics are shown in Table 2.

The estimated externality costs vary from a high cost assessment of 484.199 mills per kilowatt hour to a low cost assessment of 1.854 mills/kWh. The low cost assessment was based on an assumption that existing air pollution standards completely protect human health, a conclusion which the BPA authors considered unlikely.[27] The high cost estimate is based on a linear dose response curve with no safe threshold for exposure. The "expected" cost, 111.453 mills per kilowatt hour, was based on dose response data at projected ambient concentrations. The report notes that if larger populations are exposed to incineration emissions (BPA assumed that 75,000 persons would be exposed) then air pollution health costs would be larger.[28]

Table 2: Emissions characteristics of solid waste generic facility

	Emissions (Tons/Yr)	Tons/MWh	Increase in Morta/Morb Annual Linear dose-response
Particulates	61	9.8×10^{-4}	0.04
SO_2	220	33.5×10^{-4}	1.5
NO_2	290	44.1×10^{-4}	0
VOCs	9.6	1.46×10^{-4}	?
CO	168	25.6×10^{-4}	3.8
HCI	69	10.5×10^{-4}	0.055
H^2SO_4	5.6	0.85×10^{-4}	0.23
Lead	1.6	0.24×10^{-4}	?
Mercury	0.53	0.08×10^{-4}	0.00075
Fluoride	4.8	0.73×10^{-4}	0
Dioxin	5.1×10^{-6}	0.76×10^{-10}	0.062
POM	1.8×10^{-3}	0.27×10^{-7}	0.008
Beryllium	8.8×10^{-6}	1.34×10^{-10}	0
PCB	0.01	1.52×10^{-7}	3×10^{-5}

Source: BPA/ECO, Five Resources, Technical Appendices, Appendix C, Table C1-1. Tons/mWh calculated based on average annual output of 65,700,000 kWh. See BPA/ECO, Five Resources, Final Report at p. 4-4. The generic plant is equipped with scrubbers and particulate controls. Increases in morbidity/mortality are from Appendix C, Table C1-5, at C1-55 of the BPA/ECO Report.

Each of these figures was then reduced by the waste reduction benefit (calculated as savings from avoidance of landfill operating costs). The report does not appear to take into account any environmental effects associated with disposal of incinerator ash. The resulting damage cost figures from the BPA study are set forth in Table 3.

Emissions and control assumptions for the BPA study were based on the characteristics of a plant being constructed in 1986 at Marion County, Oregon. After the BPA study was released, EPA issued new source permitting guidelines under the federal Clean Air Act.[29] The Marion County, Oregon plant was to be equipped with scrubbers and particulate control as required by the Guidelines, but may not have included the full range of "combustion controls" that are now required to further reduce emissions of dioxin, metals, carbon monoxide, and other pollutants.

Table 3: **Levelized costs and benefits for solid waste generic facility**

(mills/kWh)

TYPE OF EFFECT	LOW	EXPECTED	HIGH
Health	.000	103.482	466.664
Visibility	1.545	6,951	15.447
Materials	.309	1.020	2.008
TOTAL COST	1.854	111.453	484.119
Waste Reduction Benefit	- 33.653	- 50.976	- 67.582
STUDY TOTAL	- 31.800	60.477	416.567

Source: BPA/ECO, Five Resources - Final Report, Table 4 - 2 at p. 4 - 6.

Additionally, EPA has recently proposed new source performance standards for waste-to-energy plants.[30] A federal court order requires these regulations to be finalized by December 31, 1990, and they will apply to any incinerator on which construction is begun after the date of the proposed rulemaking notice.[31] These regulations may or may not achieve a further tightening of standards relative to the guidelines or control assumptions for the generic plant in the BPA study.

The impacts assessed by the BPA analysis were based on a 500 ton per day plant. This may underestimate the impacts for larger plants. Some plants currently under construction or in planning are designed to burn in excess of 2000 tons of refuse per day.[32]

BPA's health risk assessment concluded that nearly 95 % of the health effects were attributable to sulfur dioxide and carbon monoxide emissions.[33] This contradicts a recent health risk assessment performed for EPA which appears to assume that virtually all health impacts of incinerator emissions are caused by trace metals and dioxin/furan emissions.[34] Further research should address the relative importance of conventional versus toxic pollutants in valuing the externalities of incinerator emissions.

EPA's 1987 risk assessment addresses carcinogenic and non-carcinogenic health effects for direct inhalation of four metals and six types of organic emissions. By far the greatest amount of cancer incidence risk was attributed to chlorinated dioxine and dibenzofuran emissions. EPA's "controlled scenario" assumed emission controls of metals and organics at levels comparable to those expected to be achieved by the proposed EPA new source performance standards. EPA estimated that the range of national cancer incidence for metals and organics was .3 - 1.5 cases per year for populations living within 50 kilometers of all projected waste combustors in the U.S. The maximum individual risk (probability that a person exposed to the highest modeled concentration of pollutants from a municipal waste combustor will develop cancer over a 70-year lifespan) was 10^{-5} to 10^{-6}.[35]

There are several severe limitations to the EPA cancer risk estimate. EPA did not look at non-inhalation exposure ("indirect exposure" from ingestion of contaminated soil or foods, and dermal contact). EPA prepared a preliminary analysis of indirect exposure but determined that the results could not be interpreted quantitatively. EPA's preliminary analysis suggests that for some persistent organics (dioxins, PCB's and chlorobenzene) the "indirect exposures to emissions deposited over long periods may be comparable to exposures through direct inhalation." Exposure of children to deposited lead and mercury was identified as needing further analysis.[36]

Some observers contend that indirect exposure routes, particularly bioaccumulation in the food chain, are more important pathways to adverse health effects than inhalation.[37] OTA's report, "Facing America's Trash", discusses the stable, fat-soluble characteristics of dioxins and furans, which allow them to accumulate and concentrate in the food chain. It describes an assessment of dioxin and furans emissions in Minnesota which concluded that 90 % of estimated incremental cancer risk from a proposed waste combustor is associated with ingestion of these substances in food.[38] The OTA report also briefly describes two other studies, one in Pennsylvania and the other in New York, addressing deposition of dioxin and furans in agricultural areas and the potential for contamination of milk from dairy cows.[39]

Another limitation is that EPA's study, like the BPA/ECO *Five Resources* study, did not address health effects of exposure to incinerator ash, either from landfill leachate, or from fugitive ash losses at the plant or in transport to land disposal.[40] For these reasons we did not attempt to use the EPA Risk Assessment Values to derive in externality costs for human health impacts.

5. A Comparison of Emissions Rates for Coal and Waste Fired Power Plants

The projected emissions for the proposed 3000 ton per day Brooklyn Navy Yard Plant are set forth in Table 4. To arrive at a pound per kilowatt hour emission estimate, we have divided the Brooklyn plant emission numbers by three to estimate the emissions for a generic state-of-the-art 1000 ton per day plant, and further divided than by a typical annual kilowatt hour production of a 1000 ton per day plant (138 million kilowatt hours, see footnote 14 *supra*). Table 5 sets forth metals emission rates in pounds per kilowatt hour.

Table 4: Emissions of toxics from Brooklyn plant

POLLUTANT	EMISSION RATE (Tons/Year)	EMISSION RATE (gm/sec)
Lead (Pb)	14.5	4.17×10^{-1}
Cadmium (Cd)	.49	1.40×10^{-2}
Chromium (Cr)	.24	6.85×10^{-3}
Copper (Cu)	.43	1.23×10^{-2}
Nickel (Ni)	.17	4.82×10^{-3}
Zinc (Zn)	27.9	8.02×10^{-1}
Arsenic (As)	6.30×10^{-2}	1.81×10^{-3}
Selenium (Se)	6.46×10^{-2}	1.86×10^{-3}
Mercury (Hg)	5.3	0.151
Beryllium (Be)	1.9×10^{-3}	5.51×10^{-5}
Flouride (FL)	93.2	2.68
Polyaromatic Hydrocarbons (PAH)	0.275	7.89×10^{-3}
Polychlorinated Biphenyl (PCB)	.178	5.13×10^{-3}
Polychlorinated Dibenzo - p - Dioxins (PCDD)	2.56×10^{-4}	7.36×10^{-5}
Polychlorinated Dibenzo - p - Furans (PCDF)	1.10×10^{-3}	3.17×10^{-5}
Tetrachlorinated Dibenzo - p - Dioxin (TCDD)	3.72×10^{-5}	1.07×10^{-5}
2378 TCDD (Isomeric Form)	2.43×10^{-6}	7.00×10^{-5}

Source:

Final Environmental Impact Statement, Proposed Resource Recovery Facility at the Brooklyn Navy Yard, Technical Appendix, Vol. III, Appendix D, Air-Quality Related Issues, June 1985. Emission rates are based on 100% of the nominal capacity, continuous operations of the fabric filter at a 0.015 gr/dscf at 12% CO_2 particulate emission.

A comparison of the figures in Table 5, which includes data on metals emissions from coal-fired plants shows that even well controlled waste-to energy plants emit lead, cadmium, copper, arsenic, nickel and mercury at levels which far exceed those of coal-fired plants, when measured in lbs/kWh. The level of chromium emissions for waste-to-energy incinerators appeared to fall within the range given for coal-fired plants. The level of zinc emissions for coal-fired plants ar not available from the sources we reviewed.

Table 5: Metals emissions - pounds per kWh

Pollutant	1000 Ton Per Day Incinerator		Scrubbed Coal-Fired Plant
	Emission Rate		
	lbs/Yr[1]	lbs./kWh[2]	lbs./kWh[3]
Lead	9666.66	70×10^{-6}	1.68×10^{-7}
Cadmium	326.66	2.4×10^{-6}	1.6×10^{-8}
Chromium	160.00	1.2×10^{-6}	1.29×10^{-6}
Copper	286.66	2.1×10^{-6}	2.4×10^{-7}
Nickel	113.33	0.8×10^{-6}	4.6×10^{-7}
Zinc	18,600	134.8×10^{-6}	-
Arsenic	42	$.3 \times 10^{-6}$	1.72×10^{-7}
Mercury	3,533	25.6×10^{-6}	7.4×10^{-8}

(1) Extrapolated from projected emissions of 3000 ton/day Brooklyn Navy Yard Plant in Table 4. The figures are 1/3 of the data reported in the Final Environmental Impact Statement, Proposed Resource Recovery Facility at Brooklyn Navy Yard, Tech. App. Vol. III, Appendix D, Air Quality Related Issues, June, 1985.

(2) Assumes annual production of 138 millon kilowatt hours.

(3) From Table 6, Chapter IV, "Environmental Costs of Electricity," based on EPA, "Estimating Air Toxic Emissions from Coal and Oil Combustion Sources", April 1989. In cases of cadmium, chromium, mercury and nickel, the EPA report gave a range of values. For this table we chose the high end of the range.

Aside from the toxics, incinerators also emit large quantities of conventional air pollution. These discharges are similar in type to emissions from coal-fired electric power plants. Table 6 describes the annual emission tonnages projected from the planned 3000 ton per day Brooklyn Navy Plant in New York City.

Table 6: Emissions of conventional pollutants

Pollutant	BROOKLYN INCINERATOR Tons/Year
Particulate Matter	161
Sulfur Dioxide	2972
Sulfuric Acid	230
Nitrogen Dioxide	2972
Hydrogen Chloride	5370
Carbon Monoxide	368
Hydrocarbons	65.7

Source:

Final Environmental Impact Statement, Brooklyn Navy Yard, supra at Table 4.

Table 7 translates these numbers into pounds emitted per kilowatt hour. It uses estimates of the annual kilowatt hour production of a 1000 ton per day plant (see footnote 16, *supra*, and accompanying text). The figures in Table 6 from the 3000 ton/day Brooklyn Plant have been divided by three to approximate the emissions from a well controlled 1000 ton per day plant. The right hand column of Table 7 presents comparable emission rates from new coal-fired plants. The latter figures are taken from Table 1, Chapter IV of "The Environmental Costs of Electricity".

Table 7: Conventional Emissions (1b/kWh)

Pollutant	Incinerators	Coal w/FGD & ESP	Coal w/IGCC
Particulates	7.77×10^{-4}	3.0×10^{-4}	1.5×10^{-3}
Sulfur Dioxide	14.35×10^{-3}	1.8×10^{-3}	$1.85 - 3.7 \times 10^{-3}$
Nitrogen Dioxide	14.35×10^{-3}	6.07×10^{-3}	6.0×10^{-4}
Hydrocarbons	$.31 \times 10^{-3}$		

This comparison shows that incinerators have higher levels of sulfur and nitrogen compound emissions per kilowatt hour than coal fired plants equipped with scrubbers. The data do not allow a direct comparison for hydrocarbon emissions.

Table 8 presents a valuation of these emissions using the cost factors developed for coal plant emissions in our book "Environmental Costs of Electricity".[41] In the table, the "same starting" point externality value developed for materials effects of utility plant sulfur dioxide emissions has been applied to incinerator SO_2 and HCL emissions., on the assumption that the materials effects of the two pollutants are similar. The right hand column gives the comparable externality cost for two types of coal-fired plants. Note that the incinerator figures do not contain a value for CO_2 emissions, while the coal plant

Table 8: Externality costs of conventional emissions from waste-to-energy plants

Pollutant	Cost Per Pound[1]	lbs/kWh[2]	Externality Cost in Cents/kWh[3]	Externality Cost Scrubbed Coal Fired Plants[4]	Externality Cost IGCC in Cents/kWh
SO_2	$2.03	.014	2.84	-	-
NO_x	.82	.002	.164	-	-
HCL	.12	.077	.924	-	-
Particulate	1.18	.00075	.089	-	-
Carbon Monoxide	-	.00085	-	-	-
Volatile Organic Hydrocarbons	-	.00015	-	-	-
CO_2	-	-	-	-	-
		TOTALS	4.019	4.5	2.8

(1) Derived from Table 2, Chapter V.B. "Sulfur Oxides", Table 1, Chapter V.C. "Nitrogen Oxides" and Table 3 Chapter V.E. "Particulates" of this report. These figures do not include acid rain effects.

(2) This figure is derived from the projected annual SO_2, HCL and NO_x emissions for the 3000 ton per day Brooklyn Navy Yard Plant. It is based on 60% capacity divided by three to estimate the emissions from a typical 1000 ton per day plant, and is then further divided by the typical annual kilowatt hour production for a 1000 megawatt plant (138 million kWh/yr.). See footnote 14, supra, and accompanying text. Kilowatt hour production figures for the Brooklyn plant were unavailable as that plant will produce exclusively steam. The Brooklyn plant emission rates were chosen because of similarity between proposed emission controls and proposed EPA new source performance standards.

(3) The externality cost for hydrogen chloride (HCL) is assumed to be the same as the materials damage cost for SO_2. We have not discovered an externality value for the HCL human health effects. See Table 2, Chapter V.B. "Sulfur Oxides" of "Environmental Costs of Electricity".

(4) From Table 1, "Environmental Costs of Electricity," Executive Summary at Page 33.

emissions from incinerators and to decide how to apply an externality figure to these emissions.

The Brooklyn plant is generally regarded as an example of a plant having controls similar to those which will be required by EPA's new source performance standards. The Brooklyn plant is not designed to produce electricity and, therefore, potential electric power outputs are not available.

Complete emissions and cost figures are not yet available for all the pollutants emitted by waste-to-energy facilities. The calculation here presented in table 8 is based solely on the acid gas emissions, and particulates. If half of the costs in Table 8 are attributable to the electric production (as opposed to the solid waste disposal function) of the plant, then this analysis would assign a social cost of 2.01 cents per kilowatt hour just for acid gas and particulate emissions. The cost does not include an estimate for health effects of HCL. By comparison, Bonneville Power Authority's "expected" externality cost for waste-to-energy plants (for all pollutant effects) is 6.0 cents per kilowatt, and its high figure is 41.6 cents per kilowatt. *See*, Table 3, *supra*.

These cost estimates are incomplete. To them must be added the so far uncalculated cancer and non-cancer health effects of the ultra-toxic pollutants emitted from incinerators, including those caused by indirect exposures. The social cost of these effects is likely to be at least as great as the damages from the conventional pollutants.

5. Methane and CO_2 Considerations

Organic municipal and industrial wastes in landfills are estimated to contribute 3-10 % of methane emissions world-wide.[42] Methane is a potent greenhouse gas.[43] It is possible that by reducing placement of solid waste in landfills, incinerators could reduce overall methane emissions. Landfills, on the other hand may tend to "lockup" organic material that will otherwise produce CO_2 in an incinerator. No studies were found which compare the relative effect of incinerators and landfills on these greenhouse gas emissions. The following general discussion may help guide future research in this area.

Methane (CH_4) production in landfills results from anaerobic degradation of organic waste. Approximately three-fourths of municipal solid waste is organic.[44] Organic wastes dioxide (CO_2) This quickly depletes the oxygen in the landfill and creates the anaerobic conditions which lead to the production of methane.[45]

Estimates of methane production are highly variable. Factors affecting the rate of methane production include moisture, pH, age of the material, volume of material, concentrations of nutrients and bacteria, and presence or absence of sewage sludge.[46] There is a one-to-two-year lag period between landfilling of wastes and the beginning of gas generation.[47] It can take from 10 to 100 years for a landfill to produce significant amounts of methane.[48]

One study estimated the rate of methane production to be from 1,000 to 7,000 cubic feet of methane/ton of solid waste (31-218 m^3/metric ton).[49] In 1988, the U.S. produced 1.48 million tons of municipal solid waste.[50] Using these figures, total U.S. methane emissions from landfills would eventually reach 2.9 - 20.7 Tg of methane per year (1 Tg = 1 teragram = $1x10^{12}$g). Using an alternative formula proposed by Bingemer and Crutzen (1987), the same amount of solid waste would generate 7 Tg of methane.[51] This rate assumes a carbon content of 22 %, that 90 % of municipal solid waste is landfilled, and a conservative methane production efficiency of 0.25 ton of methane per ton of carbon.[52]

Landfill methane emissions can, however, be beneficial. Methane can be collected from large- and medium-sized landfills to be burned in small electric power generators. One study estimates that if landfill methane emissions were collected completely and processed for energy production, they would supply 5 % of all natural gas consumption, or 1 % of U.S. energy demand.[53] Currently, only 123 of some 6584 municipal landfills collect methane to recover energy. Many more simply vent methane or burn it off with a flare.[54] Economic viability of landfill gas recovery is dependent on landfill size, local regulations, and proximity to energy users. For a discussion of recovery costs and landfill characteristics which lend themselves to methane recovery, see EMCOM Associates and Gas Recovery Inc., *Landfill Gas, Analysis of Options*, prepared for New York State Energy Office, 1981.

6. Conclusion

This preliminary analysis indicates that incinerator emissions are likely to have fairly large environmental externality costs. This information appears to conflict with the practice adopted by some state utility commissions of encouraging waste-to-energy plant construction through awards of increased rates of return, or other incentives for electricity produced. Indeed, it appears that the environmental costs of waste-to-energy plants are large enough to justify rate treatment or externality cost assignment similar to a coal plant. As noted above, incinerators emit far more heavy metal air pollution and produce greater amounts of metal contaminants in solid combustion wastes (e.g. fly ash) than do coal-fired plants (as measured in pounds lbs/kWh). Similarly, incinerators produce larger amounts of acid gas pollutants including SO_2, NO_x, and HCL. Incinerators also emit dioxins and furans that are absent from fossil fuel plant emissions.

Additional research, however, is needed before a complete social cost figure may be assigned for waste-to-energy plant emissions. Incinerators effect on overall greenhouse gas emissions is one important research need.

Footnotes and References

1 Some studies place waste-to-energy incinerators in a category of renewable energy resources. *See*, Public Citizen, *The Power of the States: A 50 State Survey of Renewable Energy*, Critical Mass Energy Project, Washington DC, June, 1990. The renewable label is probably inappropriate for waste-to-energy technology. A substantial portion of the waste stream and BTU content for incinerators is derived from fossil-fuel. Plastics constitute 12-13 % of the volume of the municipal solid waste. Office of Technology Assessment, *Facing America's Trash, What Next For Municipal Solid Waste?*, OTA-O-424, (Washington, D.C.; U.S. Government Printing Office, October, 1989), Page 242, [hereafter cited as "OTA Report"]. Incinerators also use natural gas as an auxiliary fuel at start up, shut down and at times when combustion chamber temperatures fall below specific levels. Additionally, municipal solid waste is not "renewable" in the same manner as hydroelectric, solar, wind and biomass power, as there is substantial energy and materials waste in consumer packaging.

2 As of 1987, New York State had the largest number of garbage incinerators of all states, with Florida having the largest amount of incinerator capacity, as measured in tons of refuse processed per day. U.S.E.P.A., *Municipal Waste Combustion Study, Report to Congress*, prepared by EPA Office of Solid Waste and Emergency Response, Office of Air and Radiation, and Office of Research & Development, EPA/530-SW-87-021a, June, 1987, p. 20 [hereafter cited as "*EPA Report to Congress*"); U.S. Conference of Mayors & National Resource Recovery Assn., *1989 City Currents*, Washington D.C.

3 *Draft New York State Energy Plan*, Volume V, Issue 4: Solid Waste/Energy, May, 1989, p. 6, [hereafter cited as "*Draft NYS energy Plan*"]. Although labelled "draft", these documents "form the basis for the recommendations and are part of the plan." Overview, Final New York State Energy Plan, (September, 1989), p. 1.

4 *EPA Report to Congress, supra* note 2 at pp. 20-25. For data on the number of plants that will generate electricity, see note 12 *infra*.

5 16 U.S.C. § 824a 3, (Supp. 1988); 18 C.F.R. § 292.303(a) (1987).

6 New York has required utilities to purchase power from independent power producers at a higher rate than that required by PURPA. N.Y. Public Service Law § 66-C. The law was recently upheld. *See, In The Matter of Long Lake Energy Cooperation v. Public Service Commission of the State of New York*, 148 A.D. 84, 543 N.Y.S. 755 (N.Y. App. Div. June 29, 1989), motion for leave to appeal denied Dec. 14, 1989. FERC, however, has issued a ruling prohibiting states from imposing rates that exceed avoided costs in wholesale purchases in interstate commerce. *Orange and Rockland Utilities, Inc. et al.*, 43 F.E.R.C. P61,067 docket #EL87-53-000 (April 14, 1988).

7 N.Y. Public Service Commission (NYSPSC) Case No. 88-E-246, Opinion No. 89-30, *Opinion & Order Establishing Bidding Guidelines*, (Sept. 13, 1989); NYPSC Case No. 29409, Opinion No. 88-15, *Order Concerning Bidding, Avoided Cost Pricing and Wheeling Issues* (June, 1988). *See also*, NYPSC Case No. 29409, Opinion and Order No. 88-15(A), November 2, 1988, pp. 23, 30-32.

8 *See*, Federal Energy Regulatory Commission, Notice of Proposed Rulemaking, Docket No. RM88-5-000, March 16, 1988.

9 *OTA Report, supra* note 1 at p. 242.

10 Bonneville Power Administration, *Estimating Environmental Costs and Benefits for Five Generating Resources, Final Report: Description of Their Likely Significant Environmental Effects, and the Economic Value of Those Effects*, by ECO Northwest, Ltd., Shapiro and Associates, Inc., and Seton, Johnson and Odell, Inc., March, 1986 (hereafter cited as BPA/ECO Five Resources). The authors state that a further allocation must be made where a plant provides both waste heat for factory processes and electrical generation. ESRG has developed a formula for allocating emissions impacts between electric and steam production from cogeneration plants. Its method offsets emission effects by the amount of fossil fuel use displaced by utilization of cogenerator waste heat. ESRG, *The*

Role of Hydro-Quebec Power in a Least-Cost Energy Resource Plan for Vermont, prepared for Department of Public Service, State of Vermont, January 19, 1990 at p. 9-28 to 9-30.

The need to allocate between steam and electric production, however, is diminishing for waste-to-energy plants. Most municipal waste incinerators being built today are constructed solely to generate electricity. In 1986 almost two-thirds of planned resource recovery facilities (63.9 %) were designed to produce only electricity, while 29.2 % were designed to produce both electricity and steam. Government Advisory Associates, *1986-1987 Resource Recovery Yearbook, Directory and Guide*, New York, NY, 1986, p. 36 [hereafter *Resource Recovery Yearbook*]. *See also, Draft NYS Energy Plan*, Vol. 5, Issue 4, *supra* note 3 at p. 8.

11 *Draft NYS Energy Plan, supra* note 3, at p. 8.

12 *Draft NYS Energy Plan, supra* note 3 at p. 6.

13 *Resource Recovery Yearbook, supra* note 12, at 40.

14 For comparison, Bonneville Power Authority's study determined that the expected annual electric power output of a 500 ton per day, 10 MW incineration plant would be 65,700,000 kWh. *BPA/ECO Five Resources*, supra note 12, at xiv, 2-13, 4-4.

15 The 76 % capacity factor was provided by Dr. Maarten de Kadt, Inform, Inc., New York, N.Y., personal communication February 7, 1990. The calculation in the text assumes a power production rate of 500 kilowatt hours per ton of refuse. The figure would be higher for refuse derived fuel (RDF) plants whose fuel (solid waste that is processed to remove some non-combustibles) will have a higher BTU value per ton.

16 The average tipping fees charged by mass burning plants in the U.S. was $26.64 per ton. *Resource Recovery Yearbook, supra* note 12, at p. 96. We used the more conservative figure, $39 per ton, since that is the tipping fee charged at a recently constructed landfill in New York. *See, Draft NYS Energy Plan, supra* note 3, Vol. V, May, 1989.

17 *Draft NYS Energy Plan, supra* note 3 at p. 8.

18 Personal communication from Dr. Allen Herskowitz, Natural Resource Defense Counsel, February 1, 1990.

[19] List provided by Dr. Maarten de Kadt, Inform, Inc., New York, NY, personal communication, February 7, 1990.

[20] Ujihara A. and M. Gough, *Managing Ash From Municipal Waste Incinerators, A Report*, Resources for the Future, Center for Risk Management, Washington, DC; November 1989, [hereafter cited as *Managing Ash*]. The report notes three studies which conclude that worker and public exposure risks are low. It also points out that EP toxicity tests for ash have poor reducibility and may not adequately mimic the actual conditions within different types of landfills (i.e. co-disposal facilities that take both ash and solid waste vs. monofills restricted to ash only).

[21] *Managing Ash, supra* note 22 at p. 8.

[22] Bridle, T., et al., "Evaluation of Heavy Metal Leachability from Solid Wastes", *Wat. Sci. Tech.*, Vol. 19 R, 10, pp. 1029-1026, 1987. The study was conducted by members of the Environment Canada, Environmental Protection Service, Wastewater Technology Center, Burlington, Ontario.

[23] *See*, U.S. Environmental Protection Agency, *Characterization of Municipal Waste Combustor Ashes and Leachates From Municipal Solid Waste Landfills, Monofills, and Codisposal Sites*, prepared by NUS Corp, for Office of Solid Waste and Emergency Response, EPA/530-SW-87-028A, Washington, D.C., October, 1987.

[24] *See generally, Managing Ash, supra* note 22. ("Federal laws and regulations that mandate how ash is to be managed are unclear and tests for determining whether it is a hazardous waste produce ambiguous results. This situation has served both to preserve use of incinerators and to delay federal action to improve ash management").

[25] *OTA Report, supra* note 1 at pp. 37-39, 258-259.

[26] *See* BPA/ECO *Five Resources, supra* note 12.

[27] BPA/ECO *Five Resources, supra* note 11 at p. 4-4.

[28] BPA/ECO *Five Resources, supra* note 11 at Appendix C1-55, C1-63.

[29] U.S. EPA, Office of Air Quality Planning and Standards. *Operational Guidance to State and Local Agencies for Best Available Control Technology Determination for Permitting New and Modified Municipal Waste Combustors*, June, 1987.

30 54 Federal Register 52251 (December 20, 1989), *Standards of Performance for New York Stationary Sources; Municipal Waste Combustors*. EPA also proposed standards for existing incinerators, which will require retrofit of a variety of emission controls by the mid 1990s.

31 *Id.*

32 The Hempstead, Westchester County, and Niagara Falls plants in New York State are each designed to incinerate in excess of 2200 tons per day. These were operating in 1989. The proposed Brooklyn Nany Yard plant is projected to handle 3000 tons per day by 1993. A Staten Island plant (3000 tons per day) and a Bronx plant (2000 tons per day) are being planned for construction in 1997. *Draft NYS Energy Plan*, Vol. V, Issue 4, *supra* note 3 at p. 7.

33 EPA/ECO, *Five Resources*, Technical Appendices, *supra* note 12, at p. C1-54.

34 Radian Corporation, *Municipal Waste Combustion Study: Assessment of Health Risks Associated with Municipal Waste Combustion Emissions*, prepared for U.S.E.P.A. (EPA/530-SW-87-02) Hemisphere Publishing Corporation, Washington, D.C. 1989), [hereafter cited as EPA/Radian Report.]

35 Id. at p. 2-8.

36 U.S.E.P.A., *Response to Petition for Rulemaking and Advance Notice of Proposed Rulemaking, Assessment of Municipal Waste Combustor Emissions Under the Clean Air Act*, 52 Fed Reg 25399, 25404 (July 7, 1987).

37 OTA *Report, supra* note 1, citing Ashlander, O., *The Swedish Dioxin Memorandum*, paper presented at ASCE Dioxin Symposium (NY, NY, February 10, 187); Franke, B., *Review of the Environmental Impacts of Solid Waste Incinerators Proposed for Long Island and New York City*, prepared for *Newsday* (Tacoma Park, MD); Institute for Energy and Environmental Research, Nov., 1987; Webster, T., and P. Connett, *Critical Factors in the Assessment of Food Chain Contamination by PCDD/PCDF from Incinerators*, paper presented at Dioxin '87 (Las Vegas, NV: October 4-9, 1987); World Health Organization, "*PCDD and PCDF Emissions from Incinerators for Municipal Sewage Sludge and Solid Waste - Evaluation of Human Exposure*, report on World Health Organization meeting; Naples, Mar. 17-21, 1986 (Copenhagen: Regional Office for Europe, 1987).

[38] Minnesota Pollution Control Agency, *Supplemental Health Risk Assessment Technical Work Paper, Volume I: Polychlorinated Dioxins/Polychlorinated Furans, Wisconsin County Resource Recovery Facility*, prepared by J.B. Stevens and Associates, April 1988.

[39] OTA Report, *supra* note 1, at p. 241.

[40] EPA/Radian Report, *supra* note 35 at pp. 1-3.

[41] See Table V.B.2 "Summary of Damages Due to SO_2".

[42] Bingemer, H. and P. Crutzen, "The Production of Methane from Solid Waste", *Journal of Geophysical Research* 92:2189 [hereafter cited as Bingemer]. Another estimate is that landfills in the U.S. Contribute 2-6 % of Global Methane Emissions. OTA Report to Congress, *supra* note 1 at p. 286.

[43] Bingemer, H. and P. Crutzen, as cited at *Draft Policy Options*, part IV, p. 40.

[44] OTA Report, *supra* note 1 at p. 275.

[45] *Id.*

[46] *Id.* at p. 282.

[47] U.S.E.P.A., Office of Policy, Planning and Evaluation, *Policy Options for Stabilizing Global Climate: Draft Report to Congress*, February, 1989, at p. 189 [hereinafter cited as *Draft Policy Options*].

[48] Wilkey, M., R. Zimmerman, and H. Isaacson, *Methane From Landfills: Preliminary Assessment Workbook*, Argonne National Laboratory, Department of Energy, 1982, as cited in *Draft Policy Options, supra* note 51.

[49] *Id.*

[50] U.S. EPA, *Report to Congress: Solid Waste Disposal in the United States*, Vol. II, EPA/530-SW-88-011B, Washington, D.C., October, 1988 (hereafter cited as EPA, *Solid Waste Disposal*).

[51] Bingemer, H. and P. Crutzen, *supra* note 47 at Part VII p. 189.

[52] *Id.*

[53] *OTA Report, supra* note 1 at p. 283, citing U.S. EPA, Office of Solid Waste, *Operating Criteria (Part C) Criteria for Municipal Solid Waste Landfills*, 40 CFR Part 258, EPA/530-SW-88-037, Washington DC, July, 1988.

[54] U.S. EPA, *Solid Waste Disposal*, Report to Congress, *supra* note 54.

4.8 Incorporating Externalities in Least-cost Planning: The Use of Control Costs, Standards, and Multi-Attribute Approaches

Uwe Fritsche
Öko-Institut
Darmstadt

1. Energy & Environment: A Necessary Regulatory Link

More and more, environmental aspects gain importance in sectors which are significant contributors to ecological damage. Due to the close relation of environmental impacts and energy systems, special emphasize for environmental issues is given in energy policy and planning. The complexity of the energy/environmental problem is shown in Figure 1.

As the figure indicates, the task to link (internal) costs of various energy systems with environmental impacts on several scales (local, regional, global) could be facilitated drastically if a single nominator for all aspects could be found. The principle of monetizing environmental externalities is aimed exactly at this: to express environmental costs in the same (monetary) unit as already used for internal costs.

In the US, several initiatives exist to incorporate environmental costs into the regulatory process for (utility) electricity generation, and surveys of this interesting activities (LBL 1990; NARUC 1990; Pace 1990) indicate that mainly those states lead the way which also introduced mandatory least-cost (or integrated resource) planning schemes. Seemingly, there is a trend towards total societal costs as a criterion for utility power acquisition, even if this refers only for acquisition "on the margin", i.e. new or additional capacity. Only few (and heatedly debated) concepts exist to extend total societal costing also to the existing electricity systems, e.g. by "environmental" of "full cost" dispatch (Bernow, 1990).

Surprisingly, there is no similar activity in Germany. Despite several studies prepared on externalities (see Section 2.3), the federal and state regulatory bodies have not (yet) started to include environmental costs into their doing. Only recently, the German Ministry of Economic Affairs funded a new study on externalities which can be seen as a first step (see paper of J. Keppler for this workshop).

Figure 1: The Energy/Environment Interface: Complexity

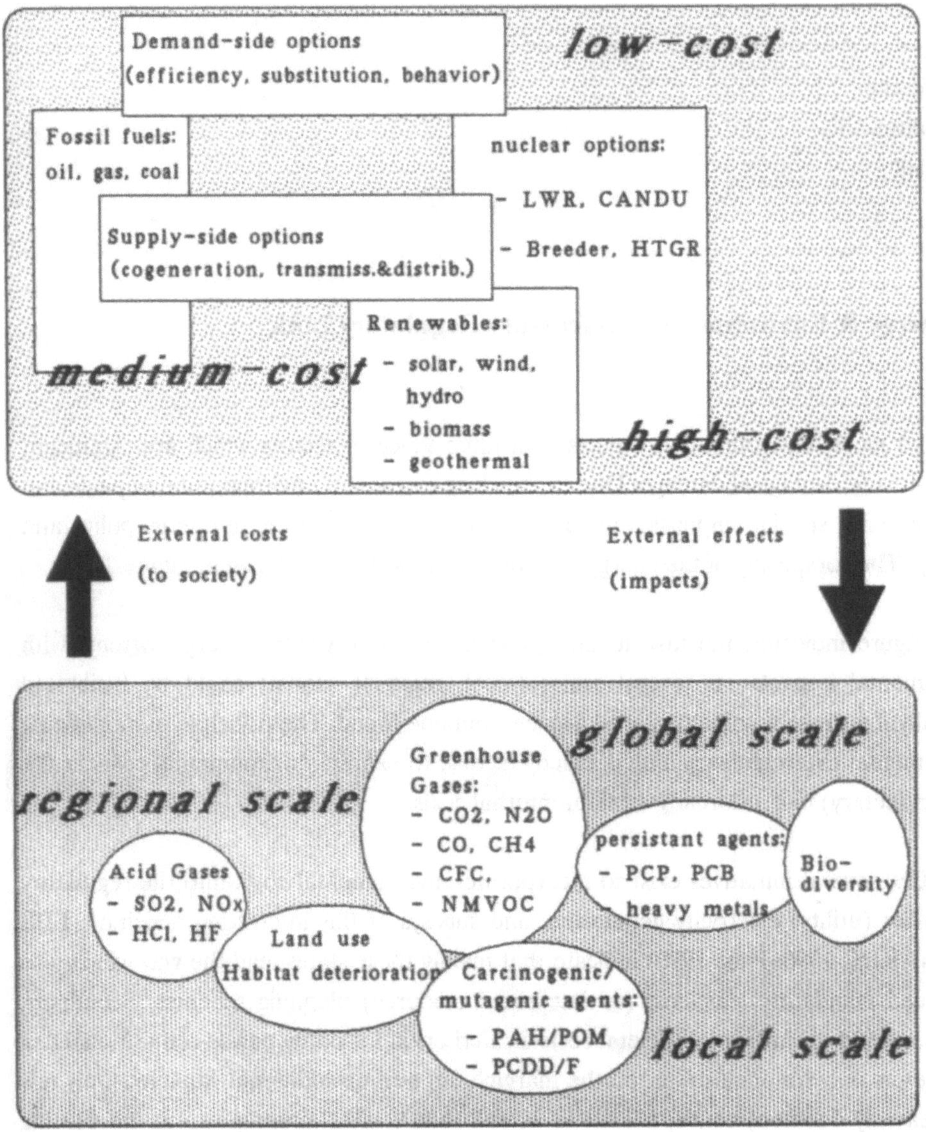

Part of the explanation for this can be seen in the somewhat stricter environmental regulations for powerplants in Germany when compared to the US emission limits (i.e. some externalities are already internalized by higher costs for control equipment), and a low demand for new capacity in the near-term. But more important are two other German circumstances: first, there is the lack of least-cost (or integrated resource) planning regulation which gives no incentive to discuss costs of power acquisition from a *societal* perspective.

Second, the German utility regulation is based on an pre-World-War II law from 1935 (!) which requires utilities to supply "cheap and secure" electricity to all customers, and does not address environmental issues at all.

Nevertheless, the concern about global warming in Germany caused some interest in least-cost planning as a means to integrate (low-polluting) demand-side options into utility acquisition strategies, and a recent study provided evidence that the LCP rationale can be implemented in Germany, too (ÖKO 1989c). Also, there is increasing support in Germany to the argument that the challenge of global warming can be addressed only with some regulatory "assistance" for the market forces (see paper of P. Hennicke for this workshop). Given the monopolistic structure of the electricity sector, this could mean also to introduce competition[1]. The incorporation of environmental costs could be an element of such "assistance", and is compatible both with the present regulation, and with least-cost orientated reforms which might introduce bidding schemes and the like in Germany, too.

2. Internalization by Monetization

In theory, market-orientated economies rely mainly on price signals as a driving force, so that the internalization of (environmental) "externalities" in monetary terms could add an environmental component to the market game without changing existing "rules".

In theory, that works well. In the real world, there is more than just prices, and one needs to take a close look to what extent externalities can be monetized before applauding to theory. A review of recent accomplishments of research in the externality area can be found in the literature (Hohmeyer 1990; ISI 1989; LBL 1990; ÖKO 1989b; Pace 1990),

1 An example for this is the new, PURPA-like German regulation regarding utility payments to non-utility generators (NUG) based on renewables. While NUG based on fossil fuels receive low payments from utilities, they have to pay from 75 % (hydro, biomass) to 90 % (solar, wind) of the customer rates to NUG.

which gives the impression of an astounding consensus regarding the magnitude of external costs of non-nuclear electricity.

2.1 Damage Versus Control?

A comparison of the various approaches which allow researchers to put dollar tags on environmental impacts quickly results in the distinction between *direct* costing methods (where health impacts, loss of crops, material damages etc. are valued in conventional economic terms), and *indirect* methods which use control costs (abatement or offsetting), revealed preferences, or also costs of marketable emission permits. Seen from the pure economic standpoint, only direct methods give the "true" costs to society, while control costs are seen as a somewhat confusing substitute. But the problems associated with direct methods when dealing with global warming gave way to a more relaxed view of the control cost concept, as long as these cost figures are derived from existing standards or realistic targets (Welsch 1989).

The fact that most studies on externalities of non-nuclear options yield similar results for air pollutants (i.e. same magnitude) can be explained easily when the sources from which the studies derive their figures are identified: the main sources for direct methods are epidemiological studies (health impacts) and field studies for other impacts (crops, materials), while indirect (non-contingent) methods rely on mainly technical data of control equipment which is used to comply with environmental standards or targets. As an early OECD study has shown for SO_2, the costs and benefits (reduced damage) of air pollutant controls are quite similar for moderate abatement goals (OECD 1981). This means that the minimum of an abatement-vs.-damage cost function is not a single point, but a range.

The most recent US study concerning externalities was prepared by the Pace University Center for Environmental Legal Studies (Pace 1990), providing an extensive discussion of methods and data for the internalization of external effects of electricity generation by monetizing environmental damages. For greenhouse gas emissions, the Pace study used control costs instead of damage costs. The following table summarizes the cost data which the Pace study found to be "reasonable".

Table 1: Values to monetize emissions based on US studies (Source: Pace 1990)

Pollutant	$/ton	DM/t
CO_2	13.6	30
SO_2	4,060	8,950
NO_x	1,640	3,615
Particulates	2,380	5,247

The results of other US work on damage costs are discussed elsewhere (LBL 1990; NARUC 1990, ÖKO 1989b), and studies concerned with indirect costing methods are presented at this workshop (see contribution of E. Caverhill for control costs, and S. Buchanan for contingent valuation).

Regarding German work in the externalities area, several recent studies cover the literature on direct methods concerned with damage costs (ISI 1989; ÖKO 1989b). In addition to this, the indirect costing method based on control costs was introduced in Germany, too (ÖKO 1989a). In the so-called GEMIS study, real pollution abatement costs were derived from German data sources (see ÖKO 1989b, Appendix for a detailed discussion). They represent the costs of required emission control measures for SO_2, NO_x, and particulates from stationary and automotive sources in (West) Germany, weighted by the relative contribution of these sources to total emissions. The costs to control CO_2 have been determined by means of fuel switching and energy conservation (low cost value), and CO_2 scrubbing (high cost value). The following table summarizes these figures.

Table 2: Values to monetize emissions based on German control costs (Source: ÖKO 1989a)

Pollutant	$/ton	DM/t
CO_2 - lower value	4.5	10
CO_2 - upper value	45	100
SO_2	2,268	5,000
NO_x	1,814	4,000
Particulates	454	1,000

These results indicate that the Pace data (see Table 1) are in good accordance with the German figures for NO_x and CO_2, while the SO_2 and particulates data are twice to five times as high. For the latter, the difference can be explained by the fact that the German data refer to total particulates (including coarse flyash), while the Pace data refer mainly to fine, sub-micron particles. If the German control costs would be based on sub-micron particulates only, the costs would be similar to the Pace data.

Because the monetized emissions which can be calculated for electricity systems are sensible regarding the inclusion of CO_2, the GEMIS study calculated three different values: first, monetized emissions based on control costs for SO_2, NO_x and particulates; second, the former plus "low" CO_2 control costs, and, finally, the same for "high" CO_2 costs. Table 3 shows results for German electricity systems, including fuel cycles, and material acquisition, too.

Table 3: Environmental costs of total emissions from German electricity systems (Source: GEMIS Model)

system	**w.o. CO_2**	**CO_2-low**	**CO_2-high**
	------------	[mills/kWh_{el}]	------------
powerplant (hard coal)	3.3	7.9	50
IGCC (hard coal)	3.8	8.0	45
cogenerator* (hard coal)	1.0	3.2	23
CC powerplant (natural gas)	1.5	3.6	23
cogenerator* (natural gas)	-1.3	-1.4	-2
cogenerator* (biogas)	-0.3	-5.6	-54
solar photovoltaics	0.2	0.3	1
wind turbine	0.1	0.1	1
micro-hydro	0.0	0.1	0
nuclear electricity**	0.3	0.6	3

* = net costs, including waste-heat utilization bonus for substituted oil-heating
** = only emissions from fuel-cycle activities, excluding costs for nuclear risks

Note that the monetized emissions include not only the operation of powerplants, but also their total fuel cycles (see discussion below).

The results suggest that *significant* environmental costs are associated with electricity generated by fossil-fuel powerplants, and that nuclear electricity offers *no* cost advantage when compared to gas-fired cogenerators, and renewables.

This holds true even when *neglecting any* nuclear-specific environmental costs (e.g. health effects). Also, the coal-fired cogen system yields similar results as the gas-fired combined-cycle powerplant, which shows that coal is not necessarily "excluded" from utility acquisition when external costs are included.

2.2 Hidden Costs: Impacts From Total Fuel Cycles

The analysis of externalities can be biased when only those impacts are considered which are directly linked to the operation of energy systems, because various activities take place before a fuel can be burned. These "upstream" activities impose also environmental impacts, some even higher than emissions from powerplant operation (e.g. CH_4 from fossil fuels, solid wastes for uranium).

Most US studies concentrate on direct impacts associated with the operation of powerplants, some include also "downstream" effects from (solid or liquid) wastes generated by the plants.

The only study which at least partially dealt with externalities from "upstream" activities like coal mining, gas pipelines etc. was prepared by PLC - now known as Resources Insight - for a Boston-based utility (PLC 1989). In the Pace Study, an appendix gives a guesstimate for total fuel cycle externalities based on the TEMIS model (Fritsche 1990).

Due to US DOE activities related to the National Energy Strategy, a better database for total fuel cycle externalities - including the manufacturing stage - will be available in the near-term, so that future US research might be able to prepare a total costing framework for energy systems. In Germany, the GEMIS study offered a comprehensive analysis of various fuel cycles (including manufacturing of energy systems), and developed a computer model which calculates the externalities for all these impacts, so that a "total costing model" is already available (see Table 3 for some results).

3. Beyond The Dollar Value: Problems In Monetization

Monetizing externalities seemingly works fine: you find an "area of consensus" for major impacts, there are computer models which even include total fuel cycles, and even (some of) the regulators agree upon such schemes. Unfortunately, there's a simple truth which cannot be ignored: monetizing externalities is possible *only for a part* of environmental impacts, as Figure 2 indicates.

On the one hand, there is a significant lack of data for several types of environmental impacts which can be monetized in principle (e.g. heavy metals, VOC, ODC), which could be overcome by more research - but extrapolating from the efforts made so far for "conventional" emissions, this will take a good while.

Figure 2: Only partial monetization is possible

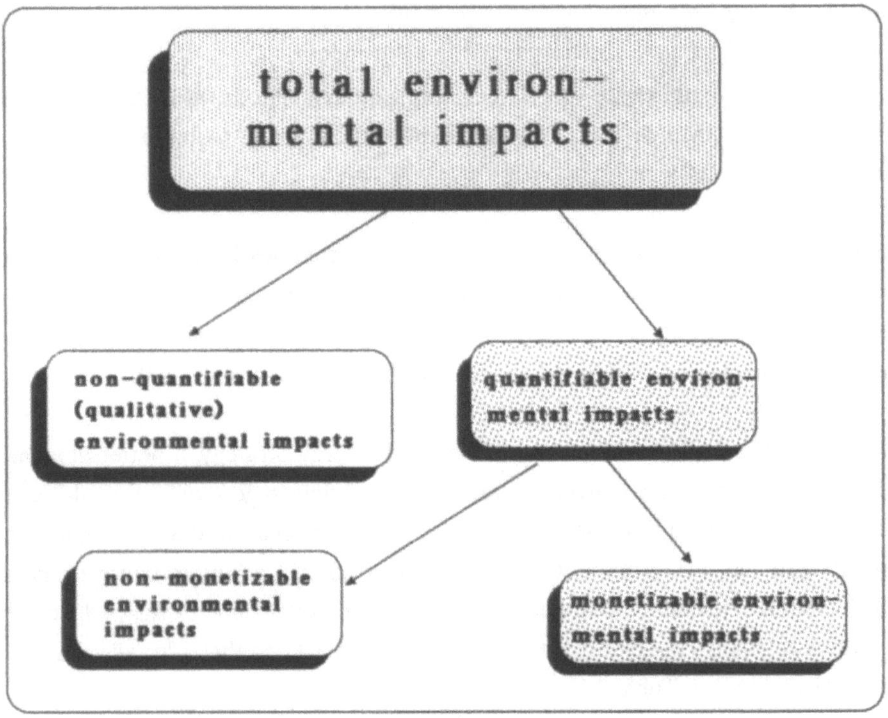

Meanwhile, monetization will be able to incorporate quantifiable aspects only partially (which is better than nothing, one might argue). On the other hand, there are impacts which cannot (at least not easily) be quantified (so that monetizing is a bit tricky), and others which cannot be monetized at all.

3.1 Climate Impacts: A "Hot" Issue

Without any doubt, the climate impacts of electricity systems are among the "core" aspects to be addressed quantitatively. Because ozone-depleting compounds are not typically associated with electricity systems, the discussion can concentrate on greenhouse gases

(GHG) and their global warming potential (GWP). The most relevant GHG for electricity systems is CO_2, although CH_4 can be prominent, too.

Monetizing CO_2 emissions is a tough issue, as the discussion in Section 2 has shown: the costs of CO_2 abatement options vary within one order of magnitude, because no definite targets for CO_2 reduction are available.

But it gets still tougher: on the one hand, econometric modelling suggests that societal costs even for small CO_2 reductions are quite high (ASE 1990), while bottom-up analyses of options (allowing to create "least-cost" packages of measures from supply curves of avoided CO_2) indicate that the net costs for significant CO_2 cuts can be very small or even zero (Enquête 1990; Krause 1990). So, finding the "true" figures to monetize CO_2 (and also other GHG) seems unachievable, especially if different targets are used in different countries or regions.

3.2 Non-quantifiable Aspects

Besides environmental aspects which can be covered by quantitative analysis and, therefore, offer the chance to be "monetizable", several other aspects of environmental relevance can only be addressed in *qualitative* terms. A discussion of such impacts is given elsewhere (AURA 1983; ÖKO 1989a; UNEP 1985).

Risks and Safety

Although the risks of occupational diseases, accidental injuries, and impacts of major hazards imposed by electricity systems are widely addressed in the literature, the outcome of all the risk analysis does not (yet?) encourage the use of quantified estimates, as the authors of a German risk study pointed out:

"The results show also that in the area of risk analysis a significant amount of additional research is still necessary before the aspect of risk as a **quantitative** criterion can be used for decision-making in energy policy" (Thöne/Kallenbach 1988, p. 793-794, translated and emphasis added by the author).

The OECD noted its COMPASS report that

"...the state of knowledge is not adequate to permit the numerical values presented in this sector for health risks associated with different electricity generation fuel cycles to be used as a basis for unambiguous conclusions regarding the relative safety of these forms of energy" (OECD, 1985, p. 130).

The disagreement in the scientific community (and abroad) culminates in the debate on (health) risks associated with nuclear power. The work of Hohmeyer and its "counter" study (IKE 1989) show that the range of possible impacts from "generic" nuclear reactor accidents covers up to 5 orders of magnitude, thereby overwhelming all other (known) external costs of nuclear. Certainly, more research could reduce this range to some extent, but the fundamental possibility of severe accidents in (not inherent-safe) nuclear powerplants, and associated risks imposed by up- and downstream activities of the nuclear fuel cycle are somewhat unique. If monetization of these impacts by a "low-high" range is rejected due to reasons of practicability, we're back to *qualitative* statements about (nuclear) risks.

Micro-Ecological Impacts and Biodiversity

In addition, the production and use of energy also imposes direct impacts on local/regional ecologies. These impacts cover a range from habitat deterioration and separation (e.g. in wilderness areas) to the interruption of migration (e.g. for fish and birds) up to the loss of especially endangered species (Plaza 1983).

The main impacts stem from the primary extraction of (fossil) fuels[2], but impacts can also be caused by energy transportation (e.g. oil spills from tankers, explosions of gas pipelines). For some of these impacts, damage or control costs can be derived from case studies (e.g. Exxon Valdez spill), but ecological variation does not permit to figure out generic costs.

Closely related are impacts on regional or even global biodiversity (Beyea 1990), for which neither quantitative estimates of cause-effect-chains are established nor are reasonable methods available to determine the societal value of decreased biodiversity: what is the cast of the spotted owl when old growth is harvested as biomass energy?

Cultural Resources

Finally (and returning to anthropocentrics), there are impacts on the cultural environment for which no reasonable costs can be determined: the spiritual intrusion on indigenous people caused by geothermal developments on Hawaii, aesthetic impacts on pristine coastal areas by off-shore oil&gas drilling, or the loss of unique cultural sites by uranium mining activities mark problems which could be considered as "marginal" when compared

2 For coal mining impacts see (Chadwick 1987), for oil and gas extraction refer to (Boesch/Rabalais 1987; Bolze 1987).

to global warming, but their societal costs could be significant, as the reactions of people in the mentioned cases have shown.

4. Towards A Multi-Attribute Approach

The previous section raised some fundamental problems of the monetization approach for internalizing externalities. Fortunately, there are *more options available* to incorporate externalities into energy planning which could be used in addition with studies where they've been (or are to be) applied.

4.1 Screening the Options: Environmental Standards and Targets

In least-cost (or integrated resource) planning schemes, there are always several options for electricity acquisition, being it supply or demand-side resources. Such portfolios can be *screened* using environmental "standards", and only those options will be considered further which are compatible with the standards.

In principle, this concept can be used for all environmental externalities, even if only qualitative evidence is available. For example, a standards could be that (even small) risks for catastrophic events are not allowed. Similarly, the qualitative impacts regarding ecological impacts can be formulated as standards, e.g. no options are considered in LCP (or IRP) which could cause interference with endangered species (see ÖKO 1989a, p. 321 ff.) A real-world example for this approach can be seen in the concept of "protected areas" for hydro development in the Pacific Northwest.

Figure 3: Multi attribute approach

NWPPC Resource Acquisition Testing
(portfolio mix under environmental constraints, Protected Area Concept for hydro development)

SCL Strategic Corporate Plan
(scenario analysis, standards, trade-off analysis)

LBL-NEES study (CO2 constraints)

Energy Enquête Commission Study for Berlin (W)
(scenario analysis)

Local CO2 Reduction Options Study for Saarbrücken (standards, CO2 constraints)

LCP/CO2 Reduction Study for Hannover & Berlin
(Standards, CO2 constraints)

In addition to screening options with respect to standards, the concept of "critical loads" can be used to derive environmental *targets*: if there is evidence (or consensus) that an environmental impact has to be reduced below a certain level, one can set up a target which has to be met by (electric) energy systems.

The most prominent example for targeting is the application of a CO_2 "cap" in least-cost planning (see paper of F. Krause for this workshop). Only those resource acquisition strategies will be acceptable which can meet the target, all other options or resource mixes are "forbidden". The targeting method will be used in Germany within two studies for municipal utilities which will explore options to integrate LCP and CO_2 reduction on the local scale.

4.2 Exploring the Options: Scenarios and Trade-Off Analysis

Related to standards and targets is a methodology developed by an US municipal utility in Seattle, WA (SCL 1990). Here, a process of iterative scenario analysis is used to identify those energy options (and mixes of options) which have the lowest impact for a bundle of (environmental and economic) goals.

First, a resource portfolio is "screened" using environmental standards developed by the utility, and each energy options is categorized with respect to its compliance with these standards. Then, a "pure" economic resource acquisition strategy (scenario) is developed which yields least (internal) costs. In addition, a least environmental impact scenario is set up using the same portfolio, but minimizing not (internal) costs but environmental impacts.

Now, the utility can clearly choose those energy options for acquisition which offer both low-costs *and* low impacts. In addition, the trade-offs between economic and environmental effects can be determined for all other options, so that the implicit costs of meeting environmental standards can be determined explicitly. Also, the magnitude of trade-offs between different goals can be used to classify the options, and to sort out those which have very low or very high trade-offs.

4.3 Supporting The Job: The Use of Software Tools

As stated before, monetizing emissions has some problems, but multi-attribute approaches which are partly non-monetary have problems, too (see Figure 4).

The most outstanding problem is that these methods are time-consuming when compared to straight-forward monetary internalization. The practicability of multi-attribute schemes can be improved when the process is supported by electronic data processing. A review of computer-based models and software tools prepared by the ÖKO-Institut indicated the possibility to use existing models[3] for the (monetary and non-monetary) incorporation of externalities in least-cost planning (ÖKO 1990). These software tools need some improvement regarding the creation of multiple-attribute graphics which display trade-offs, and the dealing with qualitative impacts, standards and targets could be improved by using advanced programming systems (e.g. expert shells) to identify those energy options which are "robust" under various constraints.

3 Besides LEAP of Tellus Institute, the TEMIS model is available (ÖKO 1989a), providing total fuel cycle ata for emissions (SO_2, NO_x, particulates, GHG), solid wastes, land use, and qualitative aspects. The English TEMIS version was funded by US DOE.

Figure 4: Problems of internalization

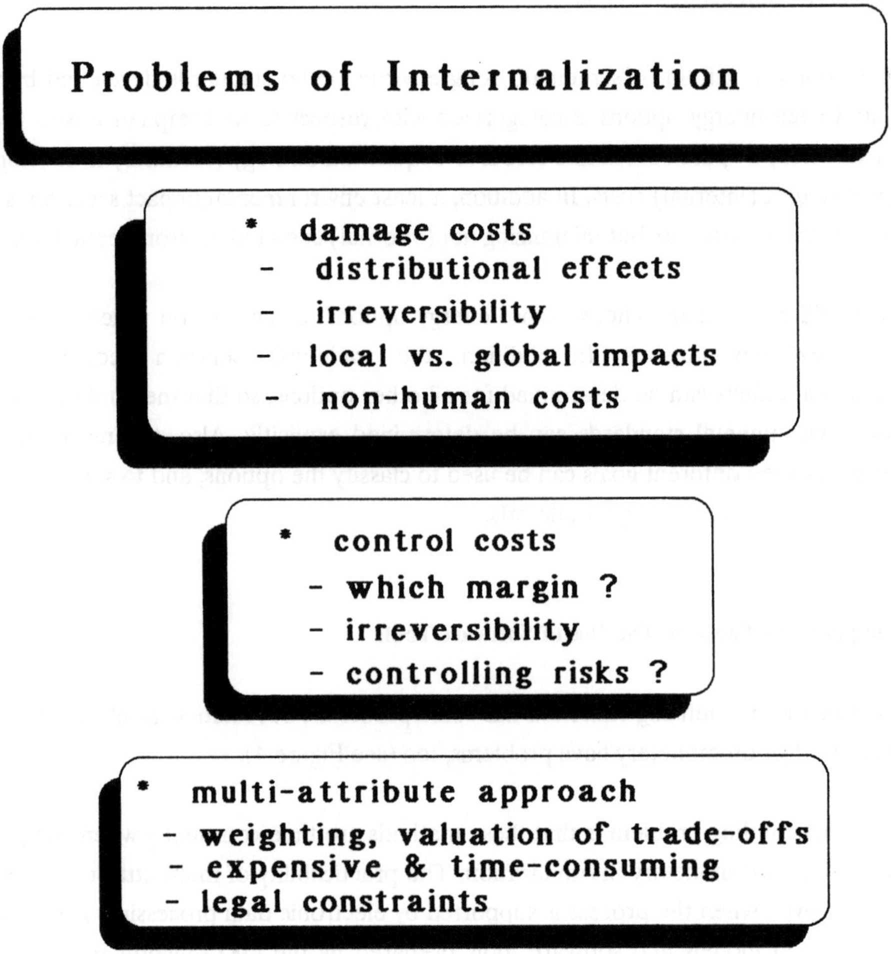

5. Conclusions

The state-of-the-art of incorporating environmental externalities allows to express some impacts in monetized terms, which has great value for utility regulation. Some data are available which allow to include total fuel-cycle externalities. Damage and control costs can be derived for air emissions, which will help to ensure proper resource acquisition with respect to human health, acid rain, and material damage caused by electricity systems.

Monetizing these externalities will also help to give correct price signals to non-utility generators (for the supply-side), and utility customers (for the demand-side). The results of studies in the US and Germany fit within a reasonable range, so that *a minimum value* of environmental externalities can be determined. But several important externalities cannot be expressed in monetary terms without a large uncertainty, and others can be expressed only in qualitative terms. Neglecting these externalities can create significant bias: "*analysts may confuse things which are countable with the things which count*" (Holdren 1982). Therefore, a more comprehensive, multi-attribute approach should be applied in LCP and bidding schemes, targets, and scenario/trade-off analysis.

It should be kept in mind that *all* approaches to internalize environmental costs have great importance also for *other, non-utility sectors* of national energy systems. Beyond regulation, these approaches can be implemented in a wide area of energy-related activities, as Figure 5 indicates.

The definition of energy efficiency standards can be improved when external cost savings associated with reduced energy use are considered, and the debate about environmental taxes or fees can draw substance from research in the externality area (although impacts of price signals have to be considered with caution due to non-market barriers). On the international scale, the incorporation of environmental costs could influence the decision-making for foreign aid in the energy sector, and could also support the "greening" of the World Bank.

Given the nature of political systems, the incorporation of externalities will take place easier in non-electricity sectors when a successful implementation of such concepts has been achieved in utility regulation. Therefore, the monetization of externalities is a promising first step which should be taken not only in the US.

On the way ahead, results from research will improve monetization schemes, and more comprehensive (multi-attribute) concepts can be used *in combination* with monetary approaches. Hopefully, US researchers, regulators, and utilities moving in this direction will be joined by German counterparts.

Figure 5: Options for implementation

Options for Implementation

- **Efficiency Standards**
- **Taxes/Fees**
- **Subsidies (e.g. World Bank)**
- **LCUP/IRP**
- **Bidding Schemes**

Literature

ASE (The Alliance to Save Energy) 1990: *Assessing the costs of climate change policies: the uses and limits of models*, Zimmerman, Mary B., Washington D.C.

AURA (American University Institute for Risk Analysis) 1983: *Assessment of comparative and non-comparative factors in alternate energy systems*, Rowe, W./Oterson, P., EUR-

Beyea, Jan 1990: *Bringing Environmental Damage Costs into the Electricity Marketplace: Gains to be Expected and Pitfalls to be Avoided*, in: NARUC 1990, p.173-200

Boesch, D.F./Rabalais, N.N. (eds.)1987: *Long-term environmental effects of off-shore oil and gas development*, London/New York

Bolze, Doreen A. 1986: *The environmental impacts of off-shore oil and gas development*, EPAD-Report #31, National Auduban Society, New York

Chadwick, M. et al. (eds.) 1987: *Environmental impacts of coal mining and utilization*, Oxford

Enquête (Enquête-Kommission "Vorsorge zum Schutz der Erdatmosphäre") 1990: *Dritter Bericht*, Drucksache 11/8030, Deutscher Bundestag, October 2, 1990, Bonn

Fritsche, Uwe 1990: *Total Emission Model for Integrated Systems (TEMIS): Preliminary Findings for the US*, in Pace 1990, p. 739-759

Holdren, John P. 1982: Energy Hazards: *What to Measure, What to Compare*, in: Technology Review, April 1982, p. 33-38 + 74-75

Hohmeyer, Olav 1990: *Stand der internationalen und nationalen Diskussion der sozialen Kosten verschiedener Energietechnologien*, in: Proceedings 7th International Sonnenforum, Frankfurt, Oct. 9.-12., Vol. III, p. 2039-2044

IKE (Institut für Kernenergetik und Energiesysteme) et al. 1989: *Externe Kosten der Stromerzeugung*, IKE/Temaplan/Ifo-Institut, prepared for VDEW, Stuttgart etc.

ISI (Fraunhofer-Institut für Systemtechnik und Innovationsforschung) 1989: *Stand des Wissens zu externen Effekten der Energienutzung*, Schlomann, B./von Deschwanden, D., prepared for the Enquête-Kommission "Vorsorge zum Schutz der Erdatmosphäre", Karlsruhe

Krause, Florentin 1990: presentation at the NARUC Externalities Conference (NARUC 1990)

LBL (Lawrence Berkeley Lab.) 1990: *Comparative Analysis of Monetary Estimates of External Environmental Costs Associated with Combustion of Fossil Fuels*, Koomey, J., LBL-28313, Berkeley/CA

NARUC (National Association of Regulatory Utility Commissioners) 1990: *Proc.Nat. Conference On Environmental Externalities*, held at Jackson Hole, WY, Sept. 1-3, 1990

OECD (Organization for Economic Co-Operation and Development) 1981: *The costs and benefits of sulfur oxides control*, Paris

1985: *Environmental Effects of Electricity Generation, Paris*

ÖKO (ÖKO-Institut) 1989a: *Umweltanalyse von Energiesystemen: Gesamt-Emissions-Modell Integrierter Systeme (GEMIS)*, Fritsche, U./Rausch, L./Simon, K.-H., prepared for the Ministry for Economy and Technology of Hessen, Darmstadt/Kassel

1989b: *Monetäre und nichtmonetäre Konzepte zur Internalisierung von Umweltaspekten im Energiebereich*, Fritsche, U./Leprich, U. prepared for the Enquête-Kommission "Vorsorge zum Schutz der Erdatmosphäre", Darmstadt/Freiburg

1989c: *Energiedienstleistungs-Unternehmen und Least-cost Planning*, Brohmann, B./Fritsche, U./Lepprich, U., prepared for the Enquête-Kommission "Vorsorge zum Schutz der Erdatmosphäre", Darmstadt/Freiburg

1990: *Environmental Manual for Power Development Phase I (Section Thermal Power Generation)*, Fritsche, U., prepared for GTZ and World Bank, Darmstadt

Pace (Pace University Center for Environmental Legal Studies) 1990: *Environmental Costs of Electricity*, Ottinger, R.L. et al., New York/London/Rome

Plaza 1983: *Impacts of energy development on wildlife: a preliminary study, EPAD* Report #13, National Audubon Society, New York

PLC Inc. 1989: *The Valuation of Externalities from Energy Production, Delivery, and Use - Fall 1989 Update*, Chernick, P./Caverhill, E., prepared for Boston Gas Company, Boston MA

Tellus (Tellus Institute) 1990a: *Environmental Externalities Measurement: Quantification, Valuation And Monetization*, Bernow, S./Biewald, B./Marron, D., presented at the 8th

NARUC Bienniel Regulatory Information Conference, Columbus, OH, September 12-14, 1990

1990b: *Full Cost Economic Dispatch: Recognizing Environmental Externalities in Electric Utility System Operation*, Bernow, S./Biewald,B./Marron,D., in: NARUC 1990, p. 151-173

Thöne, Eberhard/Kallenbach, Ulrich 1988: *Risikovergleiche für die Stromerzeugung*, in: Energiewirtschaftliche Tagesfragen vol. 38 (1988) no. 10, p. 788-794

SCL (Seattle City Light) 1990: *Strategic Corporate Plan 1990-91*, Vol. II, Seattle WA

UNEP (United Nations Environment Program) 1985: *The environmental impacts of production and use of energy*, Part IV - Phase I-III, Nairobi

Welsch, Heinz 1989: *Internalisierung externer Effekte*, EWI Köln, prepared for the Enquête-Kommission "Vorsorge zum Schutz der Erdatmosphäre", Köln

5. SUBJECT AREA 3:

IMPACTS OF THE INTERNALISATION OF EXTERNAL COSTS

5.1 Impacts of External Costs on the Competitive Position of Wind Energy in the Federal Republic of Germany

Olav Hohmeyer
Fraunhofer Institute for Systems and Innovation Research
Breslauer Str. 48
D-7500 Karlsruhe 1
Tel.: 49-721-6809-161

1. Looking beyond environmental externalities in the analysis of social or external costs of electric power generation

In the research that has been conducted at the Fraunhofer Institute on the topic of social or external costs since 1986 we have gone beyond the analysis of such costs in the environmental field because we think that there are other external costs of sizable magnitude which have to be considered as well. The costs we looked at beside environmental costs are government subsidies for energy technologies, which may constitute considerable external costs for example in the case of research and development subsidies for nuclear power, macroeconomic net effects of different energy supply strategies due to different demand and production structures leading to differing levels of induced employment or imports, and cost elements of long-term scarcity, which we think are not sufficiently represented in market prices of non-renewable energy sources. In the following all four categories will be dealt with, while numeric results on each of them will be given separately to allow comparison to other results discussed during the workshop.

In contrast to the neo-classical definition of external costs we use the term social costs in accordance with the definition by Kapp (1979) for all costs of production and consumption which are passed on to third parties or society as a result of any unrestricted economic activity. Thus, the term of social costs will be used in the following instead of the narrower term of external costs generally used in the other contributions to the workshop.

2. Basic questions to be addressed

From day-to-day experience and previous qualitative analyses we had the impression that the social costs of renewable energy sources used for electricity generation may be considerably lower than the social costs of conventional electricity generation in the Federal Republic of Germany. If this is the case, investment in technologies utilizing renewable energy sources may fall short of the optimal level in terms of a macroeconomic optimum or the best strategy for society.

To tackle this problem in the field of electric power generation three central questions need to be addressed:

1. How large is the difference between the social costs (and benefits) of renewable energy sources and those of conventional electricity generation?
2. Is this difference large enough to affect the competitive position of technologies utilizing renewable energy sources for electricity generation like wind energy or photovoltaics?
3. If so, then how does it affect the market introduction and diffusion of these technologies utilizing renewable energy sources?

As previous contributions have discussed quantifying and placing monetary values on certain social costs is difficult and faces a number of substantial uncertainties, particularly in the case of health and environmental damages. Trying to evaluate these costs we find ourselves in the situation of a researcher trying to measure an iceberg being able to see only the tip sticking out of the water. Figure 1 tries to picture the situation of our present attempts.

As a matter of fact we are looking at a number of icebergs and we see that their tips are of considerably different sizes. In a situation where we find the sizes of the visible parts substantially different, we think that these differences can be and need to be considered in decision making although we don't know the entire size of each iceberg. Not considering these (minimum) differences automatically sets them to zero.

3. Results on the order of magnitude of the difference in the social costs of different electricity generating technologies

Trying to measure minimum differences of the social costs of conventional electricity generation and electricity generated from wind energy and sunlight through photovoltaics we published some first estimates two years ago (Hohmeyer 1988). These figures have been modified a number of times according to the discussion and new findings in the field. Two later English publications have summarized these modifications (Hohmeyer 1990a and Hohmeyer 1990b). Table 1 gives the latest results and shows them in comparison to the results published in 1988.

Figure 1: Present situation of the research on the social or external costs of electricity generation

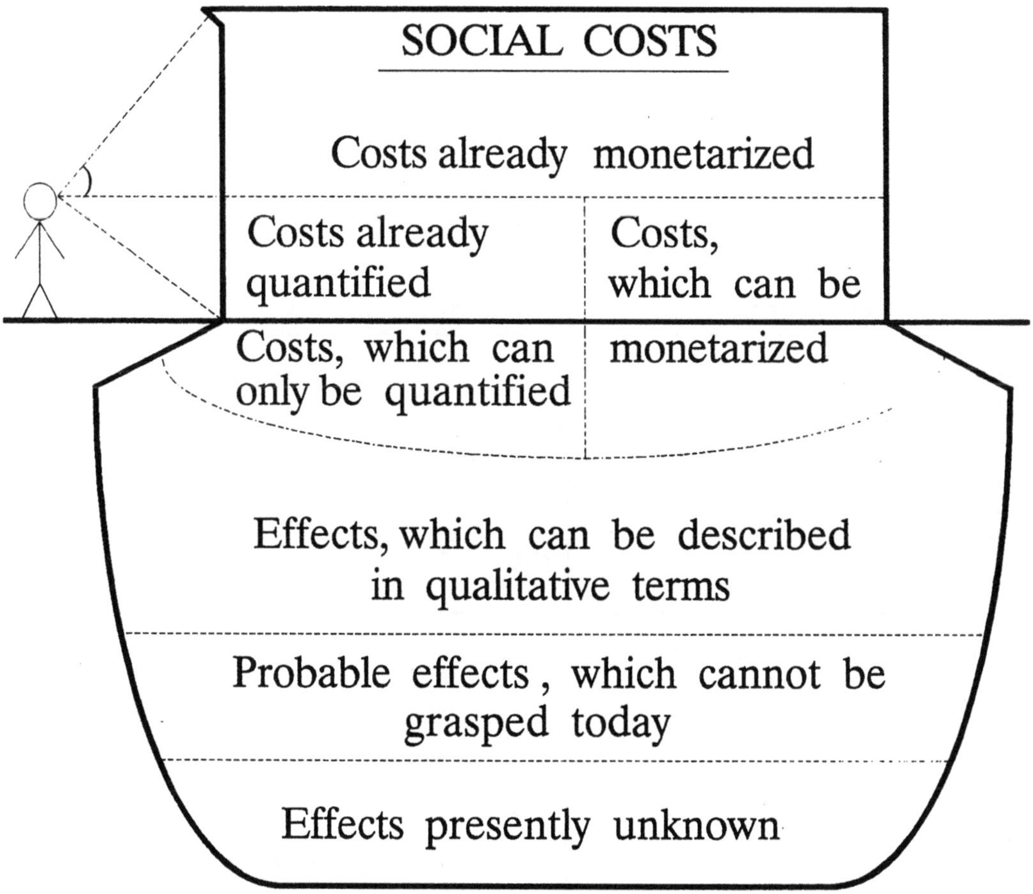

Table 1: Results of the social costs and benefits of competing technologies for electricity generation as calculated by Hohmeyer (1988, p. 8) and according to new calculations made in 1990 (see Hohmeyer 1990b, p. 2).

	Hohmeyer 1988, p. 8	New calculations 1990 incl. CO_2	
a) Gross social costs of electricity generated from fossil fuels (all figures are estimated minimal social costs)		Emissions 82	New power plants
1. Environmental effects	1.14 - 6.09	2.6 - 10.67	2.05 - 7.93
2. Depletion surcharge (1985)	2.29	0.67 - 4.71	
3. Goods and services publicly supplied	0.07	0.06	
4. Monetary subsidies (including accelerated depreciation)	0.32	0.30	
5. Public R&D transfers	0.04	0.02	
Total	3.86 - 8.81	3.65 - 15.96	3.11 - 13.03
b) Gross social costs of electricity generated in nuclear reactors, excluding breeder reactors (all figures are estimated minimal social costs)			
1. Environmental effects (human health)	1.20 - 12.00	3.48 - 21.0	
2. Depletion surcharge (1985)	5.91 - 6.23	4.88 - 47.42	
3. Goods and services publicly supplied	0.11	0.11	
4. Monetary subsidies	0.14	0.14	
5. Public R&D transfers	2.35	1.46	
Total	9.71 - 20.83	10.06 - 70.13	
c) Average gross social costs of the electricity generated in the FRG in 1984		Fossil power plants 1982	New fossil power plants 90
1. Costs due to electricity from fossil fuels (weighting factor 0.705[1])	2.87 - 6.56	2.58 - 11.25	2.19 - 9.19
2. Costs due to electricity from nuclear energy (weighting factor 0.237[2])	2.48 - 5.32	2.38 - 16.62	
Total (conventional electricity)	5.35 - 11.88	4.96 - 27.87	4.57 - 25.81
d) Net social costs of wind energy			
1. Environmental effects (noise)	(-)0.01	(-)0.01	
2. Public R&D transfers (estimate)	-0.26 - (-)0.55	-0.16 - (-)0.33	
3. Economic net effects	+0.53 - (+)0.94	+0.47 - (+)0.78	
4. Avoided social cost of present electricity generation	+5.35 - (+)11.88	+4.96-(+)27.87	(+)4.57-(+)25.81
Total social benefits rounded to three digits	+5.6 - (+)12.3	5.26 - 28.32	4.87 - 26.25
mean	(+) 8.9	16.8	15.6
e) Net social costs of solar energy (photovoltaics)			
1. Environmental effects (noise)	(-)0.44	(-)0.44	
2. Public R&D transfers (estimate)	-0.52 - (-)1.04	-0.33 - (-)0.65	
3. Economic net effects (not including 1982 figures)	+2.40 - (+)6.65	+2.35 - (+)8.35	
4. Avoided social cost of present electricity generation	+5.35 - (+)11.88	+4.96-(+)27.87	(+)4.57-(+)25.81
Total social benefits rounded to three digits	+6.8 - (+)17.1	+6.54-(+)35.13	(+)6.16- (+)33.07
mean	(+)11.9	20.8	19.6

1 Old weighting factor 0.7444

2 Old weighting factor 0.2556

Major modifications have been made in the consideration of fossil power plants by the introduction of new power plants with desulphurisation and denitrification equipment, the introduction of higher CO_2 control costs, a recalculation of nuclear accident costs based on the latest German reactor safety study, a new model to represent long-term scarcity of non-renewable energy sources, and a more realistic interest rate taken into account in the calculation of the R&D subsidies and their annual attribution. The new calculations show that the minimum values calculated originally prove to be rather robust remaining in the range of 5 to 6 Pf/kWh (3 to 4 US cents/kWh) whereas the upper values tend to be higher than originally estimated. For electricity generated from fossil fuels environmental costs account for 60 to 67 % of the difference between the social costs of fossil electricity and wind energy, while environmental costs of nuclear electricity account for 29 to 34 % of the respective difference. Of the difference in the average social costs of conventional electricity generation and electricity from wind energy the environmental effects account for 49 to 55 % constituting the largest single item of the four categories of social costs considered.

The first question on the difference between the social costs of wind energy and photovoltaics on the one side and conventional electricity generation on the other can be answered in stating that this difference is of the same order of magnitude as the internal or private production costs of conventional electricity.

4. Effect of social costs on the competitive situation and market diffusion of technologies utilizing renewable energy sources

How can one analyse the impact of considering social costs on the competitive position of a new technology versus an established technology? One way is to examine a two-product-market, as portrayed in figure 2. The costs of the established technology are increasing gradually due to rising exploration and mining costs, for example, while the costs of the new technology based on renewable energy sources are decreasing considerably over time due to technological learning.

One can show such developments empirically for conventional electricity and wind or solar energy. At the point t_0, the new energy technology reaches cost effectiveness if one considers no social costs. The substitution process can start at t_0.

Figure 2: Cost development of electricity generation over time (no social costs considered)

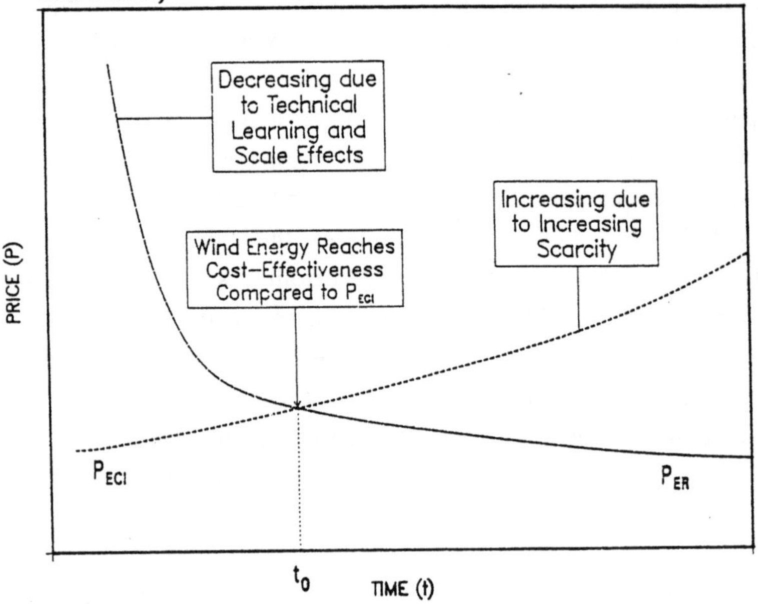

P_{ER}: Wind energy as an example for renewable energy sources

P_{ECI}: Conventional electricity (only internal costs)

Figure 3: Cost development of electricity generation over time (social costs considered)

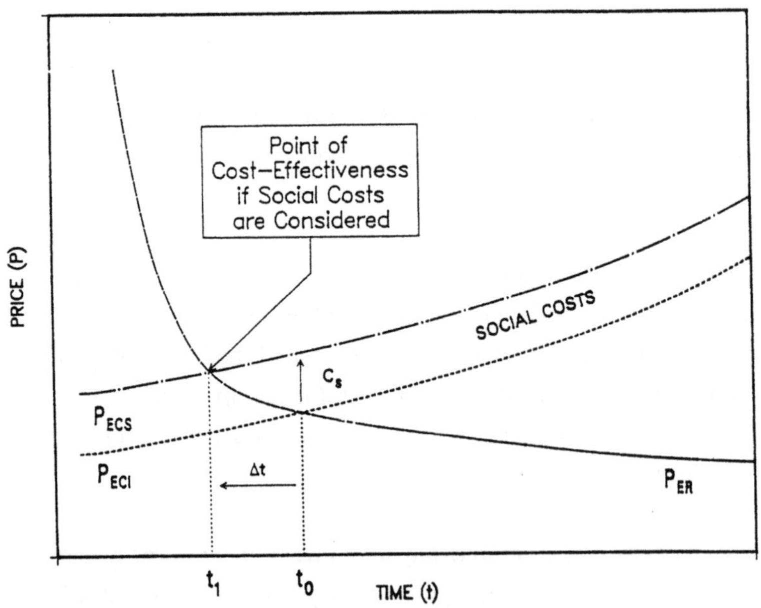

P_{ER}: Wind energy as an example for renewable energy sources

P_{ECS}: Conventional electricity including net social costs

Figure 3 shows the effect of including the net social costs. These are defined as the difference between the social costs of the conventional electricity generation and those of the new technology. A static application of the social costs for a base year (e.g. 1985) results in a parallel projection of conventional electricity's market price curve. This results in a new intersection of the renewable energy cost curve with the market price curve of conventional electricity generation and shows that the new energy technology reaches cost effectiveness at t_0 minus $Dt(t_1)$. If the social costs reach a sizable order of magnitude, then a distorted competitive situation results: The wrong price signals are given for the choice of energy technologies.

Because cost effectiveness does not lead to instant technology substitution but to a substitution - or market diffusion - process that may stretch over 20 or more years, one can picture the impact of not considering social costs as a shift of Dt in the new technology's market penetration curve as shown in figure 4. If one does not consider social costs, then the whole diffusion process is delayed by this time span as compared with the best possible diffusion time schedule for society.

The social costs quantified in table 1 are applied to the following analysis of the future competitive position and market diffusion of wind and photovoltaic solar energy. Figure 5 shows the impact of including social costs of electricity generation of the competitive situation and on the resulting market introduction of wind energy systems in the Federal Republic of Germany (comparing on the basis of new fossil power plants). An electricity price of 0.156 DM_{82}/kWh (working price) is assumed for small consumers and is applied to private production of electricity by wind energy, which substitutes electricity otherwise bought from public utilities. All assumptions are given in detail in table 2. A price of 0.065 DM_{82}/kWh is assumed for the electricity that decentralized installations sell to the grid. The analysis also assumes that WEC owners will consume only 20 percent of the electricity produced by wind energy systems and that 80 percent will be sold to the public grid. This results in an average electricity price of 0.083 DM_{82}/kWh for the conventionally generated electricity to be substituted by the wind energy systems. A real price increase of 2 percent annually for the substituted electricity has been taken into account. Figure 5(a) shows this price development as the lowest price curve for substituted electricity. For the electricity price of small wind energy systems of 50 to 100 kW nominal power, a cost curve has been derived based on the few available German wind energy cost figures for the period 1980-1986 and on the reliable Danish wind energy prices for the years 1979-1985. The data given as investment costs have been recalculated as cost per kilowatt hour assuming yearly capital costs of 9.63 percent (annuity) based on (i) a real interest rate of 5 percent, which slightly exceeds the long-term real interest rate of the past 30 years in the

Figure 4: Market diffusion of wind energy due to the handling of social costs

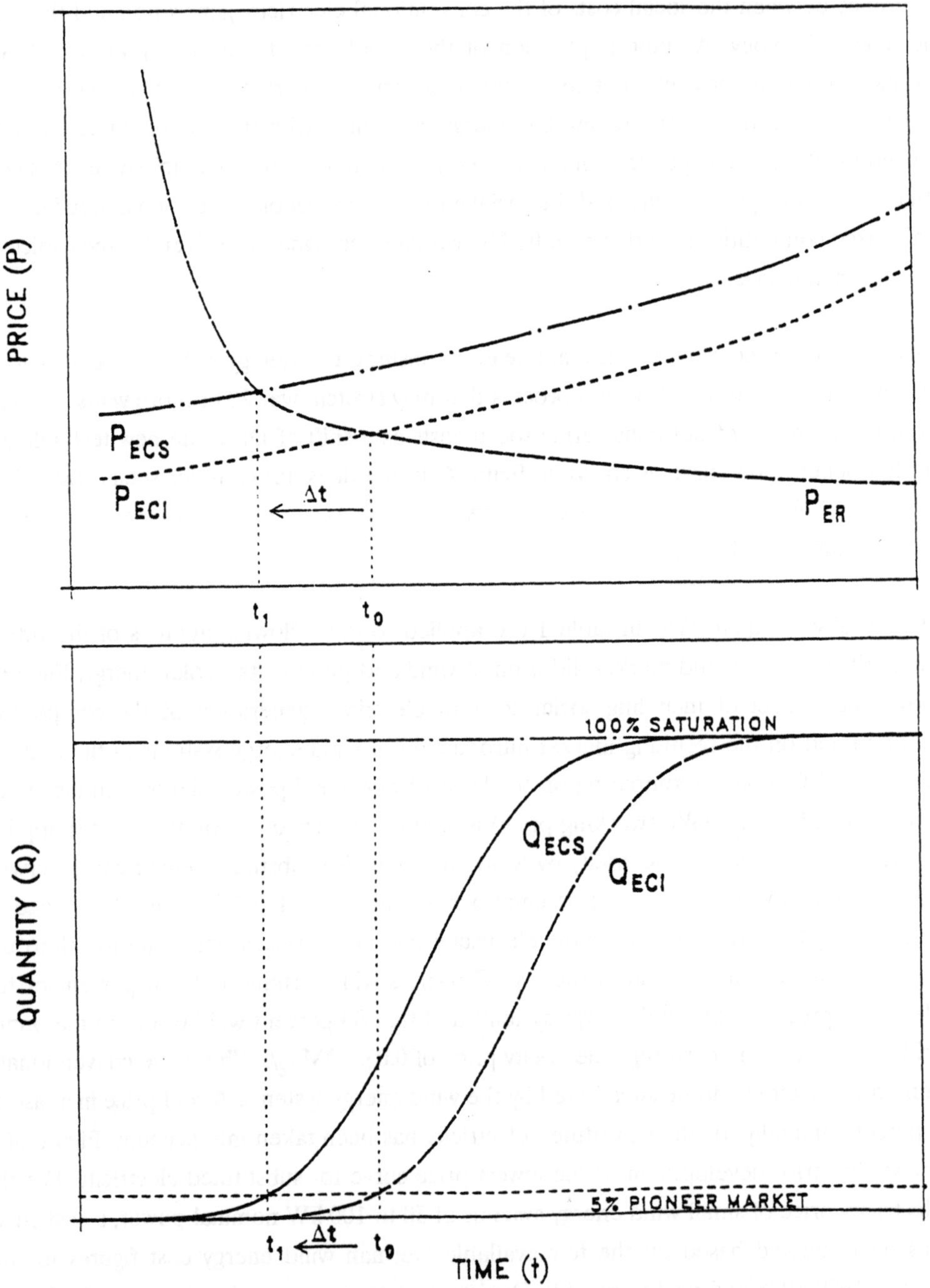

Q_{ECI}: Market diffusion curve without considering social costs

Q_{ECS}: Market diffusion curve taking into account social costs

Table 2: Assumptions Underlying the Analysis of Social Costs and the Impact on the Competitive Situation of Wind and Photovoltaics

1. General Assumptions	
1.1 Price of Substitutable Conventional Electricity (1982)	25.1 Pf_{82}/kWh
1.2 Working price (62,5 %)	15.6 Pf_{82}/kWh
1.3 Payment for Electricity Supplied to the Public Grid	6.5 Pf_{82}/kWh
1.4 Real Price Escalation of Conventionally Produced Electricity	2 %/Year
1.5 Real Interest Rate for the Financing of New Investments in Wind and Photovoltaic Machines	5 %/Year
1.6 Market Potential for Wind and Photovoltaic Machines	20 TWh/Year
1.7 "Pioneer Market" (5 % of the market potential)	1 TWh/Year
1.8 Time Period for the Diffusion Phase (5 % to 95 %)	20 Years
2. Assumptions About Wind Energy	
2.1 Share of Wind Energy Consumed by Owner	20 %
2.2 Share Sold to Utility	80 %
2.3a Compound Gain of Wind Electricity (1982)	10.2 Pf_{82}/kWh
2.3b Compound Gain of Wind Electricity Based on Working Price Assumption	8.3 Pf_{82}/kWh
2.4 Life Expectancy of Wind Energy Facilities	15 Years
2.5 Annuity	9.63 %/Year
2.6 Operating and Maintenance Cost	1.5 %Year
2.7 Wind Energy Costs in West Germany	
1980	44.8 Pf_{82}/kWh
1986	19.6 Pf_{82}/kWh
1990	15.0 Pf_{82}/kWh
2000	12.1 Pf_{82}/kWh
2010	10.2 Pf_{82}/kWh
2030	8.4 Pf_{82}/kWh
2.8 Wind Energy Costs in Denmark	
1980	12.5 Pf_{82}/kWh
1986	9.1 Pf_{82}/kWh
1990	7.6 Pf_{82}/kWh
2010	7.4 Pf_{82}/kWh
2030	7.0 Pf_{82}/kWh
3. Assumptions About Photovoltaics	
3.1 Share of Photovoltaic Energy Consumed by Owner	50 %
3.2 Share Sold to Utility	50 %
3.3a Compound Gain of Solar Current	15.8 Pf_{82}/kWh
3.3b Compound Gain of Solar Current Based on Working Price Assumption	11.1 Pf_{82}/kWh
3.4 Life Expectancy of Solar Facilities	20-30years
3.5 Annuity	8.02-6.505 %/a
3.6 Operating and Maintenance Cost	12 Pf_{82}/Wp a
3.7 Solar Energy Costs	
1982	267 Pf_{82}/kWh
1990	122 Pf_{82}/kWh
2000	62 Pf_{82}/kWh
2010	42 Pf_{82}/kWh
2020	32 Pf_{82}/kWh
2050	26 Pf_{82}/kWh

Note: Pf_{82} = Pfennig, 0.01 of a German Deutsche Mark, 1982 prices
TWh = Terawatt hour
DM = Deutsche Mark in 1982 prices

Federal Republic of Germany, and (ii) an operational life span of 15 years, which clearly is below the 20 years generally assumed.

The wind energy cost curve intersects with the market price curve of the electricity to be substituted at point A in figure 5(a). At this intersection point (about the year 2000), wind energy is competitive with the electricity to be substituted at market prices, which do not include social costs. Adding the lower range of the estimated minimum net social costs (0.049 DM/kWh) to this market price curve results in a second curve for the substituted electricity where point B (1991) is the new point of cost effectiveness for wind energy. Including these social costs shows that wind energy is competitive with the electricity to be substituted considerably earlier than market prices show. Accordingly, the market introduction of wind energy systems starts much earlier. Figure 5(b) shows this for German wind energy price conditions.

The pioneer market for wind energy systems not oriented toward cost effectiveness has been assumed to amount to about 5 percent of a technically feasible potential of about 20 TWh wind electricity produced per year in the Federal Republic of Germany. This potential translates into about 100,000 wind energy systems of 100 kW nominal power. After wind energy systems become cost effective, market penetration should begin. Market penetration should reach 95 percent of the technical potential 20 years after achieving cost effectiveness, as figure 5(b) shows.

Adding the upper range of the minimum net social costs of electricity generation (0.263 DM/kWh) to the market price of substituted electricity results in a third cost curve for the substituted electricity. This curve gives a point of cost effectiveness for wind energy systems as the intersection with the wind energy systems as the intersection with the wind energy cost curve at point C (1981). Figure 5(b) shows the change in the market penetration of wind energy systems resulting from this altered cost situation in the same way as the effect discussed above. Including social costs in the market allocation process already would have led to a substantial shift in the market penetration of wind energy systems.

The point of cost effectiveness of wind energy systems in the Federal Republic of Germany varies overall from 9 to 19 years (calculations based on new fossil power plants and existing nuclear facilities) depending on how social costs are taken into account in the near future. Thus, if social costs are not included, then a serious misallocation of resources clearly results. This causes substantial costs to society.

Figure 5: Influence of Social Costs on Starting Point of Market Penetration of Decentralized Wind Energy Systems and Future Penetration to Year 2030 (1982 prices)

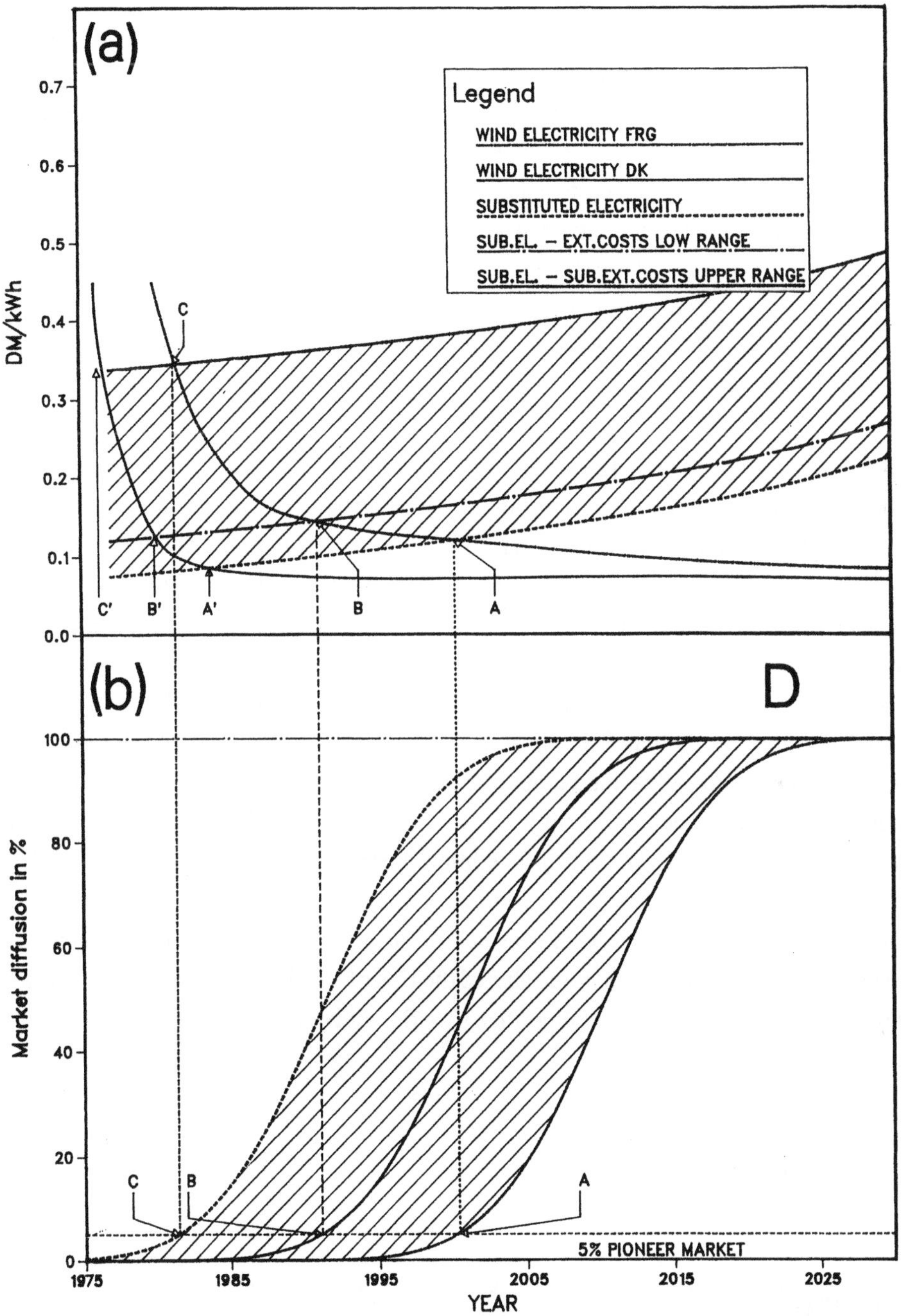

(a) Costs for Wind Energy Compared With Costs for Substituted Electricity

(b) Market Penetration of Wind Energy Based on Costs Shown Above

5. Conclusions for the Federal Republic of Germany

Because the general market pricing mechanism does not work adequately in the case of competing electricity-generating technologies, government must step in to internalize the social costs of economic processes. Theoretically, this should be done by leveling a tax against all electricity-generating processes that cause these social costs. The tax would offset the amount of these costs passed on to society. The estimated minimum cost discussed above could be a starting point for such a tax. But in practical politics, overcoming interest groups' resistance to such a tax would be difficult. One way around the practical problem of imposing a tax would be to pay a compensation for wind and photovoltaic energy systems to the extent of the social costs avoided through their operation.

In the case of wind energy, urgent action is required so as to correct the present seriously distorted market situation. The substantial social costs of electricity generation by fossil fuels and nuclear energy, or at least the positive net social benefits of wind energy, should be internalized so as to achieve energy prices that allow a better allocation of resources to the different energy systems. If this internalization is not achieved, then a sub-optimal electricity-generating system that relies too little on renewable energy sources will impose substantial social costs on society in the Federal Republic of Germany. Rapid internalization of social benefits and optimal allocation of resources in the energy market can best be achieved by paying compensation to wind energy systems for their net social benefits. At present, such compensation could range from 0.05 to 0.26 DM_{82}/kWh wind electricity produced. This would translate into approximately 150,000 to 780,000 DM_{82} per wind energy system (with 100 kW nominal power). This amount equals the estimated minimum social benefits that would result from using one such wind energy system. Even considering just the difference in the size of the tips of the different 'icebergs of social costs' makes a big difference for the competitive situation of renewables.

REFERENCES

Hohmeyer, O., *Social Costs of Energy Consumption*, Berlin, New York, Springer-Verlag, Tokyo, 1988.

Hohmeyer, O., *Soziale Kosten des Energieverbrauchs*, 2nd ed., Springer-Verlag, Berlin, 1989.

Hohmeyer, O. (1990a), "*Social Costs of Electricity Generation: Wind and Photovoltaic versus Fossil and Nuclear*", in Contemporary Policy Issues, Vd. VIII, No. 3, July 1990, 255-282.

Hohmeyer, O. (1990b), "*Latest Results of the International Discussion on the Social Costs of Energy - How Does Wind Compare Today?*", Proceedings of the European Commission's Wind Energy Conference 1990, held in September 1990 at Madrid, publication forthcoming.

Kapp, W. S., *Soziale Kosten der Marktwirtschaft*, Fischer Verlag, Frankfurt, 1979.

Pigou, A. C., *Wealth and Welfare*, London, 1912.

5.2 External Costs of Electricity Generation

R. Friedrich[+] and U. Kallenbach[++]
University of Stuttgart
[+] Institute for Energy Economics and the Rational Use of Energy
[++] Institute for Nuclear Energy and energy Systems
Pfaffenwaldring 31
D-7000 Stuttgart 80

Introduction

"External costs" of electricity generation are defined here as all negative tewchnology-related effects involved in electricity generation - including antecedent and subsequent process steps such as the erection and dismantling of power plants, the acquisition and transport of energy resources, and the disposal of wastes, which are paid for not by the producer but by the general public or other parties.

"Social costs" are all costs incurred through the generation of electricity, i.e. the total of external costs and the costs borne by the producer. Social costs do not include taxes, as these are only transferred rather than expended.

As the producer of electricity has not to compensate for external costs, he will not consider external effects in his decision process. Misallocations of scarce resources may be the consequence.

The argument of external costs has gained new impetus in connection with renewable energy technologies which are not yet competitive in terms of private economics but are supposed to produce considerably lower external costs than conventional energy sources.

So, the arising question is: will an appropriate treatment of external effects alter the competitive position of renewable energy technologies? Unfortunately, no one is able to provide a competent answer to this question, because it is not possible to quantify and monetize *all* external effects associated with different energy technologies. Consequently, the consideration of external costs can only deal with the examination, whether *quantifiable* external effects would make any decisive difference to the cost effectiveness situation of different electricity producing systems.

This paper highlights some problems when analyzing external effects with the required degree of precision; in addition, the results of an estimate of quantifiable external effects using all the information and data available at present are presented.

Forest Damage

Most of the problems incurred when quantifying external effects can be demonstrated by treating the exemplary case of damage to forests caused by environmental pollution.

The evaluations of external costs attributed to the observed damage to plants and in particular to forests requires the following principal steps:

- the determination of the contribution of a source of emissions to existing ambient concentrations of air pollutants,
- the quantification of the surplus damage caused by one additional unit of ambient air concentration of pollutants,
- the definition of a monetary equivalent of the unit damage.

However, based on present knowledge none of these steps can be accomplished with sufficient accuracy.

- Spreading and chemical transformation of pollutants can only be described in a very rough manner.
- The damage function, which gives the relation between ambient air concentration of pollutants and damage is not known.
- Commonly accepted methods to quantify damage types such as the reduced value of forest as leisure area are not available.

Nevertheless, estimates for the monetary values of forest damage have been made /2/. However, these results are not more than 'educated guesses'using the Delphi method. If the damage costs estimated in /2/ are used external costs of 0,19 Pf_{88}/kWh_{el} result for a modern coal fired power plant with a desulphurization and DENOX plant.

Effects on Human Health

Electricity generating systems may impact human health in a number of different ways. Detrimental effects stem from routine emission e.g. of toxic agents released during the operation of a coal-fired power plant as well as from accidential situations such as a major accident at a nuclear site or a coal mine disaster.

Some of these effects or health risks can be quantified with rather high accuracy, as it is the case for fatal occupational accidents in coal mines or during the construction of wind energy converters. Other effects such as respiratory diseases caused by emitted air pollutants lack quantification due to missing valid dose-response functions.

In the case of major accident at nuclear power plants it seems achievable to quantify the public dose comittment following a defined release within rather narrow bounds. However, it is still more or less disputed how often the occurence of a given radiological source term must be expected per year of operation. This is one reason why the overall risk of nuclear accidents cannot be quantified today without doubt.

Transforming quantitative risk estimates into monetary values is an even more distracting task, since commonly agreed monetary values for human lives and human health can hardly be defined. Approaches utilized in present assessments of external costs imposed by health risk of energy system do not account for the intangible values involved, e.g. the human pain and sorrow associated with illness and death.

Our estimates of external costs from health efects - except nuclear accidents - are based on a study that evaluated the different risks of energy systems /3/. For a first estimation of the risks caused by accidents in nuclear reactors, a study from the US /4/ has been used.

Utilization of Non-Renewable Resources

Non-renewable resources such as coal, oil, gas and uranium are only available in finite quantities. As they are likely to be exhausted sometime in the future, there is the question of whether the increased shortage of these goods as they are used up is allowed for in the market price in order to guarantee "optimal intertemporal allocation" as a means of

A rough calculation based on back-stop technology costs of 0.8 DM_{88}/kWh and a fossil resource availability of 200 years and with a real discounting rate of a 4 % p.a. gives - according to the Hotelling rule - a loss of benefit for future generations of approx. 0.03 Pf_{88}/kWh_{el} or approx. 0,01 Pf_{88}/kW_{th} in terms of fuel.

This value is so low that it might even be already included in the present fuel prices.

Research and Development Expenditure

The a posteriori internalization of funds already spent is not a sensible approach in terms of future investment decisions because the costs already incurred cannot be reversed. Even with a total run-down of nuclear energy the funds already spent on nuclear power stations could not be recovered, for example. For this reason, only future expenditure directly concerned with the construction, operation and dismantling of new electricity gneeration units can be included, as it are only these costs which can be influenced by today's decisions.

So an approach could be made that divides the annual R&D expenditures by the annual electricity production of a certain technique. With this approach, the external costs for research and development are 0,06 Pf_{88}/kWh_{el} for coal fired power stations and 0,45 Pf_{88}/kWh_{el} for nuclear power stations.

Problems arise, as a presise attribution of expenditures to a specific group of electricity generating systems is often difficult or - as in the case of fundamental research - impossible. Furthermore, new systems like wind and photovoltaics produce only a very small amount of electricity at present, so external costs of renewable energy systems for R&D would be extremely high; new innovative systems would be handicaped. So another approach would be to continue with the current policy and to regard R&D expenditure as a provision for the future without internalizing them as external costs.

Public Subsidies

If public subsidies are regarded as external costs, the subsequent demand for internalization implies, that the subsidies should be canceled. However, if the public subsidies are granted to compensate for an external benefit, they should not be regarded as external costs. So, the question of whether subsidies are extrenal costs changes to the question of whether subidies are justified or not. If not, they should be canceled anyway.

The public subsidies for coal, which are related to electricity generation, amount to 772 million DM_{88} in West Germany in 1988, which can be transformed into specific costs of 0,58 Pf_{88}/kWh_{el}.

External Costs of Electricity Generating Systems

Results of a first estimate of the external costs of quantifiable external effects of electricity generation from coal, nuclear power, wind and photovoltaic are shown in fig. 1 /1/.

Forest damage, damage to animals, plants and materials, health effects, accidents of nuclear power plants, public research and development expenditures, public subsidies and the depletion of natural resources have been considered. Health effects are regarded not only for power plant operation, but also for power plant construction and dismantling and the whole fuel cycle. Not considered are, amongst others, the human suffering caused by accidents and illness and the effects of climatic changes due to the release of greenhouse gases. It has to be pointed out, that the quoted data, especially the results for forest damage, health effects and the effects of accidents at nuclear power plants are based on very rough estimations of the damage functions and the monetary value of the damages.

According to the estimate presented here, the quantifiable external costs of the electricity generating systems under review represent a maximum of about 7 % of the electricity generating costs, the external costs for coal fired power plants being the highest.

Figure 1: External costs of quantifiable external effects of different power plants

IER

Cost Category	Pf_{88} / kWh			
	Coal	Nuclear	Wind	Photovoltaic
Health effects	0.18 - 0.19	0.02 - 0.06	0.02 - 0.05	0.06 - 0.09
Hyp. accidents	-	0.01 - 0.07	-	-
Environmental effects				
- forest	0.19	-	-	-
- noise	-	-	0 - 0.01	-
- other plants and animals	0.02	-	-	-
- materials	0.05 - 0.08	-	-	-
Depletion of natural resources	0 - 0.03	0 - 0.03	-	-
Public research and development	0 - 0.06	0 - 0.45	0 - 0.34	0 - 1.25
Public subsidies	0 - 0.58	0 - 0.03	-	-
Total Amount	0.44 - 1.15	0.03 - 0.64	0.02 - 0.40	0.06 - 1.34

External Costs of Selected Power Plants

Conclusions

- In principle, internalizing external costs may reduce misallocations of limited economic resources. Of course, within this process all branches of the economy should be considered, as the consideration of electricity production alone may also lead to non-optimal decisions.

- The main problem, that hampers the application of the concept of internalizing external costs, is the quantification and monetization of external effects, as methods for the quantification as well as dose-response-relationships and damage functions are uncertain, doubtful, controversial or totally missing.

- If available cost estimates are used in spite of the large uncertainties, *quantifiable* external costs result, which do not change the competitive position of electricity production from coal, nuclear power, photovoltaics and wind significantly.

- In particular, with today's knowledge it is not possible to prove, that a lower bound of quantifiable external costs exists, that would shift renewable energy systems, which are far from being economical, into a favorable competitive position. Consequently, a decision to use such renewable energy systems has still to be the result of the 'subjective' balancing of pro's and con's including the consideration of external effects that can not be quantified or monetized.

- It is not surprising, that quantifiable external costs are rather small. The reason is that identified significant external effects have always been internalized in the costs by proper instruments such as standards and regulations. For instance, the occurence of forest damage in Germany led to ordinances, that achieved a substantial decrease of emissions. These ordinances were enacted without prior knowledge of the exact relations between emissions and damage and without existence of valuable estimates of the external costs from forest damage. So, in many cases it may not be realistic to assume that the monetizing of external effects leads to the proof that effects which have been regarded as insignificant have high external costs. On the contrary, in many cases there will be the awareness that an external effect causes severe problems first and only then the problem is investigated in detail. In these cases it may be too late to wait for action, until external costs are calculated - this limits the usefulness of the external cost approach.

- In addition to the quantifiable external effects there may be effects that are not known today, and there certainly are effects, that cannot be quantified may be considerable.

- In conclusion, a full and reliable quantification of all known external effects caused by electricity production is not possible. Consequently, decisions in this field require a valuation of all - quantifiable and not quantifiable - advantages and disadvantages. Nevertheless, it is sensible to extend the quantifiable parts of the decisison process.

References

/1/ Friedrich, R.; U. Kallenbach; E. Thöne; A. Voß; H.-H. Roger; H.-D. Karl: Externe Kosten der Stromerzeugung, VWEW-Verlag, Frankfurt/M., 1990

/2/ Ewers, H.-J.; H. D. Brabänder; H.-M. Brechtel u.a.: Zur monetären Bewertung von Umweltschäden, UBA-Bericht 7, Berlin 1986

/3/ Kallenbach, U.; E. Thöne: Gesundheitsrisiken der Stromerzeugung. Verlag TÜV Rheinland, Köln, 1989

/4/ Burke, R. P.; D. C. Aldrich; N. C. Rasmussen: Economic Risks of Nuclear Power Reactor Accidents, NUREG/CR-3673, Washington DC, 1984

5.3 External Costs Involved in Generating Electrical Energy in the Eastern Part of Germany

Martin Weisheimer
Akademie der Wissenschaften der DDR
Zentralinstitut für Wirtschaftswissenschaft
Prenzlauer Promenade 149-152
Berlin 1100

In 1989, a gross total of 118,971 GWh electricity were generated in the Eastern part of Germany, the former GDR. This total is equivalent to a gross output of 7,135 MWh, or a net output of 5,851 MWh, per head of the population. The power generated came from the following sources:

Gas engine power stations	0.2 %
Run-of-river power stations	0.2 %
Pumped-storage power stations	1.2 %
Nuclear power stations	10.3 %
Steam power stations	88 %.

The steam was produced almost exclusively (approx. 85 %) from domestic brown coal resources. The percentages have been calculated on the basis of "Gesamtbilanz Energie 1989, Wirtschaftsraum DDR" (Institute for Energetics 1990).

In the five Laender comprising the Eastern part of Germany, wind energy and biogas may both take on a relevant role in the medium term, and photovoltaic power in the long term. The speed and extent to which these resources will be put to practical use depend not least on the influence exerted by internal effects and costs on the relative competitiveness of the individual source of energy in question. The prices paid to these suppliers of the national grid and other state subsidies are also relevant in this context. They in turn depend on the external costs and effects.

In conjunction with the conditions outlined above, the current emphasis of external costs lies without question on the comparison between coal-produced energy and nuclear energy. Before analyzing this in more detail, a few fundamental remarks will be necessary on the special features associated with determining prices, expenditure and costs in the economy of the former GDR.

1. External costs in a system of centralized state economy

In comparison to the FRG and other countries having a free market economy, there are generally much greater deviations in the GDR and in other Eastern European countries between the costs/expenditure to be borne by society as a whole and currently valid prices or scales of pay. There are a number of reasons for this. The main reason lies in the fact that the previous system of centralized state economy allowed for very little or no influence to be exerted by profitability calculations on decentralized decisions.

In a restricted economy, as opposed to a free market economy, state intervention predominates a priori in the form of special assistance and support measures, subsidies, financial redistribution, etc. Until now, the consequences of certain developments have been only inadequately reflected in product and performance. External costs therefore have particular significance for Eastern European countries, although they are also difficult to determine. In the literature of Eastern Europe, reference to "external" costs is uncommon. The terms "complex national costs" or "full costs and expenditure" are frequently found for the sum of direct costs included in prices and external costs not yet included in prices.

The following may be said to be generally applicable for the energy sector in a system of centralized state economy: the price level for energy sources is too low in comparison to that of the manufacturing industry. This is due to three factors.

- In the first instance, from its very nature, the method of pricing used consciously takes only incomplete account of the comprehensive, long-term need by firms for funds to finance expanded reproduction from internal earnings. Specific financing conditions stipulated by the state, such as the redistribution of amortizations, mandatory external financing with borrowed funds, interest-free state loans, low-interest public credit, etc., are inherent to this price policy as an integral part of the system.

- On the other hand, it is precisely this sector which is characterized by its high proportion of indirect costs, above all for ecological and social expenditure.

- And thirdly, the model for funding is based on the fact that natural resources, such as land and mineral deposits, are subject to little or no economic valuation, and are thus quoted as "valueless".

These distortions in practical valuation render economic analysis and comparison difficult. They give no support to substitute forms of energy by engineering, technology or science. They render energy-saving goals, and developments aimed at economizing on resources and thus protecting the environment, not particularly advantageous over other work-intensive and high-investment processes. The economic freedom of movement involved in energy-saving is excessively restricted. New sources of energy such as biogas and geothermal heat, already facing initiation problems, are further affected. Their economically justifiable exploitation potential is cited as being too low.

In a system of centralized state economy, a view of expenditure seen in relation to product and performance, costs and profit is an unreal approach, above all because the individual factors in production are subject to different levels of distortion in terms of their economic valuation. Examples of this are the "second wage-packet" to supplement a low wage or salary, the intentional overvaluation of materials and financial valuation of imports not conforming to market prices.

This current starting-point makes it essential to focus attention on practical solutions aimed at designing and exploiting market-oriented prices, and at detecting and taking account of the many direct, indirect and hidden assistance measures/subsidies provided by the state (Weisheimer 1990). It gives a different perspective to the present significance of other areas of external costs. At the same time, it helps to explain the delay in researching these problems.

The influence of macro-economic structural change on national net added value and on the long-term saving of resources is the most difficult to assess. How far, for example, can the overall economic or social order of priority of each of the energy sources be altered by such change alone?

There are also still no adequate quantitative estimations of the varied effects and expenditure involved in preventive and corrective environmental protection measures. Comprehensive research work remains to be done in the area of costs/follow-up costs for environmental damage . Those methods and practical experience presented at the workshop initiated by scientists and practical experts from the FRG and the USA will no doubt be of valuable assistance here.

2. Environmental costs of generating electricity from coal

Severe environmental damage combined with costs/expenditure by society are the result of generating electrical energy. At the centre of this damage lies air pollution, with dust, SO_2, NO_x, CO_2 and radioactivity. Combined with this are the effects air pollution simultaneously has on land and water.

- Power station operations illustrate this by the pollution load contained in their waste water, and by the claims they lay to areas of land for production and waste disposal sites (for far more than 10 million t/a of ash and particularly for radioactive material).

- In the preliminary production phases, the major ecological effects of brown coal surface mining lie in the devastation of areas of land and infrastructures exploited, and in the lowering of groundwater levels. The mining of uranium ore as a preliminary stage to utilizing nuclear energy is associated with a high level of radioactive contamination of the surroundings; in the 1,200 km^2 bismuth sector, for example, estimations of the costs for past contamination total at least DM 4 billion.

The air in the Eastern part of Germany is also heavily polluted. The SO_2 load per head of the population, for example, is 16 times higher than in the FRG; for dust, the load is double; for NO_x and CO_2, it is 1.5 times as high. The major perpetrators are without doubt to be found in the energy-producing processes. The following loads are estimated to have been emitted from brown coal power stations and from district heating power stations (approx. 18,000 MW with some 100,000 GWh) in 1989:

- dust	approx.	670	KT
- NO_x	approx.	200	KT
- SO_2	approx.	2,650	KT
- CO_2	more than	150	MT

Figures taken from Institute for Energetics (1990, p. 6)

This emission of pollutants, together with the movement of air across borders, is known to cause considerable damage to humans, animals, plants, buildings and installations, as well as influencing global climatic change. There are various possible procedures for determining the disadvantages arising for society from such developments.

On the one hand, attempts should be made to record, quantify and value the damage caused according to perpetrator. On the other hand, determining those costs required to prevent or reduce the emission of pollutants can also be of assistance. In practice, both fundamental approaches - which can of course be further differentiated - are used in combination. Both contain inherent advantages and disadvantages, and significant difficulties in terms of their practical application.

Until now, expenditure to prevent or to repair ecological damage from emissions has played no part, with the exception of certain dust filters, when examining the costs incurred for electricity on the former territory of the GDR. This is due to the fact that no desulphurization or NO_x control system is yet in operation in any of the power stations or district heating power stations. The first desulphurization plant is currently being commissioned for non-stop operation in the Klingenberg district heating power station in Berlin. It operates on the Wellman-Lord principle.

A realistic opportunity of nevertheless assessing this kind of external environmental costs for the GDR is seen to lie in the experience gained by the FRG in this field. Many reasons speak in favour of drawing such analogous conclusions, not least the fact that the limit values for emissions of pollutants are to match those stipulated in the regulations valid in the FRG by 1996 at the latest (regulations governing large furnaces, technical instruction on air pollution control, etc.), and in view of the fact that the technologies and methods for disposal will predominantly come from the West. Use can be made of those sums for additional investment and modified running costs (as a result of alterations in power consumption, materials used and profits from spin-off products) known to have been incurred in flue-gas desulphurization plants and Denox plants in the FRG. These sums can then be related to the capacities of those power stations producing electricity in the future.

The size of such power stations will no doubt be the result of economic cooperation between the two Germanies. A capacity of approx. 10,000 MW by the year 2,000 is calculated for the brown coal power stations to be up-dated. A preliminary rough calculation made on the basis of such considerations gives a resulting price increase of max. 4 Pf/kWh.

Two problem areas have admittedly been excluded from the observations made so far. These refer to CO_2 emissions on the one hand, and to past contamination arising from previous air pollution on the other hand. It is a known fact that there are yet no methods in existence anywhere in the world capable of reducing the emission of CO_2 when burning fossil fuels. As the contribution made by CO_2 gas to the greenhouse effect and to global climatic change is more than 50 %, it is a source of permanent damage to the human race. In addition, the CO_2 problem takes on a special dimension in the Eastern part of Germany, as a result of the prevailing use of brown coal. Its effects will be reduced somewhat in future, as the growing use of hard coal, oil or natural gas will lower specific emissions in relation to brown coal to 87 %, 73 % and 48 % respectively.

In order to make the external costs arising from the effects of CO_2 transparent in any way, the climatic emission charge of approx. DM 10/t CO_2 officially required in the FRG could be a first step (Cf. Berliner Zeitung, 26th September 1990, p. 3). This would mean an increase in the price of electricity produced from coal by 1.5 Pf/kWh. Such a solution nevertheless does not constitute a fundamental solution to the problem. Just as unsatisfactory would appear to be the suggestion advanced by foreign scientists, e.g. from Japan and the USA, to introduce a tax on CO_2 of approx $ 200/t CO_2 (Nordhaus 1989).

The economic assessment of past contamination from emissions is also fraught with problems. Let us take the example of dying forests. Measurements for 1989 conducted on the territory of the GDR conclude that more than 54 % of the 3 million ha forest are damaged, of which 16.4 % are subject to medium and severe damage. If we assume that acid rain due to SO_2 is with a high degree of probability the main cause, and that more than 50 % of the SO_2 load in the GDR is due to the production of electricity from coal, a rough calculation ultimately provides us with a perpetration level of approx. 25 %. In investigations into the total ecological costs associated with the power industry, that portion of such costs incurred in exploiting raw lignite (as the main source of fuel) must not be ignored. Those costs and prices cited for raw lignite currently do not take into account a significant proportion of environmental protection costs.

This applies to expenditure on follow-up investments in mining, e.g. for the devastation of infrastructures, for the temporary confiscation of useful agricultural and forest land, and for land reclamation and recultivation. Annual calculations of such costs come to DM 500-600 million, equivalent to DM 1.80-2.00/t of raw lignite mined. On top of this are the running costs, e.g. for the stabilization of mining conservation and operating areas, for compensation for real waste water charges made on draining surface mining sites and for economic hardship, and to counteract falling groundwater levels in agriculture.

A careful estimation gives a price rise of at least DM 2/t when these external costs for raw lignite are incorporated, the total costs of production currently standing at approx. DM 18.50/t. This means an increase per kWh of electricity by 4 %, or 0.32 Pf/kWh.

In conclusion, a global method for estimating the costs of environmental damage will be presented below. This method departs from the total annual sum for ecological damage, and derives from this the adequate charge due to electricity in the GDR from its corresponding average participation in perpetrating this damage. In concrete figures, this would result in the following calculation:

- Publications by environmental scientists put the total ecological damage caused to the territory of the GDR in 1988 at approx. DM 28-30 billion (Neues Deutschland).

- Energy-producing processes have a share of at least one half in the emission of most pollutants. If the weighting for the effect of electrical energy within these processes is also set at 50 %, 25 % of the same are ultimately due to electricity.

- On the basis of this calculation, at least DM 7 billion per annum in environmental damage are due to the perpetrators described here. This is equivalent to approx. 6 Pf of the costs per kWh.

Bibliography:

Berliner Zeitung
26th September 1990, p. 3

Institute for Energetics
Gesamtbilanz Energie 1989, Wirtschaftsraum DDR, Leipzig, pp. 73ff.

Neues Deutschland
24 November 1989

Nordhaus, W. D.
The economics of the greenhouse effect, contribution to the International Energy Workshop at the IIASA, June 1989

Weisheimer, M.
Preise und Subventionen im Energiesektor der DDR, Energiewirtschaftliche Tagesfragen, Düsseldorf, No. 9/1990, pp. 626-632

5.4 Facing up to the True Environmental Costs of Electric Power Generation

Stephen Wiel
Commissioner
Nevada Public Service Commission

Public awareness of mankind's impact on the environment has been growing dramatically over the past few years. Included in that awareness is a concern about global warming. With it, for good reason, comes a focus on the electric utility industry.

Electric power generation, like almost all of our activities, imposes environmental damage to our planet. Acid rain, carbon dioxide buildup, water consumption and strip mining are a few examples. Residual impacts occur even after significant pollution control requirements are imposed by various government agencies. These impacts inflict death, hospital costs, lost productivity, crop damage, disruption of wildlife habitats, deterioration of plastics, and other social costs which are not currently reflected when utilities select among competing methods of power generation.

We shoudn't ignore these impacts. If our best estimate of the real unaccounted costs of the coal-fired power plant we are about to build is two cents per kilowatt hour, we should be willing to pay up to that amount to mitigate or avoid that environmental damage. All options should be considered whether they be voluntary emission controls, the use of inherently non-polluting generation such as geothermal or solar power, or the avoidance of fuel use through improvements in energy efficiency.

The first step in internalizing residual environmental damage into our utility system operation and development is to monetize our assessment of the environmental damage which would result from any of our power generation choices. We must recognize this damage as a real cost when we prepare our long-range plans, when we commit to build or acquire a new resource and when we choose which power plant to operate on any given day.

Significant attention has been given to the assessment of environmental damage from power plants over the past two years in both the United States and Germany. Progress has been made to monetize the damage from residual pollution from various electricity generation facilities. Efforts have been underway in several parts of the United States to account for residual environmental damage in the selection of competing electricity generation resources.

This paper describes how electricity utility companies can and should make use of the results of environmental damage assessment research in selecting new resources and operating their systems.

Fifteen States Address Environmental Externalities

In August, 1989, New York became the first state to acknowledge the value in cents per kilowatt hour of residual pollution from its electric utility companies.

As of September 1990, six state commissions have modified the resource selection procedures of their electric utility companies to account for residual environmental damage (see Figure 1). Massachusetts followed the monetization path taken by New York. Three of the states have adopted a point system which avoids any expression of damage in monetary terms but implicitly achieves the same result. One state, Wisconsin, has simply assigned a penalty to fuel burning plants to account for environmental damage.

Nine other states are currently developing procedures to recognize and account for residual environmental damage from power generation in utility operations.

So far regulatory commissioners in the United States and their regulated utility companies have focused only on how environmental considerations affect long-range planning decisions and on the acquisition of long-term resources. We have not yet squarely faced several other aspects of the utility industry which will be affected by full consideration of residual environmental damage. The way we acquire short-term resources, dispatch the resources under our control, and design our rate structures will also change.

Figure 1: Status of recognizing utility environmental damage

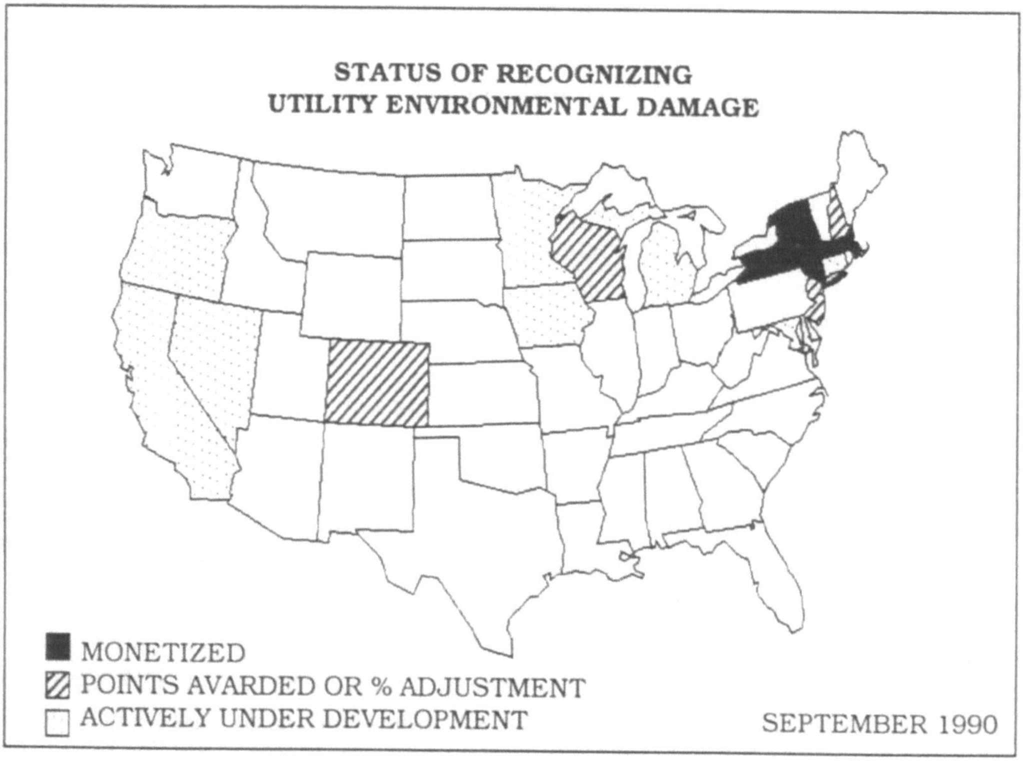

Long-Range Planning

Utility companies in the United States in their early experimentation with internalizing residual environmental externalities have tried three fundamentally different approaches.

The first is to divide the world into two classes of electricity resources; 1) environmentally benign resources such as improved energy efficiency and renewable technologies on the one hand and 2) traditional environmentally damaging resources which burn fossil fuel on the other. (The fate of nuclear power is uncertain under this approach.) Then a bonus is awarded for selection of the benign resource or a penalty is imposed upon selection of the damaging resource. While this approach is easy to apply, and in some jurisdictions may be politically satisfying, it is fundamentally flawed for its inability to distinguish amongst the relative level of damage caused by different facilities within each group.

A second approach has been to apply scores to the various competing resources to create an index of their relative desirability, weighing residual environmental damages along with costs of acquisition, operational costs, reliability and a number of other factors. This approach is also flawed for several reasons. It is administratively burdensome. It is subjective. It masks the cents-per-kilowatt-hour value of the damages which are nonehteless inherent in the results. Although the results of such weighted indices may be converted to cents-per-kilowatt-hour values after the fact, society will be better served in the long run if we calculate up front and acknowledge the cents-per-kilowatt-hour values we place on environmental damage.

The third approach, that taken by the state of New York and the state of Massachusetts, is to design values to each type of residual environmental damage. These need to be expressed as cents per kilowatt hour for a given type of power plant, but instead may be expressed as dollar damage per unit of pollutant. For example, damage from carbon dioxide emissions could be expressed as dollars per pound of carbon dioxide. Such damage values could then be converted to cents per kilowatt hour for each specific resource being considered by a utility company. This approach allows a utility company to prepare its long-range plan using residual environmental damage costs determined for each type of resource it is generically considering, yet using specifically determined costs for individual resources that are known at the time the plan is prepared. Additionally, this approach allows residual environmental damages to be considered consistently throughout all the operations of the utility company.

Utility companies in the United States are increasingly using a concept called "integrated resource planning" for conducting their long-range planning activities. Integrated resource planning involves systematically assessing all options for increasing the companies' ability to supply electricity, systematically assessing all options for reducing or managing their customers' use of electricity and integrating the demand and supply analyses in a way which accounts for the significant uncertainty in both arenas. Integrated resource planning in one form or another is required by state regulators in 25 states.

In many of these states, including my own state of Nevada, utilities look to the present worth of revenue requirements (PWRR) over the 20 year planning period as a primary selection criterion. The goal is to minimize the price of electricity over the long-run, recognizing that the price of electricity must be set to generate enough revenue to satisfy relatively sophisticated dispatch model which chooses the appropriate power plant to operate our purchase option to select during each hour of a simulated year. Adding the cost of residual environmental damage to the production cost of the electricity from each competing resource will result in the present worth of societal costs (revenue requirements plus residual damage).

The present worth of societal costs imposed by the generation of electricity (PWSC) recognizes that utility customers pay not only the amount of their utility bill for the electricity that is delivered to them, but also the largely obscure costs of residual environmental damage caused by electricity generation. While PWRR and PWSC are two among several factors which may influence the selection of a resource, clearly PWSC should be the most important of the criteria used for long-range utility resource planning.

Acquisition of Long-Term Resources

Traditionally, utility companies in the United States have built new power plants or signed long-term purchase agreements without regulatory approval and later have come to their commissions requesting increases in rates to recover the cost of these new facilities (or decreases in rates if the new source is cheaper than its existing sources.) Over the past decade, regulatory commissions have been getting involved in their utility companies' acquisitions of long-term resources. In Nevada, for example, a utility may only acquire resources identified in its commission-approved integrated resource plan. Utility companies now execute purchase power agreeements contingent upon commission approval and bring those contracts to the commission for verification that they are consistent with the long-range plan. In other states, regualatory commissions have approved bidding procedures for the selection of competing resources. In these jurisdictions, a utility's purchase agreement is considered approved if it was selected using the authorized bidding procedure.

Whichever procedure is used, the utility company must eventually justify its selection. In all cases, the PWSC and not the PWRR is the appropriate criterion for demonstration of prudence. Historically, the resources which satisfy future demand with the lowest PWRR comprise the company's referred plan. It is time to convert to minimization of PWSC.

Likewise, purchase contracts for long-term capacity and energy are made available in some states to small independent power producers which qualify under the Public Utility

Regulatory Power Act of 1978 (PURPA). PURPA is a Federal law in the United States which requires utility companies to purchase power from small independent power producers which have co-generation facilities or use renewable resources. The rates for long-term contracts under PURPA generally are based on the long-term incremental cost of the company's preferred or commission-approved plan. This is sometimes called the long-term avoided cost. The proper rate for such contracts instead should be the utility's long-term avoided societal cost (revenue requirement plus environment damage) less the specific environmental damage for the contracted resource.

Acquisition of Short-Term Resources

Internalization of residual environmental externalities should affect the acquisition of short-term resources in a manner similar to that of long-term resources. Short-term resources include economy energy purchases, capacity contracts of less than one-, or three-, or five-year duration (depending on the jurisdiction) and non-contractual purchases from qualifying facilities. These are purchases generally not covered in a utility's long-range plan and generally not requiring prior approval of the state regulatory commission.

So far, the discussion of environmental externalities in the United States has generally focused on resource planning and acquisition of long-term resources. Little attention has been given, yet, to the impact of environmental externalities on a utility company's acquisition of short-term resources or on the other issues discussed in the remainder of this paper. Still, the concept applied to long-range planning and major resource acquisition should follow through. PWSC and not PWRR should be the criterion used for the selection of all resources.

In general, utilities purchase short-term resources that have the lowest marginal cost. The marginal cost is the variable cost of producing one more kilowatt hour of electricity and is generally dominated by fuel costs. The generation of one more kilowatt hour of electricity will also create most, but perhaps not all, of the per-kilowatt-hour environmental damage considered for long-range planning purposes. This portion of the residual environmental costs must be added to the utility's marginal cost of generating or purchasing short-term power when it is selecting such a resource.

Utility companies generally establish a published short-term avoided cost which they offer they're willing to pay for non-contracted energy on an hourly basis. Typically, the rate is revised every quarter to reflect current conditions. Consistent with the arguments presented earlier, the utility company should be willing to pay its marginal societal costs (the marginal utility costs plus the marginal residual damage) less the marginal residual damage from the energy being purchased.

Dispatch of Exising Resources

The criterion used for selecting long-term and short-term acquisitions should also be used for selecting which power plant to operate or which energy to purchase each hour of each day of the year. Utilities should use the present worth of societal costs for their power plant dispatch. Conceptually, it is straightforward to modify existing dispatch models which advise dispatchers of the proper selection. Practically, it is only a matter of time before such program modifications will be made. Researchers are already experimenting with such program changes. I suspect that utility executives and regulators will want to experiment further with these revised dispatch procedures under real operational constraints to determine the actual impact on rates before feeling comfortable with full implementation of societal-cost dispatching.

Rate Design

Adding the cost of residual environmental damage from power generation to utility costs is likely to shift the actual acquisition and dispatch of resources. This will, in turn, shift the relative cost at different times for different classes of customers. Consequently, to the degree that rates are intended to be cost based, rate design is bound to change. The share of revenue requirements that any class of customers will be expected to provide in any rating time period will almost surely be different using societal costs as the selection criterion rather than revenue requirements.

If, for example, the environmental costs of base load power is greater than that of peaking power, base load power users will be expected to provide a greater share of revenue tahn they currently do. Peak users will be expected to provide less, and the ratio of on-peak to off-peak power would diminish. Using this example, large commercial customers, such as casinos and gold mines in Nevada, who use power around the clock, would likely end up with higher bills.

Cost of service studies which are used as the basis for setting rates should reflect total societal costs. Whatever the outcome, however much some customers will resent paying a higher share, the results will be more equitable from an overall societal point-of-view. And proper price signals will be sent, as they should be, to utility companies' customers.

Summary

Utility companies have only begun to experiment with internalizing residual environmental externalities. So far, they have primarily focused on long-range planning and acquisition of long-term resources. Soon, they will be addressing short-term acquisitions, hourly dispatch, and eventually, rate design.

If there is a single message that needs to be heeded as the electric utility industry evolves to a more environmentally sensitive posture, it is this: residual environmental impacts should be considered consistently throughout all operations of the utility company.

Table 1 summarizes appropriate treatment for the five major aspects of utility operations identified in this paper. Surely, many of you will discover more impacts on utility operations and will make many refinements of these five. Many of us regulators in the United States are eagerly awaiting your successful ventures.

Table 1: Summary of appropriate treatment of residual environmental damage

Utility Function	Appropriate Treatment
Long-Range Planning	Select Resources which Minimize PWSC.
Acquisition of Long-Term Resources	Acquire Resources with Lowest PWSC. (Pay up to utility's Long-Term Avoided Societal Cost Less Residual Environmental Damage of Resource Acquired.)
Acquisition of Short-Term Resources	Pay up to Marginal Societal Cost Less Marginal Residual Environmental Damage of Energy Being Purchased.
Dispatch of Existing Resources	Dispatch to Minimize PWSC.
Rate Design	Base Rates on Cost of Service Studies which Reflect Total Societal Costs.

Note: PWSC is the Present Worth of Societal Costs. It is the sum of the Utility's Revenue Requirements Plus the Residual Environmental Damage.

6. SUBJECT AREA 4:

WAYS OF INTERNALIZING EXTERNAL COSTS

THEORETICAL CONSIDERATIONS

6.1 Methods for Valuing and Incorporating Environmental Costs in Electric Resource Planning and Acquisition

Sury N. Putta
New York State Department of Public Service
Albany, New York 12223/USA

Electric power resource planning and acquisition practices in United States and many other countries consider a limited range of factors such as price, fuel diversity, reliability, etc. for achieving desirable results. This selection process often neglects the value of public goods consumed by the energy projects, and thus fails to make optimal choices for the society at large. For example, a fossil-fueled power plant consuming a greater quantity of air and water resources than a renewable energy resource may be chosen if the selection process is based mainly on the delivered price of power. To the extent that any project consumes public goods or uses public subsidies, the societal values of those goods and subsidies become external factors and neglecting their values in the decision processes leads to making suboptimal choices for the society. In the case of publicly owned and operated utility systems in which no direct private profits are involved, incorporation of external factors in the resource acquisition process is an internal decision and can be accommodated once proper values are established for externalities. In the case of privately owned but regulated utility systems such as those in the United states, incorporation of external factors in resource planning is not as easily accomplished because the market values for public goods are not readily determined and the use of the estimated values in resource selection is resisted by the private investors whose profits may be adversely affected. In the latter case, government regulators must work with the regulated utilities in finding acceptable methods for valuing public goods in electric resource planning and acquisition. For example, the utility investments on demandside management measures may be in the best interest of the society but implementation of those measures which cut utility sales and profits are resisted by the utility investors. Despite these difficulties, many regulatory commissions in the United States believe that some policy changes to the traditional scheme of regulation (in which utilities' incomes are generally based on sales and capital invested) can be made and market based policies similar to those pursued in other industrial sectors can be adopted to capture competitive market efficiencies. It is believed that given its power, free market forces will harness innovation in the utility sector to develop environmentally benign resources ultimately leading to long range cost reductions to ratepayers.

While the regulatory support for competition in the utility sector offers unprecedented opportunities for aggressive private power producers and the ratepayers to capture market efficiencies, successful implementation of competitive bidding for acquiring future resources requires the utilities and regulators to work together in devising a common measure to value price and non-price attributes for comparing electric resources. This paper discusses specific methods employed in New York state for measuring environmental values of electric power resource options and weights given to those values in the ranking methods adopted for application in the utilities' bidding programs.

Consideration of Environmental Externalities in Electric Resource Planning

Environmental externalities of electric power resources are defined as the residual environmental impacts of projects after compliance with environmental standards and regulations. Residual impacts of electric power resource options differ considerably and impose costs on the society indirectly. Considerations of the values of residual environmental discharges in electric resource planning and acquisition offers a level playing field for environmentally clean resources to compete with others and provides private power producers with some incentive to minimize total cost (including the societal costs) not just direct cost in designing their projects.

Basic principles which guide consideration of environmental externalities in electric resource planning and acquisition can be summarized as follows:

- o Environmental externalities are the residual environmental impacts of projects which meet basic environmental standardsand regulations enforced by federal, state and local environmental agencies.
- o Economic values of environmental externalities are the values placed by the society on the damages caused and the opportunities lost as a result of construction and operation of the facilities.
- o Price and social cost of environmental externalities of electric power generation are inter-related and that reducing externality costs may increase price of electric power.
- o Consideration of externalities costs in the price of electric power promotes fair competition in power generation and offers a level playing field for the bidders to participate in the utilities bidding programs for acquiring future capacity needs.
- o Estimating the cost of the damages caused by environmental discharges requires

issues which are difficult to quantify (ex: determining the damage caused by 1,000 tons of sulfur oxides and its economic value in urban areas versus in pristine areas such as national parks).

- Cost of eliminating the discharges of air and water pollutants and cost of mitigating residual land use impacts could be used as a surrogate for valuing the environmental externalities.
- Estimating mitigation costs of environmental residuals for projects which pass environmental standards and using those costs along with prices in valuing the resource options forces utility planners to consider the values of public resources in designing their project proposals.
- Energy conservation projects which reduce consumption of electric energy offer environmental benefit proportional to the value of environmental externalities associated with the amount of electric generation avoided.

Methods for Valuing Environmental Costs

Several methods were identified for valuing air pollution emissions based on (1) utility cost of controls, (2) marginal cost of control and (3) damage cost imposed by the emissions.

The utility cost of controls refers to average cost of reducing its total emissions to achieve a target level. For example, if a utility is required to reduce its sulfur dioxide emissions by 10,000 tons, it would most likely achieve that target by using least cost methods such as switching to lower sulfur fuels, installing flue gas desulfurization (FGD) controls, implementing load reduction through energy conservation programs or through a combination of several methodes. In this case the utility would value emissions reductions achieved by any other means in its region equal to the cost of achieving the same as its facilities. A method for valuing emission reductions for meeting specific target levels is discussed in New York State Energy Plan 1989. Marginal cost of control refers to the cost of reducing the last unit of emissions when a utility attains its target level of emissions. Barnow and Marron (1990) and Chernick and Caverhill (1989) provide justification for valuing emission reductions at the marginal cost in electric resource planning which are more expensive than those methods available to the utilities at the average cost. Damage cost refers to the cost of damages to public health and resources caused by the environmental discharges. Ottinger et al. (1990) and Hohmeyer (1988) provide good discussion on damage costs of environmental externalities of electric resources.

While the economic cost of air emissions may be close to the damage costs imposed, it is not clear whether society values those costs based on global damages or limit those values to damage costs to the local population and its resources. Estimating damage costs requires site specific information on local conditions such as population density, distribution of public and private resources, and meteorological data which are difficult to obtain and making judgements on the value of the affected human lives and public resources which are difficult to quantify. Use of damage costs in valuing resources presents additional problems as it would bias the decision to locate sources in less populated rural areas where the total damage costs may be lower but damage to any one person could be larger than if the source is located near an urban area. This method certainly would not offer equal protection to individuals in urban and rural areas. Given the significant difficulties for estimating damage costs and the equity problems that question their use in valuing externalities, many analysts have chosen to use control or mitigation costs as shadow prices for the externalities.

Environmental Cost of a Reference (Benchmark) Coal Plant

In approving procedures for conducting competitive bidding programs by the electric utilities, the New York State Public Service Commission (NYPSC) revealed its preference to use environmental mitigation costs at existing facilities as shadow prices for valuing societal cost of environmental discharges from future resources. The rationale for using average utility costs as shadow prices is provided in Putta (1990). New York identified the residual environmental intrusions from a reference plant - a new coal-fired generating facility meeting bare minimum legal requirements - in three major environmental categories namely air emissions, water pollution discharges and land use impacts. Average cost of controlling emissions at operating facilities for meeting some specified targets were estimated and were used as shadow prices to value the societal cost of residual emissions from new resorce options. For carbon dioxide emissions New York assigned social cost equal to the cost of reforestation necessary to sequester the residual emissions and revealed its preference to offer 20 percent of that cost as credit towards the price for sources which do not discharge CO_2 emissions and proportionately less credit for sources which emit less CO_2 than the reference plant. For water pollution discharges and land use impacts, New York examined the available studies on valuing the social costs and assigned a maximum of 0.1 cents/kwh for water discharges and 0.4 cents/kwh for land use impacts of the reference plant. The societal value of environmental externalities adopted in New York State for power resources are summarized in Table 1. In assigning the values for societal cost of environmental externalities, New York recognized the limitations and accuracy of the information used and fostered a research study to estimate values of externalities and develop methods for incorporation of more accurate values in future resource planning and acquisition.

Table 1: Environmental externalities of a base coal plant adopted in New York bidding program

Pollutant	Control Cost $/ton*	NSPS Coal Plant Emission rate lbs/MWH	Average Cost of Control cents/kwh
SO_2	832	6.0	0.25
NO_x	1832	6.0	0.55
CO_2	1[illegible].1	1820.0	0.1
TSP	333	0.3	0.005
Water Pollutants			0.1
Land Use Impacts			0.4
Total value of Environmental Externalities			1.405 ¢/kwh

* Based on application of 80% ACT and 20% LCT for SO_2
20% ACT and 80% LCT for NO_x
20% LCT for CO_2.

where ACT is Advanced Control Technology.
LCT is Low Cost Control Technology.

Methods for Incorporation of Environmental Costs in Resource Selection

While determining the economic value of environmental features is one of the most difficult problems faced by the utilities and the regulators in comparing electric resource options, the utilities are also confronted with developing a common measure to consider environmental costs along with other relevant factors - both price and non-price features - for ranking the resource options. Often, the common measure for comparison of projects is achieved by developing a scoring method in which the price and non-price factors are allocated maximum scores based on their common economic values. For example, a utility may decide to offer a maximum score of 100 points which may be awarded to a project whose price is 50 percent below the utility's bench mark price - its avoided cost. Such an offer implies that each percent of avoided cost is worth two points. the same measure is used for allocating scores to other factors based on their economic values. The relevant factors considered for valuing electric resource options generally include:

o Delivered rice of power
o Dispatchability of the project
o Reliability of technology
o Project viability
o Fuel choice and flexibility
o Environmental value

Each of these factors have some economic value to the utility proposing capacity additions. For example, if a project which is fully dispatchable, i.e., it permits the utility to vary its load according to its needs without any resrictions has a value equal to twenty percent of the utility's posted benchmark price, then the dispatchability factor will be allocated a maximum of forty points in the scoring method. Similarly, if the environmentally most desirable project is worth 25 percent of the benchmark price, the environmental factor will be allocated 50 points and the project's whose environmental value falls below the maximum will be scored according to their relative environmental values. The score allocated for the environmental factor would then be distributed for the environmental subattributes such as air emissions, water pollution discharges and land use impacts of energy resource options based on estimated environmental costs of those attributes for a benchmark resource option such as coal burning power plant that meets the minimum legal standards. A detailed discussion of weighting environmental externalities is provided in Putta (1989).

Under this system of scoring, environmentally benign sources, such as energy conservation coal plant that barely passes State environmental reviews would receive zero score for environmental attributes. The environmental score for all other projects is determined using an Environmental Scoring Form (ESF) which provides scores for all relevant environmental attributes based on their economic values. The ESF form (Table 2) adopted for use in one of New York's utility bidding programs illustrates distribution of environmental score for many environmental attributes. Using the ESF form, the score for each attribute is obtained by multiplying the number of points earned and the weight attached to them and the total of all scores becomes the environmental score of the project.

Methods for Ranking Resource Options - Implications

Selection of resource options for meeting future capacity needs of a utility greatly depends on the maximum scores allotted to the price and non-price factors and the methods employed for ranking the options. Since methods for scoring and ranking energy resources are not fully developed and tested in practice, it is important that the regulatory policies which influence utility actions be aimed at driving innovation for developing efficient methods by the utilities and refrain from prescribing solutions. Guided by these thoughts, the New York Commission offered its utilities complete freedom to design methods for scoring and ranking resource options and the utilities seized upon the opportunity to devise different methods which suit specific preferences in their franchise areas. A majority of the utilities proposed methods in which all price and non-price factors are scored and the resource options are ranked based on their composite scores. In a few cases, the utilities proposed to score only non-price factors which are ranked based on the adjusted prices. In all cases, the ranking methods developed by the utilities fall into the following three major categories:

o Arithmetic method
o Multiplicative method
o Net Present Value (NPV) of adjusted price method.

Arithmetic method:
In this method, the projects are ranked based on the sum total of scores on price and non-price factors and the projects receiving higher ranks are chosen for meeting the utilities have adopted this method as it is simple to administer and the effect of the price and non-price scores in ranking of the resources can be easily analyzed. One disadvantage of this

Table 2: Environmental Scoring Form (ESF) Case 88-E-242 Niagara Mohawk

Environmental Attributes	Weight (W)	Points (P) 0	1	2	3	4	5	Score (W x P)
Air Emissions								
Sulfur oxides (lbs/MWh)	7	>6	4.0-6.0	2.5-3.9	1.5-2.4	0.5-1.4	<0.5	
Nitrogen oxides (lbs/MWh)	16	>6	4.0-6.0	2.5-3.9	1.5-2.4	0.1-1.4	<0.1	
Carbon dioxide (lbs/MWh)	3	>1500	1050-1500	650-1049	250-649	100-249	<100	
Particulates (lbs/MWh)	1	>0.3	0.2-0.3	0.1-0.199	0.05-0.099	0.01-0.049	<0.01	
Water Effects								+
Cooling water flow (Annual intake as % of lake volume)	1	80-100	60-79	40-59	20-39	5-19	<5	
Fish protection	1	None		Operational restrictions		Fish protection provided	No public water used	
NYS water quality classification of receiving water	1	A or better	B	C+	C	D	No water use or municipal water/ wastewater utilized	
Land Effects								+
Acreage required (acres/MW)	1	0.3-0.5	0.2-0.29	0.1-0.19	0.05-0.09	0.01-0.05	<0.01	
Terrestrial	1	Unique ecological or historical value		Rural or low density suburban		Industrial area	No land used	
Visual aesthetics	1	Highly visible		Within existing developed area		Not visible from public roads		
Transmission	2	New OH >5 miles	New OH 1-5 miles	New UG >5 miles	New UG 1-5 miles	Use existing facilities	Energy conservation	
Noise (L_{eq} - background L_{90})	2	5-10			0-4.9		<0	
Solid waste disposal (lbs/MWh)	2	>300	200-300	100-199	50-99	10-49	<10	
Solid waste as fuel (% of total Btu's)	1	0	1-30	31-50	51-80	81-90	91-100	
Fuel delivery	1	New RR spur	Truck & existing RR	New pipeline	Barge	Use existing pipeline	No fuel use	
Kilometers from sensitive receptor area (per 6 NYCRR 225.1.1 (b)(4))	1	<10	10-39	40-69	70-100	>100	Energy conservation	
							Total Score (210 Max.)	=

method is that it allows projects with very poor or zero scores on certain factors to rank high based on total score. For example, a project with a zero score on environmental factor or on dispatchability factor could conceivably receive a high rank if it scores high on price and other factors. This problem is corrected by specifying threshold features which must be met by the projects before the utilities consider them in the resource ranking. Some of the threshold values could be such as (a) price shall not exceed 125 percent of the utility's avoided cost, (b) the project shall be dispatchable by the utility at least 50 percent of the time.

Multiplicative Method:
In this method, the resource options are ranked based on their product scores which are obtained by multiplying the sub scores received by the projects on price and non-price factors. This method of ranking has the effect of penalizing project designs or technologies which score low on any one factor. For example, a project which scores high on many factors but scores low on environmental factor or another non-price factor will receive a low multiplicative product and will be ranked low compared to projects which score average or above average on all other factors. This method adopted for use in one of the utilities bidding programs in New York will be analyzed for its effectiveness in attracting efficient projects when the results of the bidding will be announced in early 1991.

Adjusted Net Price Value (NPV) Method:
In this method, all resource options are ranked based on NPV of total overall adjusted cost of the proposals. First, the prices of the proposals are adjusted for energy output, availability, transmission and distribution and administrative costs. Then non-price adjustments to the price are made using the values placed by the utility on non-price factors. The non-price factors include a general category of features such as fuel diversity, reliability, risk and operational characteristics and an environmental category such as air, water and land use impacts. Projects which have best non-price features receive maximum non-price adjustment which will be subtracted from the price before the NPV of the project is calculated. The utility will decide the maximum values to be allocated for the non-price factors based on their perceived economic values in its franchise area. For example, if the economic value of environmental benefits of a benign resource option such as an energy conservation project is worth 1.4 cents per kilowatt hour of energy saved, that amount of adjustment would be made to its price in calculating its NPV. For other proposals which have less environmental benefits, only a fraction of the full 1.4 cents/kwh benefit will be provided. That fraction is calculated by dividing the project's environmental score with the maximum score allocated for the environmental factor. In using the NPV method, one New York utility valued non-price features at about 50 percent of its avoided

cost which is credited for projects receiving perfect score on non-price factors. This non-price credit which is equally divided between the general non-price features and environmental features encourages design of projects with cost effective non-price features and the resulting higher prices would not disadvantage them from competing with low-priced projects which process poor non-price features.

The effectiveness of the NPV ranking method in attracting environmentally superior projects and the difficulties in administering the method will be examined when the evaluation of the utilities bidding programs are completed and the results are announced in early 1991.

Conclusions

The advent of competitive bidding for acquiring electric resources heightened the need for considering values of environmental externalities in comparing resource options. Since the price and social cost of environmental externalities of electric power generation are related and that reducing externality costs may increase the price of electric power, comparing resource options require methods for valuing externalities and considering them along with other relevant factors. Methods for moneytizing externalities were discussed and a practical method for valuing externalities and scoring resource options in competitive bidding programs was recommended. Several ranking methods for considering the environmental score along with other relevant factors were discussed and their advantages and disadvantages in resource selection were identified.

References

Bernow, S.S. and Marron, D.B., 1990, "Valuation of Environmental Externalities for Energy Planning and Operations", Tellus Institue, Boston.

Chernick, P. and Caverhill, E., 1989. The valuation of Externalities from Energy Production, Delivery, and Use. PLC., Inc., Boston, Massachusetts.

Hohmeyer, Olav, 1988. "Social Costs of Energy Consumption," Springer-Verlag, New York.

Ottinger, R.L., et al., 1990. "Environmental Costs of Electricity," Oceanic Publications, Inc., Dobbs Ferry, New York.

Putta, S.N., 1989. "Competition in Electric Generation - Consideration of Environmental Externalities," New England Governor's Conference, Boston, Massachusetts (October 1989).

Putta, S.N., 1990. "Consideration of Values of Environmental Externalities in Electric Capacity Planning and Acquisition," Workshop on Externalities, Albany, New York (April 1990).

6.2 Role of Control Costs in Developing Climate Change Policy

Dr. Ajay K. Sanghi[1,2]
New York State Energy Office
Two Rockefeller Plaza
Albany, New York 12223

Application of Cost-Benefit Analysis to Climate Change Policy

There is consensus among scientists that escalating concentrations of atmospheric greenhouse gases are likely to result in increased mean global temperatures over the next century. However, there is considerable uncertainty with respect to the magnitude of this warming trend, its timing, and the severity of its impacts (DOE 1990; EPA 1989; IPCC 1990). Most of the current debate is focused on the degree and timing of the warming trend, not on whether or not it will occur. It is likely that intense debate on the impacts, political pressure is mounting for some carbon mitigating policy action to be implemented in the near future.

Like all other desirable things in life, potential solutions to the global warming issue must compete for the use of scarce economic resources. The reduction of greenhouse gas emissions therefore presents itself as an allocation problem, which could be suitable for the application of cost/benefit analysis. Within this framework, the relevant cost would be the control cost (i.e. the net expenditures associated with the implementation of any direct or indirect measures which reduce emissions of greenhouse gases). The relevant benefit would be the damage from potential temperature increases which is avoided by implementing the control measures. Because the abatement of greenhouse gases requires the diversion of scarce economic resources from other productive activities, there is also a substantial opportunity cost associated with such expenditures. The opportunity cost of emissions abatement thus consists of the value of foregone production. This opportunity cost could also be manifested in the short-run by inflation, unemployment, closure of industrial plants, and a general reduction in economic growth (Jorgenson and Wilcoxen 1989). For this reason, the amount of expenditures for global warming control programs cannot be limitless.

Although the cost-benefit approach is required for determining optimal control targets and for ensuring that scarce economic resources are being properly utilized, this methodology is based on the presumption that the value of benefits (damage costs) can be accurately determined. However, the benefits which result from mitigating any type of environmental pollution are extremely difficult to quantify with a reasonable degree of certainty. The sheer magnitude of the potential impacts of global warming exacerbates this problem, and it is generally conceded that the dollar value of these damages would be beyond estimation due to uncertainties associated with impacts of global warming and the catastrophic nature of potential impacts. Consequently, it may be a futile exercise to attempt to calculate the damage costs inflicted if, for example, Manhattan becomes submerged in the Atlantic or the mid-west breadbasket of the U.S. becomes a dustbowl.

The problems associated with attempting to achieve estimates of damage costs with reasonable certainty place severe limits on the usefulness of cost-benefit analysis, thereby precluding its application in most environmental contexts. The result has been that emission reduction targets and environmental quality standards in the U.S. have been determined by public setiment and political negotiation, rather than by optimization levels established by cost-benefit analysis. The Acid Rain provisions of the pending Clean Air Act amendments provide a clear example of this process, as policies are likely to be adopted which will curtail emissions of acid rain precursors before the specific value of acid rain damage is established through further research. Because the physical processes and economic linkages associated with global warming are far more complex than those associated with acid rain, it is very unlikely that conclusive cost/benefit analysis of carbon reduction policies can be performed within the near future. For this reason, policy responses to the global warming issue may follow a path similar to that established by responses to the acid rain issue.

Use of Control Costs

The response of energy policy makers to environmental regulation has always been carried out in terms of the cost of compliance (control). In traditional utility rate of return regulation, public utility commissions seek to ensure that compliance with environmental regulations is achieved by the utilities at the least cost. When the price of electricity paid by ratepayers is subsequently increased to include the cost of compliance with mandated emissions rates, etc., the effect is to internalize a portion of the environmental damage from the regulated pollutants may continue even after applicable standards have been met.

This occurs because pollution control mandates do not generally require a complete elimination of the pollutant. For this reason, compliance with emissions control mandates does not internalize many of the costs associated with the residual emissions.

These costs generally arise from the use of resources for which no market exists (i.e. externalities). When externalities are present, the prices of the goods and services produced by an action do not reflect the full marginal cost of the resources allocated to their production. Because these external costs are as much a part of the opportunity cost of making a good available as are wages and materials costs, the monetization and subsequent internalization of externalities results in an adjustment to the opportunity cost of using the resource as valued by the market. Such an adjustment to price leads to improvements in resource allocation and economic efficiency (Sanghi, 1990)[3].

Carbon emissions resulting from the combustion of fossil fuels impose externality costs on society in the form of global warming. Climate change policy requires that these externality costs be monetized and incorporated in the resource selection process if the threat of global warming is to be reduced. Given that there are presently no legislative mandates for reducing carbon emissions, the value of carbon externalities may be estimated from the cost of carbon reduction. The monetized value of carbon externalities, consistent with a given reduction goal, is the fee or tax which must be levied on each ton of carbon emission, so as to result in the desired reduction of carbon emissions.

The least cost method of achieving a target reduction would entail moving along the marginal control cost curve. However, such a result will be difficult to attain through market mechanisms. Therefore, government policy should ensure development of mechanisms and institutions which will result in implementation of the least cost method. One such mechanism is to charge a fee equal to the monetized externality values. The proceeds from the fee would be deposited in an environmental trust fund, which would be used to finance pollution mitigation measures. An externality value based on the weighted average cost of the required mitigation measures to achieve a given target reduction is consistent with this approach.[4]

It is also important to note that political acceptance of the monetized value of carbon externalities requires that the externality value based on conservative estimates and that it be kept small initially. Monetized externality values based on the average cost of the carbon mitigation measures needed to achieve a targeted amount of carbon reduction are consistent with this concern.

New York State Initiatives

Many studies have been undertaken at both state and national levels to analyze policy options for reducing emissions of greenhouse gases (CEC 1989; Oregon 1990; Spectrum 1990; DOE 1990; EPA 1989; Morris et al. 1990). The major instrument for addressing global warming policy in New York State has been the State Energy Plan, released in September 1989 (New York State 1989).

The New York State Energy Plan (SEP), prepared jointly by New York's State Energy Office (NYSEO), Department of Environmental Conservation (NYDEC) and Public Service Commission (NYPSC), represents the first time that energy planning, environmental and regulatory agencies at the State level have worked together to integrate energy planning with environmental policy objectives. Following are some of the major policy actions which were recommended in the SEP. These policy actions will reduce carbon emissions, as well as contribute to other planning objectives related to environmental quality and energy security:

- Pursue aggressive electric utility DSM programs which would achieve statewide annual electricity savings of 15 percent by 2008.

- Encourage increase of Federal Corporate Average Fuel Economy (CAFE) standards to reach 42 mpg by year 2000.

- Improve energy efficiency by reducing consumption in state-owned buildings (20 percent by 2000), increasing building efficiency codes, increasing appliance efficiency standards and extending and enhancing the Home Insulation and Energy Conservation Act (HEICA).

- Increase the proportion of electricity generation fueled by natural gas (e.g. promote fuel-switching from oil and coal).

- Accelerate the research, development and demonstration of renewable resource technologies for end-use and electricity generation applications.

- Develop a power plant efficiency standard for life extension projects.

The SEP also analyzed the environmental externality costs associated with the combustion of fossil fuels. It recommended explicit monetary externality values for SO_2, NO_x, and CO_2

emissions for consideration in electric utility competitive bidding programs to be carried out under NYPSC guidelines. The NYPSC has adopted the SEP values for SO_2 and NO_x[5], but has adopted an alternative value for CO_2. The SEP recommended an externality value of 0.64¢/kwh (levelized nominal) for carbon emissions from a coal plant (NYSEO 1989a), but the NYPSC subsequently used a "compromise" value of 0.1¢/kwh in the initial round of utility competitive bidding programs (NYPSC 1989). New York's incorporation of an explicit monetary environmental externality value which recognizes the potential damage from CO_2 emissions is the first in the nation.

To facilitate the development of more specific policy actions which would reduce the threat of global warming, the SEP directed the New York State Energy Office, in collaboration with the other agencies, to examine the actions necessary to achieve a 20 percent reduction in CO_2 emissions from 1988 levels by 2008. As a part of this study, the NYSEO is also evaluating the impacts of a federal carbon tax on New York's energy, environmental and economic projections. Although the final report of this study is not expected to be released until later this year, some preliminary results are discussed in the following sections.

Cost of Controlling Carbon Emissions: New York State Experience

To achieve a 20 percent reduction in New York carbon emissions from 1988 levels by 2008 would require a carbon reduction target of about 37 million tons from the SEP Business-as-Usual scenario. This carbon reduction target includes electricity generation, transportation and residential, commercial and industrial sectors. The target is calculated as follows: Estimated 2008 SEP Business-as-Usual carbon emissions (87 million tons) minus estimated 1988 carbon emissions (63 million tons) plus 20 percent of 1988 emissions (13 million tons) equals the target reduction (37 million tons).

Figure 1 presents the costs and carbon reduction potential for selected carbon reduction measures to be applicable in New York State. For each individual measure, this graph indicates the estimated cost in real levelized $/ton of carbon reduced (1990$) as well as the estimated number of tons of carbon reduction. This list of measures should be considered as partial and subject to continual revision and expansion, as it represents only the measures which have been identified and quantified at this time. Many additional measures may be developed and become practicable over the next 20 years, especially if a strong public policy of reducing carbon emissions is adopted early on. It should also be pointed

Figure 1: Life cycle costs of carbon mitigation measures and potential carbon reduction in 2008

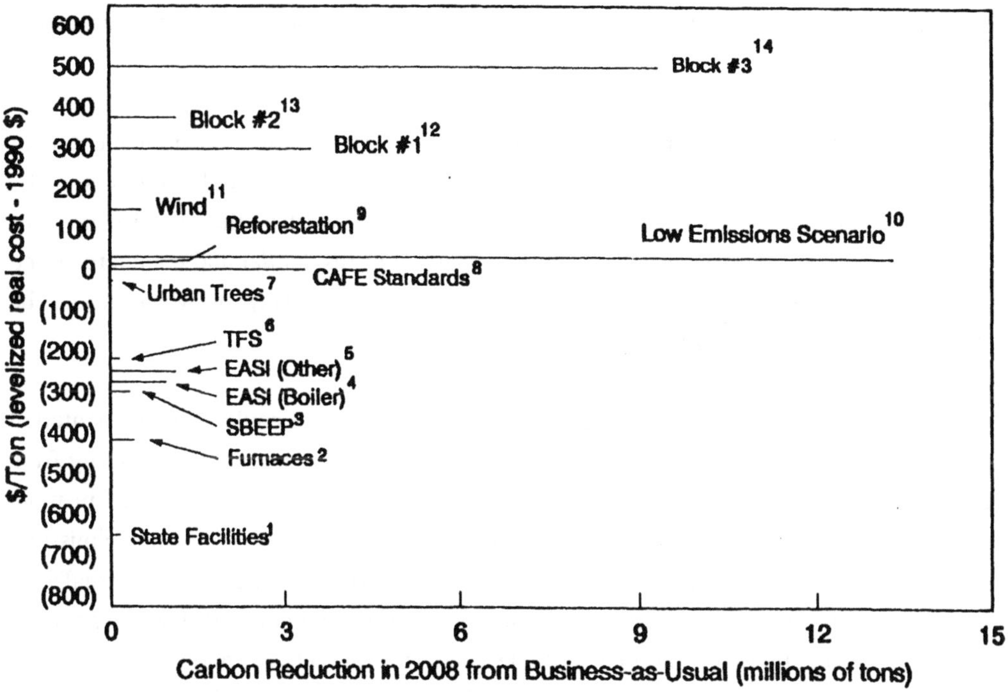

NOTES: (All levelized $/ton costs include capital costs plus fuel savings, if applicable, over the life of the measure)

1/ State Facilities - Assumes 20% reduction in energy use by 2000.
2/ Furnaces - Improvement of residential furnace efficiency from 81% to 91%.
3/ SBEEP - Small Business Energy Efficiency Program provides on-site surveys geared to the needs of small commercial businesses and not-for-profit organizations.
4/ EASI (Boiler) - Energy Advisory Service to Industry program provides on-site surveys to identify energy saving improvements. Available to small and medium sized commercial and industrial firms.
5/ EASI (Other) - See #4 above; separate programs available for boilers and all other systems except boilers.
6/ TFS - Technical Feasibility Studies program offers matching funds for analyses of energy conservation projects. Available to small and medium industrial, commercial and agribusiness firms.
7/ Urban Trees - Planting trees to provide shade, thereby reducing air conditioning use.
8/ CAFE Standards - Assumes average fuel mileage increases from 28 mpg in 1988 to 42 mpg in 2008.
9/ Reforestation - Four measures: tropical reforestation, upgrading of NY forests, reforestation on NY public land, reforestation on NY private land.
10/ Low Emissions Scenario - Electricity supply scenario which employs increased natural gas use, reduced coal and oil use, and increased DSM - i.e., reduced need for new capacity. Includes difference in revenue requirement from Reference Scenario and customer share of DSM costs.
11/ Wind - Assumes that wind generation capacity (1200 MW operating at 15% capacity factor) reduces the required energy production at existing coal plants.
12/ Block #1 - Could include any number of new technologies which affect supply or demand; cost and impact estimated by proxy. Assumes that all new utility-owned generation capacity required by 2008 (2,500 MWs of AFB coal is required by Lower Emissions Scenario) is replaced by hybrid generation units consisting of photovoltaics (PV) backed up by high-efficiency gas combustion turbines (the back-up is required because the PV's would operate at only 30% capacity factor, while the AFB replaced would operate at about 80%). Gas price assumed to double because of supply constraint.
13/ Block #2 - Estimated by proxy (see note #12). Assumes that 25% of existing coal capacity is retired early and replaced by the hybrid PV and gas turbine units described in Block 1.
14/ Block #3 - Assumes that remainder of carbon reduction required to reach 20% objective could be achieved at $500/ton. Assumes that societal demand for carbon reduction facilitates development of new technologies and associated carbon reductions. Reductions could occur in electricity generation (supply or demand side), transportation, residential, commercial or industrial sectors. As the cost for this block is conceptually a weighted average of many individual measures, some of these measures (e.g. biomass generation, additional DSM or reforestation, etc.) could come in at costs much lower than $500/ton while others (e.g. PV with storage back-up, use of hydrogen fuel, CO_2 Scrubbers, etc.) could be higher. Also, a portion of this block of carbon reduction could result from reduced energy use due to price increases.

out that a carbon tax is not considered as a control measure in this context due to the wide range of effects it may have, depending on the tax rate and the use of the revenues collected. For this reason, the potential effects of a carbon tax are discussed separately in the next section.

Figure 1 identifies seven carbon reduction measures which could be implemented at negative net costs and one measure, increased Corporate Average Fuel Economy (CAFE) standards, which could be implemented at zero net cost. Most of the negative net cost measures represent improvements in energy efficiency in residential and commercial and industrial applications (e.g. high-efficiency furnaces, etc.), many of which would be implemented through programs administered by the NYSEO.

Among the positive net cost measures, two are identified which could be implemented at relatively low costs. The first is New York's potential share of carbon sequestering from reforestation (2 million tons at an average cost of $19/ton), which is described in more detail in Figure 2. The second low cost measure is referred to as the Lower Emissions Electricity Supply Scenario (13 million tons at $32/ton). This measure is actually an integrated group of measures comprised primarily of increased demand-side management (i.e., reduced generation requirements); fuel switching from oil to gas and from coal to gas; and early retirement of certain coal plants. Over 8 of the 13 million tons of carbon reduction attributed to the Lower Emissions Electricity Supply (LEES) Scenario are the direct result of the implementation of DSM measures above and beyond the DSM measures assumed in the Business-as-Usual-Scenario. This additional block of DSM consists of about 30,000 gwh of avoided electricity energy and is equal to the 15 percent energy reduction in 2008 which is recommended in the SEP. Examples of measures included in this block for the residential sector are compact fluorescent lamps, heat pump water heaters, very high efficiency refrigerators, high-efficiency new construction, weatherization measures and submetering of master-metered apartments. For the commercial and industrial sectors, examples are improved lighting technology, HVAC and refrigeration improvements, high-efficiency and adjustable speed electric motors and building shell improvements.

It should be noted that above mentioned DSM measures are in addition to approximately 2,000 gwh of energy savings which are due to proposed utility DSM programs (per DSM Plans as required by the NYPSC). These DSM measures are also above and beyond the energy savings resulting from the assumptions for increased energy efficiency which are embedded in the energy demand forecast. Examples of energy efficiency improvements included in the energy demand forecast are enhanced residential and commercial building

Figure 2: Reforestation: Carbon reduction supply curve

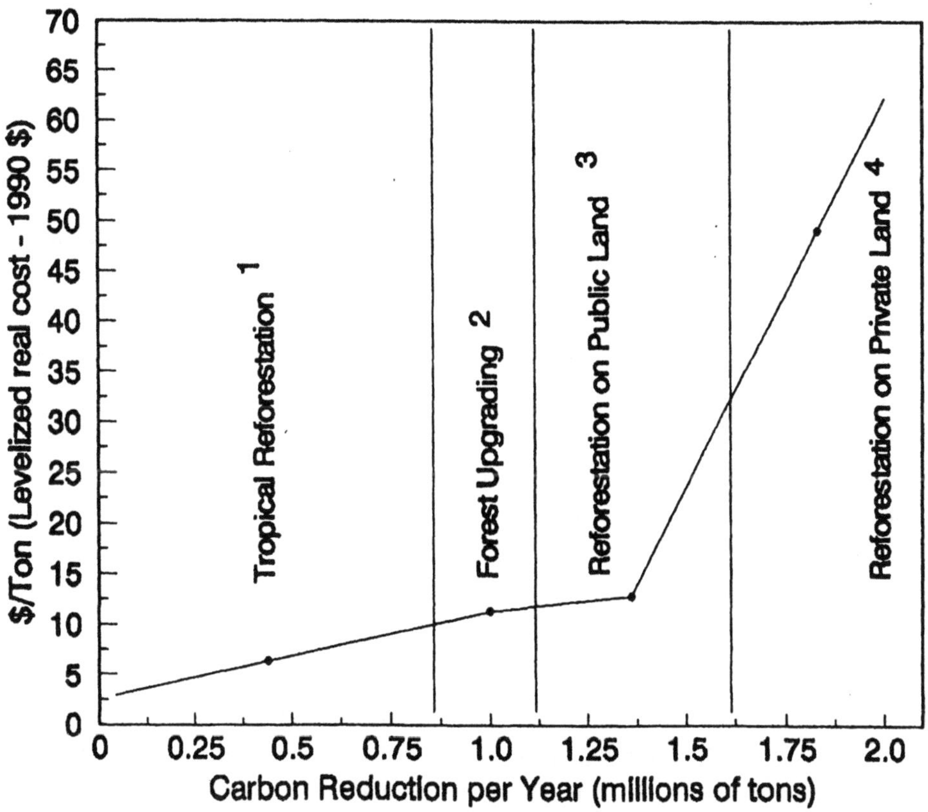

NOTES:

1/ Tropical Reforestation ($6/Ton) - Costs assumed by New York are somewhat higher than those reported for the carbon sequestration project in Guatemala initiated by Applied Energy Systems (AES) to offset carbon emissions from a new generation plant in Connecticut ($2.50/ton). Increase reflects the extreme level of uncertainty associated with this type of project. AES estimate included no labor costs and no allowances for losses from pests, natural disasters, social or political unrest or possible institutional barriers (WRI 1989, Pace 1990). The NY value represents the midpoint between the AES value (assumed to be the lower bound) and an estimated upper bound based on estimated planting costs in NY and an annual carbon uptake rate for a tropical forest of 6.2 tons/hectare. The amount of land targeted represents NY's estimated share of this low-cost reforestation. It is expected that world demand for carbon reduction will drive up the cost of tropical carbon sequestering, so it is unlikely that NY will be able to purchase significantly larger amounts of carbon sequestering at this price.

2/ Forest Upgrading ($11/ton) - Understocked New York forests can contribute to carbon sequestration by the implementation of management practices which facilitate more rapid regeneration (brush clearing, etc.), thus increasing the carbon uptake for these areas. Available land based on U.S. Forest Service data for NY (USFS 1984, 56).

3/ Reforestation on Public Land ($13/ton) - Targets highway corridors, etc. Assumes no land costs; planting costs assumed to be $688/hectare. Costs partially offset by the economic value of the timber after 40 years of growth ($1250/hectare - 1990$). Average long-term carbon uptake rate of 2.4 tons/hectare/year assumed is considerably lower than the rates generally used in other published studies. The higher uptake rates used by other studies are more representative of tropical forests or short rotation plantations.

4/ Reforestation on Private Land ($49/ton) - Based on the same assumptions as public land except that a lease payment for the land is added based on an average market value of $1250 per hectare.

codes, increased appliance efficiency standards, and general adoption of more efficient technologies which are cost-effective, such as high-efficiency fluorescent lamps.

Figure 3 shows the least-cost mix of measures and the carbon reduction contribution of each which would be required to achieve carbon reductions up to 20 percent in 2008. This figure indicates that achievement of a 20 percent objective would require an additional 14 million tons of carbon reduction above and beyond the 23 million tons potentially provided by the lower cost measures discussed above. This additional carbon reduction, identified as Blocks 1, 2 and 3, is likely to be achieved at high costs ($300-$500 per ton). These blocks could be comprised of any number of supply-side or demand-side technologies which may become technically feasible and commercially available in the future. Examples in the electricity generation sector could be increased use of biomass, photovoltaics, fuel cells, ocean thermal, tidal and wave energy; hydrogen fuel and CO_2 scrubbers. In the residential and commercial sectors, additional measures might include increased use of solar space and water heat and end-use photovoltaics. New technologies in the industrial sector could reduce carbon emissions from manufacturing processes. In the transportation sector, a major restructuring could include such technologies as electric cars and a mass transit system based on high-speed trains.

The specific costs and amounts of carbon reduction which could result from individual measures in Blocks 1, 2 and 3 are highly uncertain. For this reason, the $/ton costs and carbon reduction potential of these blocks were estimated by proxy. The technology selected as the proxy measure for Blocks 1 and 2 is a hybrid configuration of central station photovoltaic electricity generation units. This technology was selected because it is presently developed at least to a point where theoretical cost estimates are available (EPA 1989; ICF 1990; PC/SECC 1988; SERI 1989) and because the potential carbon impacts are large enough to contribute substantially toward the reduction targets. Block 1 (3.5 million tons at $300/ton) is based on the assumption that all of the new utility-owned generation capacity required by 2008[6] is replaced by hybrid generation units consisting of photovoltaics backed-up by high-efficiency gas combustion turbines[7] (Williams 1989). Block 2 (1.2 million tons at $375/ton) is based on the assumption that 25 percent of existing coal capacity is retired early and replaced by the hybrid photovoltaic and gas turbine units. Block 3 (9.3 million tons) is based on the assumption that the remainder of carbon reduction which is required to reach the 20 percent objective could be achieved at a cost of $500/ton. This cost assumes that societal demand for carbon reduction would facilitate the development of new technologies and associated cost reductions. Also, a portion of this block of carbon reduction could result from reduced energy use caused by increases in the price of energy. Because the cost for this block is conceptually a weighted average of many individual measures, some of these measures (e.g. biomass generation, additional DSM or

Figure 3: Least-cost mix of measures and carbon reduction contribution of each needed to achieve up to 20 % carbon reduction in 2008

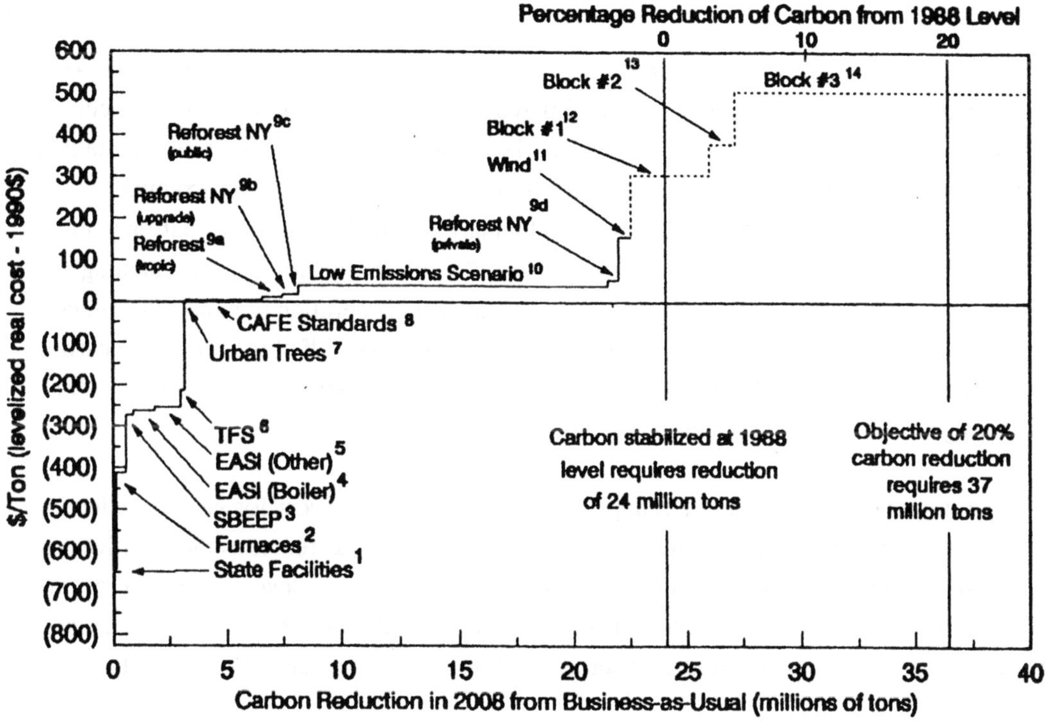

NOTES: Numerical data on which Figure 3 is based are provided in tabular form below.
A definition for each measure may be found in the footnotes to Figure 1.

#	Measures (in order of cost)	Levelized Real Cost of Carbon Removed 1990 $/ton	Reduction in Carbon from 2008 Business-as-Usual million tons
1	State Facilities	-650	0.17
2	Furnaces	-417	0.42
3	SBEEP	-277	0.35
4	EASI (Boiler)	-267	0.96
5	EASI (Other)	-259	1.14
6	TFS	-219	0.18
7	Urban Trees	-9	0.05
8	CAFE Standards	0	3.35
9a	Reforest (tropical)	6	0.86
9b	Reforest NY (upgrade)	11	0.25
9c	Reforest NY (public)	13	0.48
10	Low Emissions Elec. Supply Scenario	32	13.32
9d	Reforest NY (private)	49	0.48
11	Wind	150	0.55
12	Block #1	300	3.46
13	Block #2	375	1.17
14	Block #3	500	9.33
	Weighted Average =	156	Total = 36.51

reforestation, etc.) could come in at costs much lower than \$500/ton, while others (e.g. photovoltaic generation with storage back-up, use of hydrogen fuel, CO_2 scrubbers, etc.) could be higher.

Regardless of the mix of measures which is assumed to become available in Blocks 1 through 3, it is clearly evident that a 20 percent reduction target would be difficult to achieve within the short time frame of 18-20 years. It would require substantial restructuring of our energy systems and patterns of energy use. As an example of the severity of the impacts of this objective, preliminary calculations indicate that only about half of the carbon reduction required by Block 3 could be achieved by eliminating the remainder of coal generation (75 percent of existing coal generation remains after Block 1 and Block 2 assumptions) or, alternatively, by eliminating one half of the remaining carbon emissions from automobiles (by assuming that CAFE standards are increased to about 75 mpg or that a high-efficiency mass transit system is developed).

Figures 4 and 5 show the net change in total energy service cost for carbon reduction objectives up to 20 percent. Total energy service cost initially decreases as the negative cost measures are implemented and then increases gradually as low cost measures are implemented. With the addition of high-cost measures, total energy service cost increases rapidly up to the point where the 20 percent carbon reduction objective is achieved.

Figures 4 and 5 indicate that carbon emissions could be stabilized at the 1988 level (zero percent reduction objective) with no net increase in total energy service cost. To achieve 5 percent, 10 percent and 20 percent carbon reduction objectives, total energy service cost would increase by about 3 percent, 8 percent and 17 percent, respectively, above the Business-as-Usual scenario. Figure 5 indicates that these selected carbon reduction objectives could be achieved at weighted average costs of about \$44/ton for 5 percent, \$84/ton for 10 percent and \$156/ton for 20 percent reduction. Thse results emphasize the fact that the costs of carbon reduction are extremely sensitive to the number of tons of reduction required by the selected objective. For example, as shown in Figure 5, moving from a 5 percent objective to a 10 percent objective would require only 4 million additional tons of reduction (31 million vs. 27 million), but this increment would more than double the total cost of the reduction from \$1.2 billion to \$2.6 billion. This occurs because of the high costs of the carbon reduction measures at the margin.

The cost calculations presented in Figures 1 through 5 are based on the assumption that all expenditures (and savings) are allocated toward carbon reduction and that consequently no value is credited for the benefits which result from reductions in other emissions, such as achieving a given carbon reduction objective would, in effect, be decreased if the measure costs were adjusted to reflect the value of the SO_2 and NO_x reductions. The magnitude of this adjustment, however,is very sensitive to the externality values assigned to the SO_2 and NO_x reductions. For example, if SO_2 and NO_x reductions were valued at the externality costs estimated in the SEP,[8] the cost of achieving a 5 percent carbon reduction objective would decrease by about 20 percent. Alternatively, this cost would be offset entirely if SO_2 and NOx reductions were valued at the externality costs estimated in the Pace study.[9]

In exploring the measures required to reach a 20 percent objective, the analysis presented above indicates that this target would be difficult to achieve within a 20 year planning horizon due to high costs and the uncertainties associated with undeveloped technologies. In contrast, a 5 to 10 percent reduction target appears to be more technologically feasible and is likely to be achieved at a much lower unit cost of reduction.

It should be noted that achievement of a specific carbon reduction objective is sensitive to the rate of growth in energy demand which is assumed in the Business-as-Usual Scenario. For example, if the rate of annual energy demand growth in all sectors is assumed to drop by 0.25 %, this would result in a carbon reduction in 2008 of about 6 million tons. Under such a reduced energy demand forecast, the 20 percent reduction target is likely to be met more easily than under the original energy demand forecast.[10]

Federal Carbon Tax

This section evaluates the carbon abatement potential of a federal carbon tax and estimates the tax rates which would be required to achieve up to 20 percent reduction of carbon emissions in New York State. Required carbon tax rates are developed under two alternative sets of assumptions regarding the use of the tax revenue collected. In the first case, it is assumed that tax revenues are used for deficit reduction or for other purposes not related to carbon reduction. In the second, revenues are used to fund the carbon reduction measures discussed in the previous section. In either case, the carbon tax rate represents the monetized value of carbon externalities, assuming that this monetized value would be actually collected as an emission fee.

The impact of a federal carbon tax was estimated by tracing the following effects on the energy system: (1) increases in the prices of fossil fuels and changes in their relative prices, which induce interfuel substitution; (2) reductions of energy demand as a result of

Figure 4 - Net Change in Total Energy Service Cost in 2008 to Achieve up to 20% Carbon Reduction (1990 $)

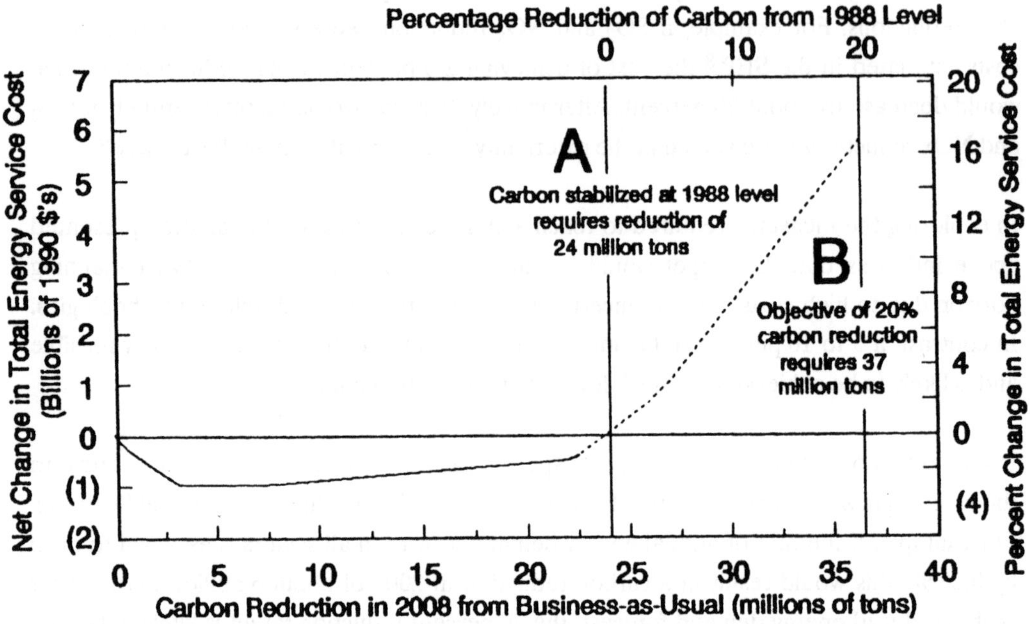

NOTE: Total Energy Service Cost (for Business-as-Usual) includes revenue requirement for providing electric service for the Reference Scenario (capital, O&M and fuel) plus fuel costs for the transportation sector and the residential, commercial and industrial sectors. The Net Change in Total Energy Service Cost (due to carbon reduction) includes: 1) the change in utility revenue requirements between the Reference and Low Emissions Scenarios (including utility DSM expenditures); 2) costs incurred by DSM participants; and 3) capital costs less fuel savings resulting from carbon reduction measures in the transportation sector and residential, commercial and industrial sectors.

Figure 5 - Levelized Average Real $/ton Cost to Achieve Selected Societal Carbon Reduction Objectives

Societal Goal: Carbon Reduction from 1988 Level		Net Change in Total Energy Service Cost		Levelized Real Cost
percent	millions of tons	billions of 1990$	percent	$/ton 1990$
0	24	0	0	0
5	27	1.2	3	44
10	31	2.6	8	84
15	34	4.3	13	126
20	37	5.7	17	156

increases in energy cost; (3) increasesin the cost-effectiveness of energy efficiency measures and their substitution for fossil fuels; and (4) potential accelerated adoption of new energy technologies with lower carbon emissions (primarily renewable resources).

Most of the carbon tax proposals in the U.S. assume that the tax revenues collected are to be used for deficit reduction or for other purposes which have long-term benefits other than reducing carbon emissions. In such a case the carbon reductions are solely a result of consumer response to increasesin fuel prices due to a carbon tax. Figure 6 presents the estimated impacts on New York State of a federal carbon tax when revenues are used for deficit reduction (as opposed to funding carbon abatement measures). To achieve a 20 percent carbon reduction objective would require a federal tax rate of $447 per ton emitted, which means that the average carbon tax paid (revenue collected) per ton of carbon reduced would be $1152 per ton.[11] Alternatively, a 5 percent objective would require a tax rate of $217 per ton emitted, or average revenue collected per ton of carbon reduced of $1065 per ton.

Distinction should be made between the carbon tax rate and the average carbon tax revenue collected per ton of carbon reduced (abatement cost). Although both are expressed in $/ton of carbon, Figure 6 shows that, for any given percentage reduction of carbon, the $/ton carbon tax rate is much lower than the $/ton average tax revenue collected. By definition, the carbon tax rate would be applied to all the tons of carbon emitted, while the tax revenue per ton of carbon reduced is based only on the tons of carbon removed. Therefore, the carbon tax rate would be much lower than the tax revenue per ton of carbon removed.

In the case where tax revenues are earmarked for carbon abatement programs, the average abatement cost drops significantly from the case where revneues are used to reduce the deficit. Calculation of the carbon abatement cost and the required carbon tax rate under a policy which uses the tax revenues to finance carbon abatement measures is illustrated in Figure 7 for the 5 percent reduction target (27 million tons). It is important to note that a carbon tax strongly impacts the relative prices of alternative enrgy sources and on average makes energy use more costly. It is estimated that the income and substitution effects due to the change in relative fuel prices result in about one million tons of reduction in carbon emissions.[12] As shown in Figure 7, the abatement measures required to reduce the remaining 26 million tons would cost approximately $590 million. The expenditure represents an average abatement cost of $22/ton.

Figure 7 indicates that the carbon tax rate required to finance $590 million worth of abatement expenditures would be on the order of $8 per ton (5 percent objective). Carbon

Figure 6 - Estimated Impacts of a Federal Carbon Tax on New York State When Revenues Are Used for Deficit Reduction

Carbon Reduction From 1988 Level		Federal Carbon Tax Rate (1990$)	Federal Carbon Tax Collected from NY State (1990$)	Revenue Per Ton of Carbon Reduced (Abatement Cost) (1990$)
Percentage	Tonnage /a			
	(Mil. Ton)	($/Ton)	($ Mil.)	$/Ton)
5%	14	$217	$14,762	$1,065
10%	17	$273	$17,976	$1,056
15%	20	$366	$22,993	$1,141
20%	24	$447	$26,849	$1,152

a/ Assumes the Lower Emissions Electricity Supply Scenario is adopted, irrespective of the carbon tax. Therefore, the required tonnage reductions from the carbon tax are net of the 13 million ton reduction associated with the LEES Scenario.

Figure 7 - Estimation of Average Abatement Cost for a 5 Percent Reduction When Carbon Tax Revenues Are Used to Finance Carbon Mitigation Measures

	Carbon Tax (1990$)		Tons of Carbon Abated		Abatement Costs (1990$)		
Measures	Rate	Total Revenue	Total for the Measure	Carbon Cumulative	Unit Cost	Total for the Measure	Cumulative
	($/Ton)	($ Mil.)	(Mil. T.)	(Mil. T.)	($/Ton)	($ Mil.)	($ Mil.)
Carbon Tax	$8	$590	1.2	1.2/a			
Negative Cost Measures			6.6	7.8	($147)	($975)	($975)
Reforestation			2.1	9.9	$6 - $49	$38	($937)
LEES Scenario			13.3	23.2	$32	$432	($505)
Wind			0.6	23.8	$150	$83	($422)
Block #1			3.4	27.2	$300	$1,012	$590

Average							$22

a/ Resulting from demand reduction and fuel substitution

tax rates and corresponding carbon abatement costs for reduction objectives up to 20 percent are presented in Figure 8.

Figure 8: Estimated impacts of a federal carbon tax on New York State when revenues are used to finance carbon mitigation measures

Carbon Reduction From 1988 Level		Federal Carbon Tax Rate (1990$)	Federal Carbon Tax Collected from NY State (1990$)	Revenue Per Ton of Carbon Reduced (Abatement Cost) (1990$)
Percentage	Tonnage			
	(Mil. Ton)	($/Ton)	($ Mil.)	($/Ton)
5%	27	$8	$590	$22
10%	30	$20	$1,529	$50
15%	33	$35	$2,619	$78
20%	37	$50	$3,752	$103

A comparison of Figure 6 to Figure 8 indicates that by using the tax revenues to finance mitigation measures, less tax revenues are required to achieve a given reduction target than if tax revenues are used for deficit reduction. For instance, about $15 billion in tax revenues are needed to achieve a 5 percent reduction if the revenues are used to reduce the deficit, compared to the slightly more than half a billion dollars ($590 million) of tax revenues which are required if the revenues are used to fund carbon abatement measures. Consequently, the average abatement cost decreases from $1,065 per ton to $22 per ton, and the required tax rate drops from $217 per ton to $8 per ton.[13]

Incorporation of the externality value of carbon emissions in energy planning required, at minimum, that this value should be added to the cost of production of alternative energy resources prior to the selection of the least cost option. Any serious attempt to mitigate the threat of global climate change will eventually require the actual collection of the monetized value of carbon externalities as an emission fee.[14] Under such circumstances the externality value of carbon emissions would be equivalent to the carbon tax rate required to achieve a given reduction objective, assuming that the tax revenue would be utilized to pay for mitigation measures. Figure 9 presents the tax rates (monetized value of carbon externalities) required to achieve carbon reduction objectives up to 20 percent. Tax rates are shown as both $/ton values and as equivalent cents per Kwh for various energy technologies. These values could be used as adders to production costs to reflect the externality cost of carbon emissions.

Figure 9: Externality value of carbon emissions for selected generating technologies

Societal Goal: Carbon Reduction from 1988 level	Externality Value		Selected Generating Technologies				
	Real	Nominal	Coal: FGD or AFB	Natural Gas Combined Cycle	Coal Cogen-eration	Natural Gas Cogen-eration	DSM
Percentage	$/Ton of Carbon		Cents per Kwh (Levelized Nominal)				
5%	$8	$12	0.33	0.15	0.24	0.10	0.00
10%	$20	$29	0.82	0.38	0.59	0.26	0.00
15%	$35	$51	1.44	0.67	1.04	0.46	0.00
20%	$50	$73	2.06	0.96	1.48	0.66	0.00

The monetized externality values shown in Figure 9 are expressed in levelized nominal terms rather than in the levelized real 1990$ used throughout this paper, to be consistent with the methodology used in New York's lon-run avoided cost proceedings and competitive bidding programs. This figure indicates that, for a 5 percent objective, the monetized externality value of carbon emitted is about $12 per ton (levelized nominal value starting in 1990). This value is equivalent to 0.33 cents per kwh for an FGD coal plant. This value increases to 2.06 cents per kwh for a reduction objective of 20 percent.

In summary, a much lower tax rate will be required to achieve a given carbon reduction objective if the revenues are used to finance carbon abatement measures than if the revenues are used for deficit reduction. Alternatively, a modest tax will generate significant revenues, but by itself will not be effective as a carbon reduction tool.

Conclusions

- Carbon emissions could be stabilized at their 1988 levels, without any net increase in energy service cost, largely by pursuing the policies set forth in the State Energy Plan and by implementing low cost mitigation measures.

- The goal of a 20 percent reduction in carbon emissions by the year 2008 would be difficult to achieve due to high costs and the uncertainties associated with technologies which are presently undeveloped. Furthermore, a commitment to this target would require a large-scale restructuring of our energy systems and patterns of energy use. In contrast, a 5 percent to 10 percent reduction target appears to be more technologically feasible and is likely to be achieved at a much lower unit cost of reduction.

- Using a 5 percent reduction objective, an externality value of $12 per ton of carbon emitted is estimated (levelized nominal value starting in 1990). This value may be initially used in energy planning in New York. In the future, however, a higher externality value based on a higher reduction goal could be used.

- A federal carbon tax would be a powerful tool for either raising revenue or reducing carbon emissions. A modest carbon tax rate could raise a large amount of revenue which could be used to fund carbon mitigation measures. However, if a carbon tax is used solely as a revenue raiser (i.e., the tax revenue is used for deficit reduction or reducing other taxes), its impact on carbon reduction would be minimal.

- Reduction of carbon emissions is likely to accrue a significant amount of other environmental benefits, such as reduction of SO_2 and NO_x emissions. For a 5 percent reduction objective, between 20 percent and 100 percent of the cost of mitigating carbon emissions could be allocated to environmental benefits other than those resulting from reduction in global warming.

References

CEC (California Energy Commission). *The Impacts of Global Warming on California: Interim Report*. Sacramento, CA. August 1989.

Cohen, Sam; Eto, Joe; Goldman, Chuck; Beldock, John; and Crandall, Geoff. "Environmental Externalities: What State Regulators Are Doing." *The Electricity Journal*, July 1990.

EPA (Environmental Protection Agency), Office of Policy, Planning and Evaluation. *Policy Options for Stabilizing Global Climate*. Draft Report to Congress. Feb. 1989.

DOE. *Energy and Climate Change*. Report to the DOE Multi-Laboratory Climate Change Committe. Lewis Publishers, Inc. Chelsea, MI. Feb. 1990.

ICF Resources, Inc. "Preliminary Technology Cost Estimates of Measures Available to Reduce U.S. Greenhouse Gas Emissions by 2010." Report Submitted to EPA. August 1990.

IPCC (Intergovernmental Panel on Climate Change). "Policy Makers' Summary of the Scientific Assessment of Climate Change." Report to the IPCC from Working Group I. June 1990.

Jorgenson, Dale and Wilcoxen, Peter. "Environmental Regulation and U.S. Economic Growth." Harvard Institute of Economic Research. October 1989.

Koomey, Jonathan. "Comparative Analysis of Monetary Estimates of External Costs Associated with Combustion of Fossil Fuels." Lawrence Berkeley Laboratory. March 1990.

Manne, Alan and Richels, Richard. "CO_2 Emission Limits: An Economic Cost Analysis for the USA." *The Energy Journal*, April 1990.

Spectrum Economics, Inc. "Economic Impacts of Greenhouse Gas Reduction Plan." Report Prepared for the California Coordinating Council. August 1990.

Morris, Samuel; Solomon, Barry; Hill, Douglas; Lee, John; and Goldstein, Gary. *A Least Cost Energy Analysis of U.S. CO2 Reduction Options*. Report Prepared for Climate Change Division of EPA. 1990.

New York State. *New York State Energy Plan*. Sept. 1989.

NYSEO (New York State Energy Office). "Environmental Externality Issue Report." Published in *Draft New York State Energy Plan*, Vol. VI, May 1989a.

NYSEO (New York State Energy Office). "Air Impacts/Electricity." Published in *Draft New York State Energy Plan*, Vol. VI, May 1989b.

NYPSC (New York Public Service Commission). Case 88-E-240. "Order Issuing a Final Environmental Impact Statement and Adopting Staff's Response to Agency Comments." Issued May 16, 1989.

Oregon. *Report to the Governor and Legislature*. Oregon Task Force on Global Warming. June 1990.

Ottinger, Richard. "Getting the True Cost of Electric Power." *The Electricity Journal*, July 1990.

Pace University, Center for Environmental Legal Studies. *Environmental Costs of Electricity*. Report Prepared for the New York State Energy Research and Development Authority and U.S. Dept. of Energy. Oceana Publications. New York. 1990.

PC/SECC (Public Citizen and Safe Energy Communication Council). "Turning Down the Heat: Solutions to Global Warming." Sept. 1988.

Sanghi, Ajay. "The Role of Externalities in Utility Bidding Programs." Paper Presented at the Rutgers University Graduate School of Management Ninth Annual Conference, Advanced Workshop in Regulation and Public Utility Economics. New Paltz, NY. May 1990.

SERI (Solar Energy Research Institue). "Photovoltaics for Large-Scale Applications." Prepared by Ken Zweibel. Paper Presented at 1989 Energy and the Environment Conference. Denver, CO.

USFS (U.S. Forest Service). "An Analysis of New York's Timber Resources." Prepared by Thomas Considine. 1984.

Williams, Robert. "Low-Cost Strategies for Coping with CO_2 Emissions Limits" Center for Energy and Environmental Studies. Princeton University. December 1989.

WRI (World Resources Institute). "Forestry as a Response to Global Warming: An Analysis of the Guatemala Agroforestry and Carbon Sequestration Project." Prepared by Mark Trexler, Paul Faeth and John Kramer. June 1989.

Footnotes

1/The views and opinions expressed in this paper are solely those of the author and do not necessarily state or reflect the views, opinions or policies of the New York State Energy Office.

2/The author wishes to acknowledge Karl Michael, Jianguo Wang and Tony Joseph for their help in preparing this paper.

3/The issue of environmental externalities (i.e. damages associated with residual emissions after meeting all applicable environmental regulations) has recently received a great amount of attention in energy planning and particularly in utility competitive bidding programs. To date, over half of the states in the U.A. are considering some form of evaluation of environmental externalities in energy resource planning (Cohen et al. 1990; Koomey 1990; Pace 1990).

4/The highest marginal cost of achieving a given target reduction represents the "implied value" which society has implicitly placed on the last desired unit of reduction. However, basing the externality value on the implied value would impose an unnecessarily high cost on society.

5/The SO_2 and NO_x values adopted by the NYPSC are those reported in NYSEO 1989a (Preliminary Draft, Feb. 1989). These values were subsequently revised prior to the publishing of SEP.

6/2500 MWs of AFB coal capacity is required by the Lower Emissions Electricity Supply Scenario.

7/The back-up is required because the photovoltaics would operate at a maximum capacity factor of only 30 percent, while the AFB plant replaced would operate at about 80 percent.

8/SO_2 valued at \$416/ton (1990\$) using 50/50 mix of low cost (fuel switching) and high cost (FGD) measures. NO_x valued at \$1697/ton (1990\$) using 70/30 mix of low cost (low-NO_x burners) and high cost (SCR) measures. For this analysis, the SEP values were adjusted from nominal to real 1990\$ (NYSEO 1989a, 11: NYSEO 1989b, 4).

9/SO_2 valued at $4060/ton; NO_x valued at $1640/ton; real levelized 1990$; assumes "starting point" values, which were judged by authors to represent "typical values", based on a range of estimates from available literature (Pace 1990, 213, 230).

10/Under the reduced energy demand forecast (growth rate reduced by 0.25 %), the 20 percent reduction target could be achieved by implementing the same amount of carbon removal which is required to achieve a 10 percent reduction target under the original energy demand forecast.

11/As a point of comparison, Manne and Richels (1990) have estimated that the carbon tax rate required to achieve a 20 percent reduction in the U.S. rises to about $580 per ton of carbon by the year 2020, and then recedes to about $250 per ton of carbon in the long run.

12/While these economic adjustments may not occur without some costs to society, these costs are not included in Figure 7.

13/It is assumed that the Lower Emissions Electricity Supply (LEES) Scenario is adopted, irrespective of the carbon tax. If the carbon tonnage reduction associated with the LEES Scenario (13 million tons) were to be achieved through the carbon tax, the estimates reported in Figure 6 would increase substantially. For example, a 5 percent reduction target would require a carbon tax rate on the order of $550/ton, resulting in about $30 billion in tax revenues collected.

14/The proposal advanced by former Maine Commissioner David Moskovitz provides an example of how an emission fee could be implemented. He has proposed that utilities would be charged a fee equal to the quantified environmental costs of each resource selected. The proceeds would be deposited in a Pollution Mitigation Fund, which would be used to finance pollution mitigation measures (Ottinger 1990, 22).

6.3 Cooperative Research, Development, and Demonstration Opportunities in Environmental Externalities and Energy Planning

John A. Beldock[1], Kenneth G. Moore[2], and Robert L. San Martin[3]

Introduction

Environmental and economic externalities and energy planning are likely to be inextricably linked from now on. Significant progress has been made over the past several years by various stokeholders in establishing a basis for factoring externalities into energy decisions and operations. In particular, the States and State regulatory authorities have been instrumental in shaping the initial externalities research agenda and in moving to apply the concepts in practice. This has been accomplished while recognizing the need to balance the legitimate interests of utilities, consumers, and other interested parties in the research and development process.

It can be argued that States have played a key leadership role in the process and believe that it reflects the appropriate, essential manner by which energy is produced, consumed, and regulated in the United States. Recognize as well that energy production and use have important regional, national, and international implications which should receive research attention. Accordingly, this paper presents - for discussion purposes - a potential set of programmatic goals and objectives, an analytic framework, and a set of analytical research needs to help shape a results-oriented, multi-year national research effort.

---

1 Dr. John A. Beldock is a staff scientist in the U.S. Department of Energy's Office of Planning and Assessment in the Office of Conservation and Renewable Energy.

2 Kenneth G. Moore is the Acting Director of the U.S. Department of Energy's Office of Planning and Assessment in the Office of Conservation and Renewable Energy.

3 Dr. Robert L. San Martin is the U.S. Department of Energy's Deputy Assistant Secretary

The constructive efforts of States and others including international practices are paving the way for innovative methods that will result in environmentally sound energy planning and operations. Federal efforts in the U.S. play a meaningful support role in the process of developing environmental approaches to planning and facilitating their implementation and evaluation. This paper examines examples of these cooperative support avenues and the frontier opportunities for further analytical research and advancement.

In each stage of the externalities research, development, and demonstration process, there are lessons to be learned from tangential research that examines the forces affecting why new ideas are adopted and/or rejected in social systems. This paper applies some of the key findings in innovation diffusion research to the challenges ahead in the field of externalities.

The concepts and issues raised in this paper are not intended to be exhaustive or definitive. However, this is an opportune time for this important national and international arena.

Goals, Objectives, and Strategies

The ultimate goal of integrated resource planning is to aid in the development of technology-neutral, market "value-based" planning and operations. Several underlying purposes help define what is meant by this objective: 1) to ensure that markets are free to make optimum energy, economic, and environmental investment decisions, considering both supply and demand options, 2) to promote the richest possible set of reliable energy supply and demand options, available at reasonable cost, and 3) to encourage that full fuel cycle costs are considered in investment decisions. In the utility sector, these objectives are best pursued in conjunction with power marketing authorities, States, FERC, utilities, independent energy producers, trade organizations, research organizations, consumer/citizen advocacy groups, and other interested parties.

Realizing these goals and objectives requires examination of potential programmatic objectives and the progress made to date toward achieving these ends. Research to date suggests that future electric supply and demand options can achieve:

- Low initial and life-cycle costs,
- Improved competition among technology choices including both supply and demand options,

- o Market-based pricing which reflects total fuel cycle costs and benefits,
- o Environmentally sound and systematically reliable operations, and
- o Improved opportunities for utility returns from investments on the demand-side and on environmentally attractive technologies.

Current progres in the field has reached the point where it is opportune to review the analytic foundations that have supported the development of externalities work to date and to frame the types of future analyses which may be required.

An offspring of our Least-Cost Utility Planning Program is the Integrated Resource Planning Program that has as one of its missions the forestering of rational methods for estimating externalities accounting methods. Promising options exist for allowing supply-and-demand-side options to compete on an equal basis. In addition, utility and utility planning experience suggests that one can simultaneously pursue the multiple objectives of integraed resource planning, including:

- o Cost-Effective Power and Capacity,
- o Reliability,
- o Environmentally Sound Operations,
- o Industry Profits And Equitable Prices.

Bottlenecks and infrastructure impediments often slow and discourage progress towards cost-effective investments on both the demand and supply sides of the equation. However, there has been some significant progress. With help from the States, future options include reliable, cost-effective utility systems. These systems will be managed efficiently by regulations designed to allow the market to determine the best all-around investments - investments where energy, environmental and economic goals are balanced.

A number of cooperative projects involving research and development of environmental externalities methods and analysis are underway or have been completed. These include the Pace University study on the Environmental Costs of Electricity (Pace, 1990), DOE/U.S. Environmental Protection Agency (EPA) study with the Lawrence Berkeley Laboratory on the probable rate impacts of incorporating externalities in New England, the MSB & Associates' Pennsylvania, New Jersey, Maryland (PJM) Power Pool/externalities study, and the NARUC survey of State PUC activities which incorporate environmental externalities into electric utility planning and regulation. The Pace University work describes a significantly large universe of externality definitions and approaches. The study at LBL has modeled the probable effects of externalities programs on electricity rates for a Northeast utility, with the intent of extending the analysis to forecast the effects of programs in other service areas. The MSB work is addressing regional inertie questions and the effects of externalities programs on power pools. And finally, the NARUC survey describes the ongoing State PUC activities in this arena and the information needs that PUCs consider vital to furthering the process. There are a significant number of other efforts also contributing to this body of information which forms the cornerstones of the multi-stage research process. Figure 1 depicts the stages in the environmental externalities research process.

Research Stages

Research on environmental externalities appears to be progressing through a series of stages, each contributing to the accounting for environmental concerns in energy planning. A rough sketch of these stages follows:

o **Categorization of Externalities**, specifically directed at the environmental and socio-economic effects of energy use at all stages of the fuel cycle;

o **Measurement, Description, and Data Base Development and Refinement** of the types and magnitudes of environmental impacts of energy generation, delivery, use, and by-product disposal;

o **Development and Refinement of Accounting Methods** for characterizing environmental effects, including the derivation of economic costs or comparable factors within the context of resource planning;

o **Analysis of Institutional Issues** in the context of dynamic regulatory and utility structures, involving local, State, Federal, and international boundaries (including analysis of jurisdictional issues, market characterizations, and other limiting factors);

o **Monitoring, Program Evaluation, and Refinement** to assess and incorporate the lessons learned from on-line environmental accounting programs, leading to improved applications.

Figure 1: Environmental externalities research stages

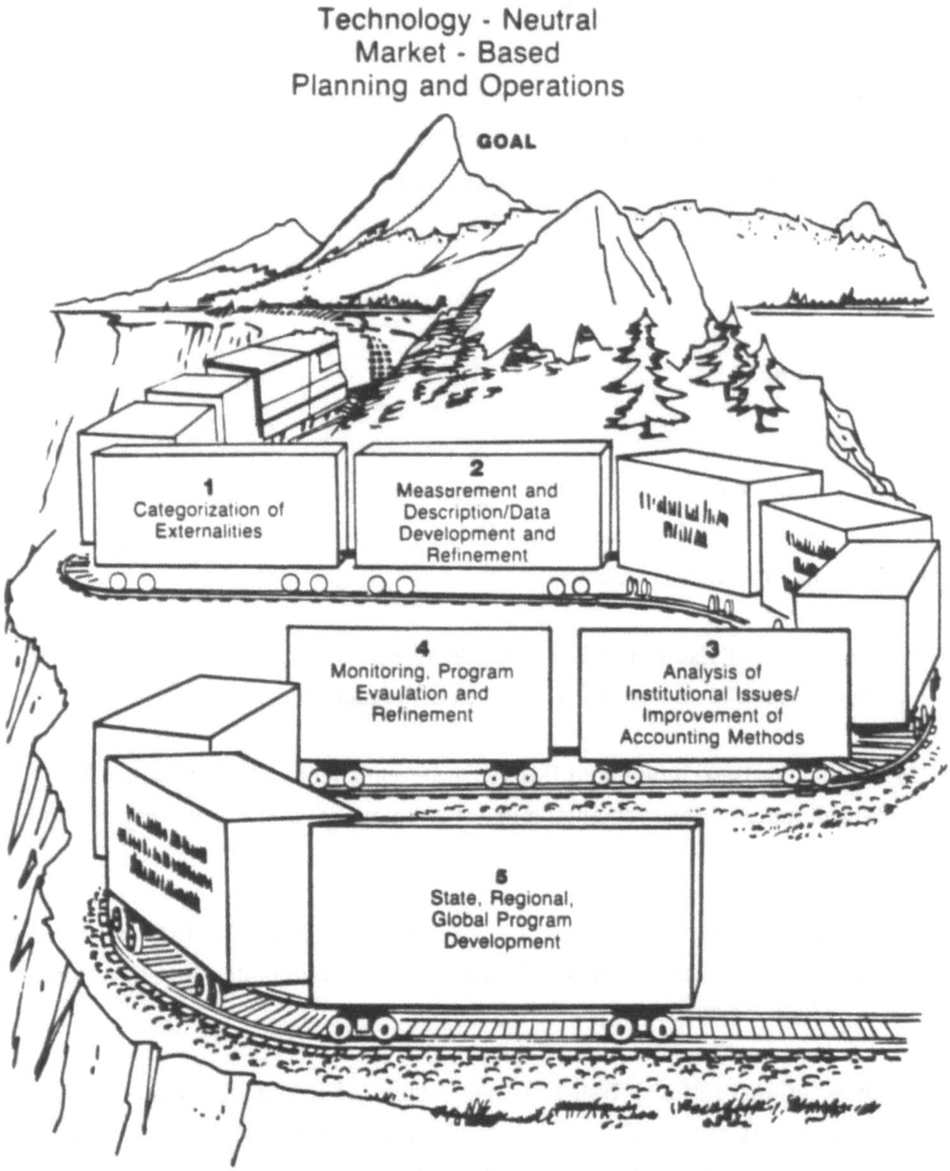

o **State, Regional, Global Program Development** involving the widespread implementation and subsequent continued improvement of externalities accounting programs that take into account State, regional, and international priorities.

These are overlapping research stages, involving a significant number of feedback loops, perhaps best approached as a process rather than as a single research path. Accurate data and rational methods are an integral part of the development of viable accounting methods. So too, are the development of implementation strategies and analysis of institutional issues key inputs to the process of developing technology neutral market-based planning and operations. Along similar lines, the results of monitoring and evaluating on-line programs are essential inputs in the effort to develop accepted and standardized methods. To be successful, each of the stages must be supplemented by strong technology transfer and information sharing activities.

Figure 2 compares an externalities research process with a similar description for technology research and development. Many of the stages are comparable, and the feedback loops are as important to one process as they are to the other.

Figure 2: Comparison of research and development processes

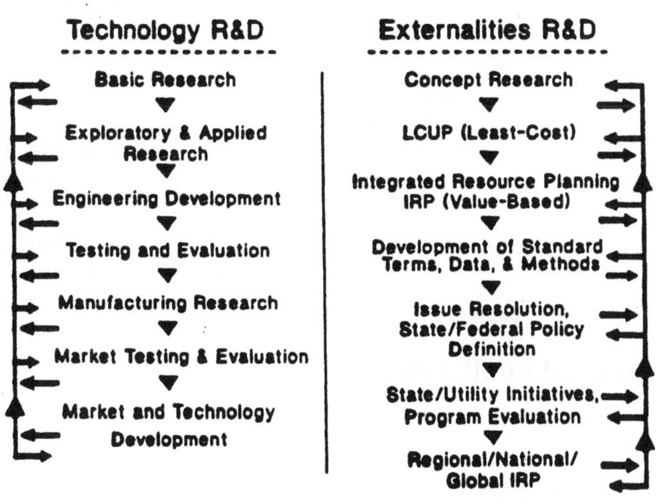

Figure 3 describes some of the interrelationships in the environmental analysis relevant to energy efficiency and renewable energy activities, portraying the need to integrate significant pieces of the puzzle. Ongoing and future analysis of emission factors (units of pollutants or impacts per unit of energy use) for each energy sector (utilities, transportation, industry, and buildings) will provide the basis for describing the process tree or total fuel cycle effects of energy use. The emission factors analysis, fuel cycle characterizations, and analysis of the costs of regulatory compliance for energy systems are likely to serve as the foundational work for refined environmental externalities accounting methods. Bringing together these presently-less-than-optimally-integrated-activities will allow for accurate analysis of the options for mitigating the environmental concerns related to conventional and alternative technologies. Significant integration efforts lay ahead in building the necessary bridges between past, current, and future analytical efforts.

Figure 3: Integration needs in environmental analysis relevant to energy efficiency and renewable energy

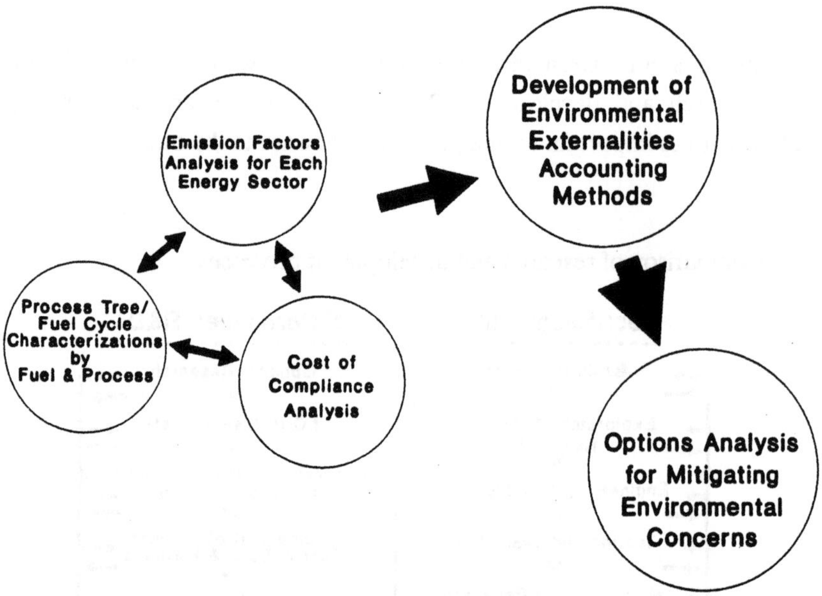

Lessons Learned and Limitations

In the U.S., the States have taken the innovative leadership role in developing and implementing externalities accounting methods. This leadership role offers us many opportunities to enhance these efforts through cooperative research, development, demonstation, and program evaluation projects. One of the most interesting aspects of the state approaches developed to date is that no single approach has emerged as the be-all-end-all method. This leaves open the opportunity for further improvements and

Where some states have resources to develop (state-level) accounting methods or approaches, the Federal government and other organizations have resources that can broaden the scope of these efforts. Federal resources can be used to develop regional and then national externalities approaches, transferring elsewhere the lessons learned in specific States.

Sharing experiences will be an important ingredient in developing and implementing viable accounting methods. New York, Wisconsin, Maine, California, and others have learned that including all the major players early in program development makes them more likely to work with new environmental accounting methods and the regulators. Problems are more prone to occur without consensus building. This is a well established finding in the literature (Rogers and Rogers, 1976). If utilities are afforded the opportunity to help develop the accounting methods for incorporating environmental externalities in planning decisions, they are more likely to work with the PUCs when regulators institute the new procedures. At present, this is evident where control cost estimates underpin most of the currently deployed externalities methods, and utilities are in one of the best positions to provide cost of control values. For example, New York utilities played a pivotal role in developing the methods now on record in that State. These findings also suggest working closely with utilities creates an opportunity to be sensitive to the barriers that utilities face with investments, such as those in demand-side management.

Uncertainty and Opportunity: With Some Progress, Where Are the New Frontiers?

Analysis of Jurisdictional Uncertainties

The scope of anylysis activities completed and underway suggests there are a number of unanswered questions. One of the subject areas with unanswered questions involves jurisdictional issues of how to establish definitional and system boundaries and who will be responsible for regulating, monitoring, and enforcing the accounting procedures for externalities programs.

Definitional uncertainties pinpoint one of the key questions in environmental externalities research - what factors can be legitimately considered externalities? Given the extraordinary difficulties in defining the boundaries associated with externalities, one must ask whether to direct future analytic resources toward expanding program breadth or depth. Posed differently, how does one determine whether primary, secondary, or tertiary order effects must be analyzed in depth? For example, it is a currently accepted first order conclusion that demand-side efficiency options: a) are environmentally attractive, b) are currently utilized below their level of economic efficiency, c) face major institutional barriers (capital hurdles) with both utility investors and facilities operators, building owners, and other potential adopters, and d) have positive influences on local, regional, and national economic productivity and employment. How far must one go down the fuel cycle in validating these conclusions?

Integrated resource planning initiatives at the State and Federal levels will likely only be widely successful, if there is a process for data gathering, estimation, and analysis, addressing issues such as:

- o At what level of market adoption does enrgy efficiency begin to exhibit neutral or even negative energy, environmental, or economic returns, compared with the other alternatives on the margin?

- o At what levels of market adoption would energy efficiency pass through the so-called "market innovator" stage of the supply curve and command a reduced incentive as compared to a market entry premium?

- o At what level of detail does one monitor program effects in order to best characterize the effects on utility load profiles and dispatching regimes?

- o For each of the above data elements, what levels of accuracy are required for future decision making at the level of the Federal government, the State PUC, the individual utility, the utility investment community, and consumer/ratepayer interests?

- o How far down the total fuel cycle continuum does one go in analyzing secondary or tertiary environmental impacts associated with the manufacture of energy efficiency and renewable energy applications (primary level manufacturer/assembler, secondary suppliers, etc.)?

If externalities are defined as everything from electromagnetic fields (EMF) to mining emissions and safety concerns, then considerable definition, measurement, and up-keep work is necessary. Additional definitional questions include how to deal with: land-use and drinking water externalities, species diversity uncertainties, and potentially positive externalities such as the economic efficiency resulting from supply- and demand-side

System boundary uncertainties is another area without consensus on the treatment of environmental effects that do not obey local, State, regional, or international boundaries. Which State, regional, national and/or international environmental approaches offer the potential for addressing regional and national utility emission challenges? At the first level of analysis, one can attempt to distinguish between urban and rural externalities, as the costs and benefits of energy fuel cycles often appear inequitably distributed when considered from these two perspectives.

And lastly with respect to system boundary uncertainties, who should be responsible for **regulating, monitoring, and enforcing** the accounting procedures for externalities? Should State PUCs be those responsible for implementing regional programs? What should the FERC do at the national level with regard to incorporating externalities into wholesale power transactions? or into public power ratemaking? How does one consider combining the efforts of States at one level and those of the Federal government at another? Are PUCs and the Federal government those that should be addressing externalities or should methods be implemented and enforced at some other level? These are the issues at the heart of the needed research and analysis on incorporation strategies.

In trying to find **the effective medium between local, State, regional, and Federal levels**, the lessons learned in the U.S. appliance industry offer salient advice to those in search of cooperative avenues. California took the innovative lead with minimum product energy efficiency standards for appliances. Other states followed the West's lead, resulting in a fragmented market. Energy efficiency improvements were achieved, but at a cost to the industry, in analysis, R&D, and marketing, suggested by some to be, significantly higher than the cost of meeting standards set at consistent national levels. However, without the innovative lead of the States, and the impact of potential regulations on the market, it may have been more difficult to reach a consensus. One of the key questions is how to learn from this or other examples and how to best use the innovative force of the States, supplementing it with Federal support? This is an opportunity to work cooperatively with the states and the regional power councils. The utility industry has the challenge to: a) think the externalities problem through and b) get out in front and help develop the viable accounting methods to assure the procedures instituted will be acceptable. The integrated resource planning participants should view the utility industry as the cooperating colleague. It is necessary to collectively improve the understanding of the utility business environment and to represent information on externalities programs in terms that the industry can a) understand and b) use to make decisions. Part of this task involves understanding issues such as the limits on utilities with demand-side management investments (e.g. profitability, see for example Moskovitz, 1990) and the impacts of potential Federal policies (e.g. taxation of rebates).

The dialogue should continue and be strenghtened. Consideration of the information and decisison needs of IOUs and public power companies in shaping the future analytic agenda for integrated resource planning and externalities is a key to productive accomplishment.

Normalizing the Available Methods

One of the next key steps is to distill the innovative methods of the States into a package that calls on the best attributes from each of the approaches, yet is also consistent in the treatment of resources and effects that have very different values as a function of locale. The Pace University study team (1990), Koomey (1989), Chernick and Caverhill (1989), the Ohio PUC (1990), and the NARUC survey (1990) move in the direction of consistently describing the presently on-line and developing approaches. But some fundamental questions remain unanswered. First, how best to account for State and regional distinctions in fuel and energy supply, load variations, etc.? Second, how best to account for the differing priority placed on capacity across State and regional boundaries? Third, how best to account for the differing values that States and regions may place on distinct types of environments and services? At the next level, what are the distributional effects of a 2 cents/kwh surcharge in a state where electricity is priced at 8 cents/kwh and in one where the price is 14 cents/kwh? There is much to be tested before one can consider developing a normalized method. On the bright side some methods are in practice and have started the process.

The Effect of Legislative Activity

Further uncertainties and analysis opportunities exist in the area of legislative activities, specifically with regard to the new Clean Air Amendments. These uncertainties are significant, beginning at the level of cost of compliance. Based on a system of offsets and credits, the Federal government is proposing a market control plan to reduce the emissions of acid rain precursors. Additional analysis is needed to determine the effects of the new legislation on the costs and benefits incurred by utilities, stockholders, ratepayers, other interest groups, and by the environment. Some of the big questions are:

1 how will the new legislation affect state level initiatives, including on-line externalities programs?

2) how will the costs of compliance for the Clean Air Act affect utility resource mixes?

3) how to distinguish between the externalities that are considered in the new legislation and those considered in the on-line State initiatives, to avoid double counting?

4) how will the legislation affect the value and development of up-and-coming technologies such as photovoltaics, wind, municipal solid waste incineration, and demand- and supply-side energy efficiency improvement options?

5) how will the conventional non-fossil fuel options, such as nuclear and hydroelectric technologies, be affected, and

6) from the externalities accounting perspective, what does the new legislation, based principally on a control cost view of the world, hold in store for a field of research and testing activity that appears to be moving in the direction of damage costs?

Damage and Health Costs as the Elusive Objective

In the Federal legislative arena and in the state activity thus far, control cost estimates have served as the surrogate, in the absence of any more accurate estimates, for the value of the environmental effects of generation technologies. The value of the damages to the environment imposed by technologies at each stage of the fuel cycle has yet to be accurately incorporated into the mainstream of externalities accounting methods. There are several reasons for the absence of accurate damage cost estimates. One of these is that one might expect different interest groups to value the effects of energy and the services that energy provides differently. The simplest example involves the way in which bass and trout fisherman value the damage created by the construction of a hydroelectric facility. Needless to say, the revealed preference approach to accounting for externalities leaves open the possibility of diverse responses.

The Challenges of Program Evaluation

With on-line externalities programs in several States and close to fifty percent of the public utility commissions in the nation considering environmental programs, there is a rapidly increasing need to develop measures to evaluate program performance. There are a number of fundamental questions to be answered in approaching the evaluation of externalities program? This question is best defined by the program objectives. Consider the possibility of achieving the ultimate goal, technology-neutral, market "value-based" planning and operations. To phrase the question differently, if a State were to evaluate an externalities program, what would it need to know? Some of the program evaluation criteria might include:

a. **Load Profile Effects?** (ie. Are programs moving load profiles away from baseload capacity units towards less reliable intermediate load facilities?)

b. **Supply Mix Effects?** (ie. What supply mix shifts will result from externalities programs, including coal to natural gas, coal to nuclear, nuclear and/or coal to other supply sources?)

c. **Demand-Side Effects?** (ie. Are utilities and customers participating at the desired rate of adoption for demand-side management in States where initiatives have been launched? Are programs causing demand-side management to reach levels of economic efficiency? What impediments remain (eg. first versus life-cycle cost)?)

d. **Supply- and Demand-Side Effects** (ie. What are the new supply-side options on the margin in states where externality initiatives are active? and are DSM incentives in reality competing at the margin with the renewable energy technologies?)

e. **Economic Indicators** (ie. What effects will programs have on energy prices and the distribution of risks between ratepayers, investors, taxpayers, and others?)

f. **Business As Usual Forecasts?** (ie. What would happen under normal operating conditions in the absence of externalities programs?)

g. **Ecosystem Effects?** (ie. How do we best measure the results of externalities programs on the environment? Which indicators should be those on which to concentrate?)

h. **Distributional Effects?** (ie. How will the programs affect different interest groups, including utilities, ratepayers, investors, taxpayers, and others?)

i. **Social System Effects?** (ie. What effects will programs have on development, social programs and services, elderly and handicapped citizens, and other societal concerns?)

j. **Utility Investment Effects?** (ie. Where will utility investment portfolios shift in the future as a result of programs, towards diversification, lower risk investments, environmental management, etc.)

k. **Engineering/Research Effects?** (ie. Where will research and development efforts concentrate as a result of programs? and where should one expect to see innovative advances?)

l. **Communication Effects?** (ie. Is the value of externalities and the effects of programs being expressed in such a way as to influence consumer and utility behavior?)

These are some of the questions that must be answered collectively in evaluating environmental externalities programs. All the answers are not known, but it is essential to continue a program aimed at obtaining accurate responses. The information acquisition process for resolving these uncertainties starts with the field data from the States and others, followed by the ongoing synthesis and integrative analysis of many organizations.

Diffusion of Innovations

Accounting for the significant set of uncertainties above will do more than merely produce paper studies. Adequately addressing these uncertainties will also assist in making environmental externalities programs acceptable to a wide range of interest groups. Innovation diffusion research dedicates itself to analysis of the communication of innovations such as externalities initiatives. The principal objective of analysis into the diffusion of innovations is to better understand the variables that influence the acceptance (adoption) or rejection of innovations (new ideas, technologies, etc.) in social systems. Results of the thousands of innovation studies - are best characterized by Everett Rogers in Diffusion of Innovations (1983). Rogers examined 3,085 studies related to the diffusion of innovations, and derived a set of generalizations about the forces influencing innovation adoption.

In simple terms, Rogers finds that the adoption of new ideas (such as externalities programs) is influenced by (1) the attributes of the innovations themselves (ie. both the inherent attributes of externalities programs and the attributes that the programs are perceived to have) and (2) the characteristics of the systems in which the new ideas are presented, including communication channels, characteristics of the decision makers, and activities of those whose opinions influence the decisions of others (opinion leaders). In the context of environmental externalities programs, the key findings from the innovation diffusion generalizations (and the answers to the program evaluation questions posed above) can be applied as follows:

Compatability

The Degree to Which Externalities Programs are Perceived by Decision-Makers as Compatible with existing infrastructures in which energy planning and policy decisions are made is likely to positive influence the degree to which the programs are adopted.

Complexity

The Complexity of Externalities Programs is likely to adversely affect the degree to which the programs are accepted. If decision-makers are unable to understand the methods, they are unlikely to support their inclusion.

Relative Advantage

The Degree to Which Decision-Makers Perceive the Relative Advantage of Externalities Programs, relative to the methods they are intended to replace (in this case methods where certain environmental externalities were implied through omission to be equal to zero), is likely to positively influence the adoption of the new programs.

Perceived Risk

The Degree of Perceived Risk Posed by Externalities Programs from the Perspective of Decision-Makers is likely to adversely affect the adoption of new approaches. Key questions include: who pays?, who is at risk?, and who benefits?

Observability/Trialability

The Degree to Which Decision-Makers Can See Successful Externalities Programs in Place and Test-Out the New Methods is likely to positively influence the adoption of the innovative approaches.

Along with the perceived attributes of innovations above, Rogers finds that several other factors influence the adoption of new ideas, including 1) the characteristics of the decision-makers themselves (objectives, etc.), 2) the communication channels which link the system in which innovations are introduced, and 3) the opinion leaders to whom decision-makers look for credible and visible references when faced with new ideas.

Analysis and description of the uncertainties associated with the effects of externalities programs will make the task of externalities champions easier, by addressing the forces that affect process and institutional change.

Conclusions

Recall that one of our ultimate objectives in the U.S. DOE Integrated Resource Planning Program is the development of technology-neutral, market-value based planning and operations. To accomplish this, we need to ensure markets are free to make optimum economic and environmental investment decisions, promote availability of a rich set of reliable energy supply and demand options at reasonable costs, and encourage consideration of full fuel cycle costs and benefits.

There are several ways to achieve these objectives, and a framework is in place for examining the alternatives. Initial findings suggest cost-effective power and capacity, reliability, environmentally sound operations, and utility profitability at equitable prices are not mutually exclusive objectives. Moreover, it must be recognized that the answers to remaining uncertainties can only be generated through a collective process, where local, State, regional, and Federal efforts are pooled together. Different parts of the problem will require different analytical and policy treatments. States offer some of the best test-bed opportunities for solving practical problems involving externalities., yet the Federal government may be able to complement State activity with integrative research, analysis and legislative support.

There is consensus that any bottlenecks in the infrastructure often slow and discourage progress towards cost-effective investments on both the demand and supply sides of the equation. To mitigate this state of affairs a number of cooperative projects are underway aimed at assessing the options for refining regulatory structures to allow for proper accounting of externalities. It is also suggested that there are as yet untapped opportunities in each stage of the research, development, and demonstration process.

One of the most important observations to date is that States have again taken the lead in innovation and in developing and implementing accounting methods. The prowess of the States provides a number of opportunities to support these efforts through cooperative research, development, demonstration, and program evaluation projects.

A large number of uncertainties have been identified, and it is suggested that these can, and should, be viewed as opportunities. Included in the set of uncertainties are: definitional and boundary uncertainties, regulation, monitoring, and enforcement uncertainties, uncertainties as to how best to normalize amongst the on-line State methods, uncertainties regarding the effects of new legislative activities at the Federal level, questions as to how best to capture the value of damage costs, and finally, opportunities to define the most appropriate criteria for evaluating programs.

Externalities programs can capitalize on the findings of innovation diffusion research by: 1) making the programs user friendly, easy to understand, compatible with the existing systems, visible, and capable of being tested on a trial basis, 2) designing systems where the relative advantages of programs are clear compared with those being replaced, 3) communicating program risks upfront, 4) making sure that decision-making opinion leaders are well-informed and supportive of the programs, and 5) being sensitive to the characteristics of the decision-makers and the systems in which externalities programs are under development.

The good news appears to outweigh the bad news on first inspection. There is now consensus among a large number of interested parties that the environmental costs and benefits of energy options can no longer be ignored or assumed through omission to be zero. Given this consensus, the key is to cooperatively keep the analytical research, development, and demonstration balls in the air so as to promote the most effective environmental externalities programs.

References

Berry, L. and E. Hirst. 1989. *Recent Accomplishments of the U.S. Department of Energy's Least-Cost Utility Planning Program*. ORNL/CON-288. Oak Ridge National Laboratory. Oak Ridge, TN.

Cavanaugh, R. 1986. "Least-Cost Planning Imperatives for Electric Utilities and Their Regulators." The Harvard Environmental Law Review, 10, no.2: 299-344. Cambridge, MA.

Center for Clean Air Policy. 1989. *An Efficient Approach to Reducing Acid Rain*, Prepared by the Center for Clean Air Policy. Washington D.C.

Chernick, P. And E. Caverhill. 1989. *The Valuation of Externalities from Energy Production, Delivery, and Use*. PLC, Inc. A Report prepared for the Boston Gas Company. Boston, MA.

Congressional Budget Office.1989. *The Role of Technology and Conservation in Controlling Acid Rain*. Staff Working Paper, Congressional Budget Office. Washington, D.C.

Dudek, D. 1988. *Offsetting New CO2 Emissions*, Environmental Defense Fund. New York, NY.

Cohen, S., J. Eto, C. Goldman, J. Beldock, and G. Crandall. 1990. "A Survey of State Public Utility Commission Activities to Incorporate Environmental Externalities into Electricity Utility Planning and Regulation." National Association of Regulatory Utility Commissioners, Washinton, D.C.

---------------What States are Doing About Externalities. *The Electricity Journal*. July 1990. 24-35.

Geller, H., M. Ledbetter, E. Miller, and P. Miller. 1987. *Acid Rain and Electricity Conservation*. American Council for an Energy Efficient Economy and the Energy Conservation Coalition. Washington, D.C.

Goldman, C., E. Hirst, and F. Krause. 1989. *Least-Cost Utility Planning in the Utility Sector: Progress and Challenges*. Lawrence Berkeley and Oak Ridge National Laboratiorie. LBL-27130, ORNL/CON-284. Berkeley, CA. and Oak Ridge, TN.

Hohmeyer, O. 1988. *Social Costs of Energy Consumption: The External Effects of Electricity Generation in the Federal Republic of Germany*. Springer-Verlag. Berlin. FRG.

Koomey, J. 1990. "Comparative Analysis of Monetary Estimates of External Costs Associated with the Combustion of Fossil Fuels." Lawrence Berkeley Laboratory. Berkeley, CA.

Krause, F., W. Bach, and J. Koomey. 1989. *Energy Policy in the Greenhouse. Volume One: From Warming Fate to Warming Limit: Benchmarks for a Global Climate Convention.* Published by the International Project for Sustainable Energy Paths in Conjunction with the European Environmental Bureau. El Cerrito, CA.

Moskovitz, D. 1989. *Profits & Progress Through Least-Cost Planning*. National Association of Regulatory Utility Commissioners. Washington, D.C.

New York State Energy Office. 1989. *Environmental Externalities Issue Report*. New York State Energy Office, Division of Policy Analysis and Planning. Albany, N.Y.

Pace University Center for Environmental Legal Studies. 1990. *Environmental Costs of Electricity*. Oceana Publications. Dobbs Ferry, N.Y.

Public Utilities Commission of Ohio. 1990. *Including Societal Benefits in Rates for Cogeneration and Small Power Production*: Prepared by the Public Utilities Commission of Ohio with the Assistance of Battelle Memorial Institute and the Cincinnati Gas and Eectric Company. Columbus, OH.

Rogers, Everett M., *Diffusion of Innovations*, Third Edition, The Free Press, a Division of MacMillin Company, Inc. New York, NY, 1983.

Rogers, Everett M. and Rekha Agarwala Rogers (eds.), *Communications in Organizations*, New York, Free Press.

Rosenfeld, A. and R. Mowris. 1989. *Energy Efficiency: The Best Way to Save Money and Slow Global Warming*. Testimony for the Hearing on an Integrated National Energy Efficiency Strategy, U.S. House of Representatives, Committee on Energy and Commerce, Subcommittee on Energy and Power. Berkeley, CA.

United States Department of Energy. 1989. *A Preliminary Analysis of U.S. CO2 Emissions Reduction Potential from Energy Conservation and the Substitution of Natural Gas for Coal in the Period to 2010*. DOE/NBB-0085, prepared for the U.S. DOE Office of Energy Research, Office of Basic Energy Sciences, Carbon Dioxide Research Division. Washington, D.C.

U.S. Department of Energy. 1988. *Energy Technologies & the Environment: Environment Information Handbook*. DOE/EH-0077. Washington, D.C.

U.S. Environmental Protection Agency. 1989. *Policy Options for Stabilizing Global Climate: Draft Report to Congress*, Office of Policy, Planning, and Evaluation. Washington, D.C.

Wiel, S. 1989. "Making Electric Efficiency Profitable." Public Utilities Fortnightly, July 6, 1989. Arlington, VA.

6.4 Environmental Externalities in Utility Planning

Donald C. Peterson, Jr. and Daniel M. Violette
RCG/Hagler, Bailly, Inc.
P.O. Drawer O
Boulder, CO 80306

1.0 Introduction

Environmental externalities are gaining new importance in utility resource planning. This paper proposes a framework for addressing environmental externalities within a utility integrated planning process. This framework was previously proposed by Violette and Peterson (1990) and Violette, et al. (1990), and has also been suggested by other researchers, including: Bernow and Marron (1990), Cropper and Oates (1990), Ottinger (1990), and Brick and Edgar (1990). For purposes of this paper, this approach is termed the "environmental constraint" method.

Prior to presenting this framework, the general "environmental externalities" problem is discussed. In characterizing the externalities problem, some "simple truths" that have appeared in recent literature will be examined, and while not necessarily showing them to be untrue, the discussion will hopefully illustrate that these "truths" are not so simple. The purpose of this examination is to help clarify policy issues facing regulators, and assumptions implicit in decisions to incorporate environmental externalities in utility resource planning. While we believe it is important for regulators to explicitly address residual environmental externalities, we are not as sanguine as some about what has become the "conventional wisdom" on the issue. We believe that regulators should broaden the scope of analysis and policy tools being considered for incorporating externalities as part of the planning process.

2.0 Background

State Public Utility commissions in the U.S. have been pressing ahead with provisions to quantify environmental externalities in utility least-cost planning proceedings. By one count 24 jurisdictions have at least started to formally consider incorporating externalities in utility planning (Cohen, et al., 1990). Three principal arguments have been advanced for this, and are well stated in Ottinger, et al. (1990). These arguments are:

- Allowing utilities to only meet applicable governmental laws and regulations requiring emission limits or rate standards ignores externality costs of emissions which exceed regulated levels. These emissions will have some impact on the environment, and should be taken into account in utility planning.

- It is the regulators' purview to require these costs be accounted for in a least-cost plan, since it is their responsibility to protect consumer wwelfare. By including environmental externality costs, regulators are simply assuming their traditional role of assuring prudent resource allocation by utilities.

- Granting that any monetary valuation or other quantified weighting of environmental externalities is extremely uncertain, commissions must take the general position that approximate damage estimates are better than implicitly valuing these externalities at zero. This position is probably best expressed by the Oregon Commission in a recent decision (Order No. 89-507) where it states "when the certainty of external costs is known, but the amount of the costs is not, zero is the least desirable and least accurate cost to apply."

There are several different approaches for incorporating environmental externalities in utility planning. The approach selected will depend, in part, on regulatory objectives. Several approaches have been suggested for explicitly incorporating externalities in utility least-cost planning. The two methods most often cited are: the cost of control approach; and the direct valuation of damages using economic methods.

A third method that does not involve a monetization of the externalities, but instead uses a set of weighting criteria has also been suggested. Several fairly complex weighting schemes have been devised. See, for example, the California Energy Commissions's ER-90 Report and Destribats, et al. (1990) for a matrix weighting system proposed by New England Electric System.

The first approach, based on avoided costs of control assumes that if lawmakers require controls that, at the margin cost $500 per ton, then an additional one ton reduction in emissions is worth $500 to society. Thus, the cost of additional reductions is assumed to be equal to the marginal control costs. This assumes that lawmakers are acting as society's representatives and are applying reasoned economic judgement. Recent U.S. Clean Air Act Analysis shows that cleanup of acid rain will require expenditures of up to $1,500 a ton to reduce SO_2 emissions. This assumes that an additional one ton reduction is worth $1,500.

The second approach, damage cost estimation, attempts to identify specific resources at risk, environmental injury to sensitive receptors, and the value to society associated with injured environmental services. Direct assessment of environmental damages is the conceptually preferred approach, as there is no relationship between control costs and the value of natural resource injury. The often-cited difficulty of direct assessment is due to complexities in the required analyses and the uncertainties associated with our knowledge of pollutant fate and transport, ecosystem structure and function, and exposure-response functions relating pollutant concentrations to ecosystem, human health, and man-made materials damage. In addition, estimating monetized damage values using techniques available to environmental economists is highly uncertain and may provide values difficult to defend in public forums.

Opinion is sharply divided regarding the appropriateness of these two approaches for estimating externalities. Each method has its proponents, its strenghts and weaknesses, and requires numerous assumptions. These have been widely discussed in recent examinations of environmental economics.

Before further developing discussions of incorporating environmental externalities in utility planning, it is important to recognize that a significant policy question is how to appropriately price and/or control *residual* emissions, i.e., emissions after all applicable regulations have been met. In this context, residual emissions can be broadly interpreted to include air, water, solid waste as well as land use and aesthetic externalities. The policy question, at its most basic is, whether the future time path of residual emissions is set to a socially optimal level. A corollary to this question may be expressed as: since different options for providing energy services result in different residual emissions, should integrated utility plans that produce lower residual emissions be given some type of credit relative to a plan that produces higher residual emissions? In comparing planning options, the argument is that options, such as conservation, that produce fewer detrimental externalities should be provided credits relative to other supply options.

These considerations may be summarized in two questions:

One - Has society paid enough? That is, is the current level of residual emissions such that incurring additional costs to achieve greater reductions in residual emissions (for specific pollutants) are not worth the benefits they achieve?

Two - If additional rsidual emissions reductions are needed, what mix of policy tools can best achieve this reduction?

Related to the second question, it should be noted that there is an extensive literature in environmental economics on choice of policy instruments. A policy maker can choose either pricing or quantity setting policy tools. If regulators set prices for externalities, then polluters (i.e., the market) determine the quantity in response to the set prices. If regulators set quantities, then the market sets an implicit price. Cropper and Oates (1990) show that with perfect knowledge either prices or quantities can be used to achieve the desired result. With imperfect information, uncertainty and regulatory constraints, arguments may be made that quantity tools are preferable.

Policy tools considered to date by commissions have been almost exclusively pricing mechanisms within the resource plan.

Cropper and Oates (1990) provide an excellent discussion of this issue in their recent survey of environmental economics. They show that under uncertainty, the choice of price or quantity becomes more complex. Weitzman (1974) established the theorem that the preferred policy tool depends on the relative steepness of the marginal damage and cost curves. When the damage curve is steep relative to the cost curve, a quantity standard is preferred. When the cost curve is steeper, a price instrument is preferred. Roberts and Spence (1976) further developed the Weitzman theorem by examining combinations of price and quantity instruments and showed that, in the presence of imperfect information and uncertainty about costs and benefits, these policies used in combination can provide welfare gains. Further research on policy instruments in a regulatory policy setting (Bohm and Russell, 1985) illustrates advantages of quantity instruments when growth and uncertainty are present. This literature has not entered into this specific policy debate on incorporating externalities in utility planning, but may be directly relevant.

3.0 Some "Simple Truths" That Are Not So Simple

Background presented above on the general approaches used to incorporate environmental externalities allows several popular statements to be explored. The purpose in examining these statements is to place estimation of environmental externalities in a proper regualtory perspective. This may help identify some important regulatory perspective. This may help identify some important policy implications in any decision to incorporate externalities in utility planning. Four specific statements will be explored.

- Simple Truth # 1: Full compliance with all applicable environmental laws by utilities is insufficient because residual emissions and environmental impacts remain; *and* that this effectively assigns a value of zero to these residual emissions.

- Simple Truth # 2: It is important to use marginal control costs, i.e., the most expensive control costs used to obtain the last ton of reduction to place a value on environmental externalities when using the avoided cost approach.

- Simple Truth # 3: It is always cheaper to control new sources than existing sources, thus the correct policy focus is on new sources rather than existing sources.

- Simple Truth # 4: That a value of zero is the least desirable estimate to apply when environmental externalities are known to exist with certainty, but the amount of physical injury and economic damage is unknown.

3.1 Discussion of these "Simple Truths"

The statement that current regulation implicitly assigns a value of zero to residual emissions reads as a call to action for all those who are environmentally aware, because no reasonable person would support placing a zero value on resulting environmental injuries and economic damages. However, in establishing a policy standard, damages associated with additional emissions reductions are usually depicted by a declining curve; that is, the first ton reduction may have a high value to society, but each subsequent ton reduction has a lower value.

An often stated objective in setting a pollution standard is to equate the marginal benefits of control with the marginal costs. This is illustrated as point P in Figure 1. This figure roughly approximates the political compromise that was struck on the acid rain provisions of the proposed Clean Air Act amendments - Senate Bill 1630. The bill calls for a cap of approximately 9 million tons per year on SO_2 emissions. It is estimated that marginal control costs will be approximately \$1,500 per ton of SO_2 removed for units with the highest cost of control.

Figure 1: Comparison of marginal environmental damages with marginal control costs

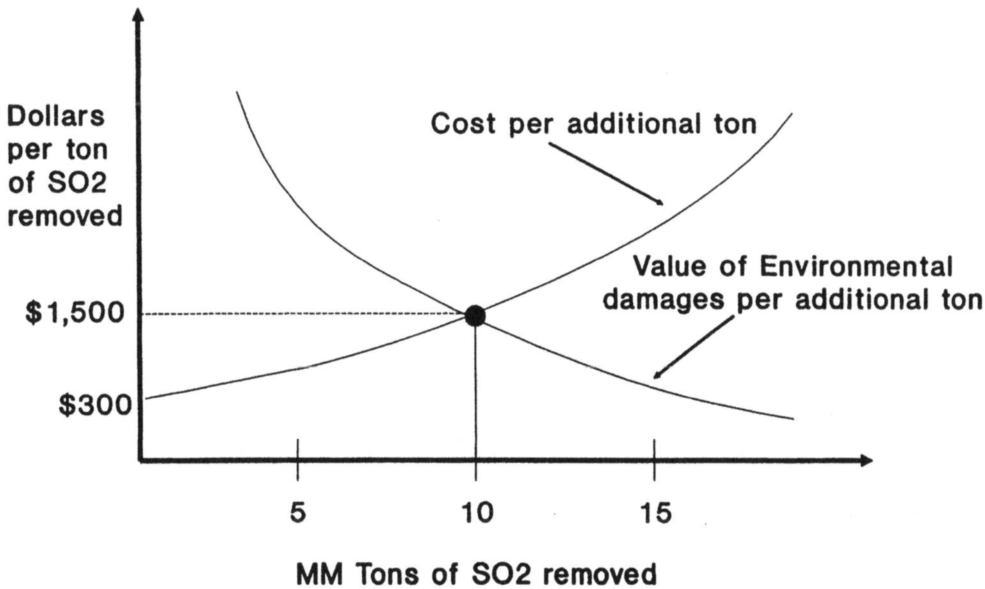

Given the information shown in Figure 1, is it true that the residual damages are assigned a zero value? The answer is no. They are simply assigned a value lower than the costs of reducing emissions to a lower level. The control actions that take place (i.e., those that are reported) are only those "transactions" where the value of the reduction outweighs the cost. Analyses of markets and estimates of consumer surplus for market commodities show that there is an observed price (or value) associated with the purchase only if the benefits of these transactions outweigh the cost. If not, then the only value observed is zero since no transaction has occurred.

In absence of an actual market, it may be assumed that lawmakers analyze pollution abatement costs and attempt to find a market clearing price (i.e., cost of removal) and/or quantity. If omniscient, then lawmakers could take into account all control strategies and factors that contribute to, or reduce, emissions over time. This would include, in the acid rain example, all energy conservation options and the socially economic construction of coal fired power plants. Thus, lawmakers and federal regulators may have already assumed specific energy use levels and different mixes of fossil use for both utility and non-utility generation in their deliberations. To argue that additional costs should be incurred to reduce emissions, then we must assume that the lawmakers have erred, and that their bias is towards under control of the pollutant in question. It is interesting to note that advocates of the avoided control cost of valuing residual emissions support their position by appealing to the argument that lawmakers make decisions that at least approximate equating marginal benefits and costs (Chernick and Caverhill, 1989).

The second simple truth addresses the use of marginal cost values in the avoided cost approach. This aproach identifies the highest marginal cost reduction strategy required by legislation. Assuming that lawmakers are rational economic agents, this "marginal" value places a lower bound on the costs that society assigns to an additional increment of emissions control. Again, appealing to Figure 1, it is clear that this statement is not necessarily valid. If lawmakers are rational, then the highest control cost (i.e., the marginal cost) is actually the upper bound on the amount that society would be willing to pay for an additional ton reduction. If we believed that higher control costs would yield emissions reductions producing benefits exceeding the costs, then the standards would have been set more tightly. Marginal control costs may be a reasonable approximation for the costs of small changes in emissions around point P in Figure 1, but it is difficult to argue that it is an upper bound.

While most analysts have argued that the avoided control cost approach likely will underestimate true externality values compared to direct damage assessment, we believe this issue to be indeterminate, depending on the amount of reduction already obtained in a region and the corresponding pollutant damage.

The third "simple truth" is that it is easier to control new sources rather than existing sources. This has stemmed largely from the assumption that all controls are similar to Best Available Control Technologies (BACT), such as flue gas desulfurization devices or scrubbers. Scrubbers have extensive space, venting, piping, and solid waste removal requirements that are more easily incorporated in new construction. However, studies of economic responses to the proposed U.S. acid rain legislation have shown that many of the most economic compliance actions are those at existing plants with high emissions rates. These compliance options involve operational rather capital intensive actions such as fuel switching, dispatching, or even retiring older, dirty plants. Again, if the real policy goal is to reduce residual emissions in the most cost-effective manner, regulators may want to consider options within the least-cost planning process where tradeoffs between externalities produced by new sources can be offset by emissions reductions at existing sources. This type of integrated compliance plan could produce the same, or reduced, level of residual emissions at a lower cost to ratepayers.

The fourth simple truth states that if residual emissions and environmental externalities are known to occur with certainty; then, the least desirable estimate to apply to external costs is zero. While this statement has appeared in many papers, is it really an unarguable truth? The implication of this statement is that any action taken by regulators to incorporate a value for emissions within an integrated planning perspective will increase rather than decrease social welfare. While regulatory actions in addressing residual externalities may well be appropriate, is this level of confidence deserved?

For example, consider the estimated SO_2 emissions through year 2030 without new acid rain regulation shown in Figure 2. Projected emissions, given provisions of Senate Bill 1630, are shown in Figure 3. Given that both Figures project a SO_2 emissions decline and that additional reductions from an already lower baseline emissions level may Produce fewer benefits per ton, is it obviously true that using a marginal cost of control of $1,500 to reduce an additional ton of *residual* SO_2 emissions, will improve social welfare over another policy that does not place a price adder on emitting sources? Consider if the "true" social value was $500 per ton. Is a price adder of $1,500 better than the status quo? Of course, the above considerations are quite different for currently unregulated pollutants.

3.2 How Do the "Simple Truths" Influence Policies to Incorporate Environmental Externalities in Utility Planning

Assessing policies for incorporating environmental externalities in the utility planning process remains one of the most important issues currently facing both utilities and regulators. What is clear from the exploration of the "simple truths" is that incorporating externalities in utility planning is complicated and that a number of apparently simple arguments may incorporate broader range of controversy than might first be apparent. When these assumptions are accounted for, it is not quite so easy to dismiss arguments utility analysts have proposed that previous investments in pollution control already incorporate externality values. The position is that electric power companies have already invested billions of dollars meeting pollution control, land-use and other environmental and safety requirements, and that this already internalizes many externalities.

Figure 2

Sulfur Dioxide Emissions for the U.S.

Base Case Scenario

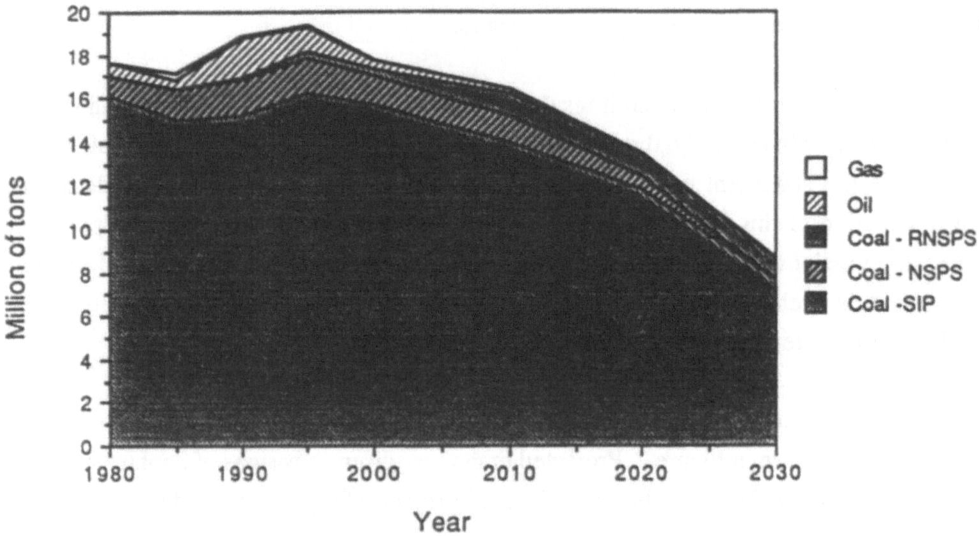

Figure 3

Sulfur Dioxide Emissions for the U.S.

Control Case Scenario

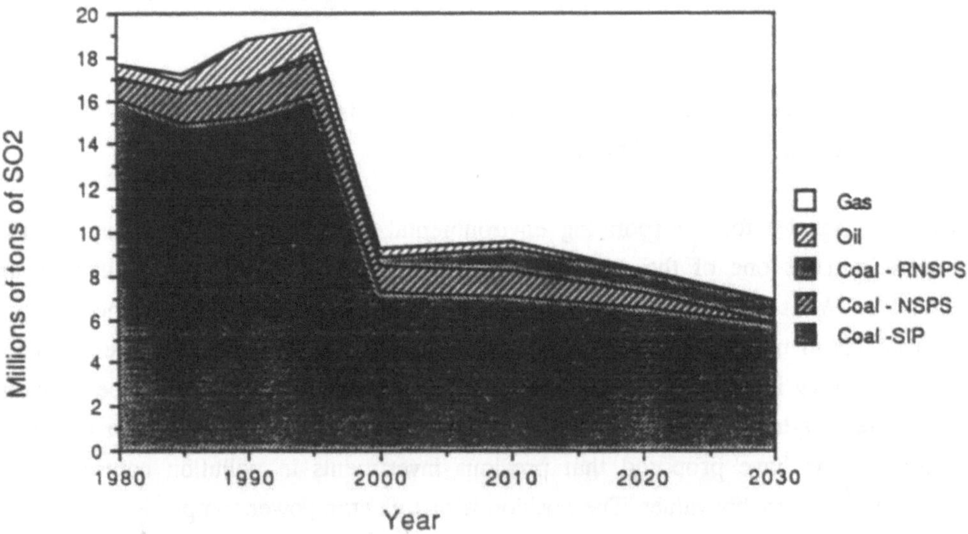

Results shown are from Advanced Utility Simulation Model (ANSM) Analysis by RCG/Hagler, Bailly, Inc. (1990) for the National Acid Precipitation Assessment Program and U.S. EPA.

When setting externality policies, one of the factors that should be considered is the projected path of emissions *over time* in a specific region. The critical importance of the time dimension is illustrated by M. Hazilla and R. Kopp (1990). They state "specifically, regulations affect interteporal microeconomic decisions and cause social cost to increase over time. Failure to quantify and account for these intertemporal phenomena in a benefit-cost analysis will likely lead to errors of unknown magnitude. On the basis of our findings of significant intertemporal impacts, currently practiced methods are unacceptable." While they advocate a general equilibrium framework for benefit-cost analyses, the basic message of their work does not unreasonably complicate the types of factors regulators should consider.

Other factors that should be considered include: the stringency of current regulations on each externality, the current and projected levels of residual emissions in a region, and the regional political licensing and siting process. For example, the public siting process can be effective addressing aesthetic, amenity and land use externalities associated with a new power facility. This process may ensure that multiple sites for new sources are considered. Certain clean supply options, such as energy conservation, that avoid this siting review may reduce atmospheric emissions but shift some risks closer to urban areas, exposing more individuals to potentially harmful pollutants than a power plant in a rural area.

In conclusion, environmental impacts of energy use and production decisions are important and complex. Regulators should understand the environmental implications of any integrated least-cost plan proposed by a utility, at least, in terms of the change in emissions and wastes expected to be produced by the plan. A decision to place high values on residual emisions through price or cost adders should be carefully reviewed and fully account for the projected costs and reductions expected from existing emissions regulations. At a minimum, regulators should develop a check list of externalities and require utilities to quantify the change in quantities of emissions, pollutants, waste products and other externalities. This list could be available for public review and comment. Many commissions may want to go a step further and directly incorporate quantitative factors within the integrated planning models and screening processes.

4.0 Options for Incorporating Environmental Externalities in Utility Planning

Most emissions that have taken quantitative action on incorporating environmental externalities have used pricing mechanisms that influence resource selection or bidding processes. As previously discussed, direct pricing policy tools available include the use of avoided compliance cost estimates and direct damage estimates. These both have weaknesses: principally, that engineering costs do not bear any necessary relationship to the value of avoided environmental damages (particularly on a site-by site basis), and direct damage estimates often require unavailable data.

Another set of policy options available to commissions involve the use of quantity tools. As was discussed earlier, there is an extensive literature addressing the relative merits of choosing emissions pricing or quantity setting tools. To date, however, the discussion regarding environmental externalities has focused on pricing methods. It is possible that practical aspects of regulatory processes, as well as economic theory, may support use of quantity constraints.

The regulatory policy or environmental constraint option, involves the use of quantity policy tools rather than pricing tools. In this approach, commissions would set either emissions levels or externality quantity targets (e.g., a zreo increase in air emissions) and then impose this as a constraint on the planning process. A number of authors have recently discussed and proposed this type of approach including Bernow and Marron (1990) in a case study done for Vermont; Violette and Peterson (1990); and Brick and Edger (1990).

The environmental constraint approach has a number of potential advantages over direct pricing tools. As discussed earlier, quantity policy tools can offer a number of advantages over pricing tools when there is uncertainty and imperfect information. While the avoided cost of control approach is based on national data that give the same value to emissions reductions in Texas as it does in New York, the environmental constraint method allows for a separate, unique determination of emissions quantity (i.e., Level) for cash jurisdiction. This method also avoids the complexities and uncertainties inherent in a direct damage assessment. In addition, environmental constraints explicitly show the tradeoffs between environmental quality improvements and increased revenue requirements. Integrated planning methods can be used to calculate the costs of meeting different environmental quantity targets by developing an estimate of the revenue requirements under each proposed policy. Regulators, intervenors, utilities and ratepayers can then offer comments on whether they believe this expenditure to be appropriate, given the environmental constraints that will be achieved.

Explicit consideration of environmental costs and the comparison of these costs to environmental improvements, even if performed in a subjective policy context, is an important responsibility of the regulatory community. Of course, to the extent that direct monetary estimates of avoided environmental damages are available, as well as any other costing information, this comparison can be made easier and more concrete. The decision that regulators will face is clear:

> Given that regulators will have a reasonably accurate estimate of the costs of achieving target environmental constraints, the policy judgement they will have to make is whether, given the evidence, they believe it is likely that the benefits of these environmental improvements will exceed the costs.

Violette and Peterson (1990) point out that if regulators decide the benefits are not likely to outweigh the costs, then a new set of environmental constraints can be considered and costs of meeting those constraints estimated. In this way, "sensitivity studies could be performed using different residual emissions levels or limits, and the change in revenue requirements could be calculated to show regulators how much more it would actually cost to further reduce or increase emissions" (p.7).

The obvious problem with the environmental constraint method is the complexity of setting acceptable quantities of pollutants and externalities. While difficult, this process is expected to be easier than developing prices for each individual externality.

In discussing the general constraint approach, Bernow and Marron (1990) state that "it is recognized that the valuation of environmental impacts is so fraught with difficulty that, at this time, public policy on the environment should precede and motivate technical analysis. For that reason, emissions reductions goals should be seen as an input to the electric evaluation, rather than as an output." While generally agreeing with this statement, it should be pointed out that the monetary damage valuations still may be useful in setting the quantity limits/targets and in assessing the likelihood that the benefits of meeting the targets will outweigh the increased revenue requirements.

Brick and Edger (1990) point out that the use of planning constraints is not at all new to utilities, stating "this is not a new twist to utility planning; quite the contrary. Operting reserve margins are routinely treated as planning constraints. Appropriate reserve levels are derived from assumptions about the value of unserved energy and system reliability standards based on past practice and conventional wisdom" (p. 58). A similar approach could work in devising and meeting environmental constraints. Bernow and Marron also provide a practical application of this method in their Vermont case study which used a zero increase in air emissions as their residual emissions constraint.

5.0 Summary and Conclusions

One purpose of this paper was to present quantity constraint policy tools as another option in the set of planning tools available for incorporating environmental externalities in integrated resource planning. Emissions quantity targets have potential to provide lower cost strategies for reducing residual emissions. Again, concern with incorporating environmental externalities in planning is based on residual emissions and damages that occur, even after all regulations have been met. Therefore, the overall goal should be the cost-effective, socially optimal reduction of these residual emissions. An overall constraint on emissions would allow utilities to devise least cost plans that meet that constraint. This could include adjustments to existing units, control strategies for new units, bidding processes, non-utility generation, load management, and conservation options. By letting utilities decide how to meet the targets, a wider array of options could be made available and less expensive strategies developed.

Regulators must consider numerous factors when determining how to address environmental externalities in a planning context. The literature on this subject is in its early stages of development and the thinking in this area will surely evolve as these issues continue to receive attention. While it is important not to become so overwhelmed with all of the potential complexities and interdependencies implied by environmental externalities that action is impossible, it is equaly important not to oversimplify the issues involved.

6.0 References

Bernow, S. and D. Marron. 1990. "The Treatment of Environmental Impacts in Electric Resource Evaluation: A Case Study in Vermont," Tellus Institute, Boston, MA, January 22, 1990.

Bohm, P. and C.F. Russell. 1985. "Comparative Analysis of Alternative Policy Instruments," in *Handbook of Natural Resource and Energy Economics*. Eds. A. Kneese and J. Sweeny. Amsterdam: North Holland, 1985, pp. 395-460.

Brick, S. and G. Edgar. 1990. "Blunting Risk with Caution: the Next Step for Least-Cost Planning," *The Electricity Journal*, Volume 3, Number 6, July 1990.

Chernick, P. and E. Caverhill. 1989. "The Valuation of Externalities from Energy Production, Delivery, and Use," from a Report to the Boston Gas Company, PLC, Inc., December 22, 1989.

Cohen, S.D., J.H. Eto, C.A. Goldman, J. Beldock, G. Crandall. 1990. "Environmental Externalities: What State Regulators are Doing," *The Electricity Journal*, Volume 3, Number 6, July 1990.

Cropper, M.L. and W.E. Oates. 1990. "Environmental Economics: A Survey," Discussion Paper QE90-12, Resources for the Future, Washington, D.C., January 1990.

Destribats, A.F., J.B. Lowell, D.S. White. 1990. "Dispatches from the Front: New Concepts in Integrated Planning," presented at EPRI Innovations in Pricing and Planning Conference, Milwaukee, Wisconsin, May 3, 1990.

Hazilla, M. and R.J. Kopp. 1990. "Social Cost of Environmental Quality Regulations: A General Equilibrium Analysis," *Journal of Political Economy*, Volume 98, Number 4, August 1990.

Ottinger, R.L. 1990. "Getting at the True Cost of Electric Power," *The Electricity Journal*, Volume 3, Number 6, July 1990.

Ottinger, R.L., D.R. Wooley, N.A. Robinson, D.R. Hodas, and S.E. Babb. 1990. "Environmental Costs of Electricity," prepared for New York State Energy Research Development Authority, Pace University Center for Environmental Legal Studies.

Roberts, M.J. and M. Spence. 1976. "Effluent Charges and Licenses Under Uncertainty," J. Public. Econ., April/May, *5*, pp. 193-208.

Tobin, J. 1958. "Estimation of Relationships for Limited Dependent Variables, *Econometrica*. 26:24-36, 1958.

Weitzman, M.L. 1974. "Prices vs. Quantities," *Review of Economic Studies*, October, pp. 477-491.

Violette, D.M. and D.C. Peterson, Jr. 1990. "Utility Planning and Evaluating Environmental Externalities," presented at Decision Support Methods for the Electric Power Industry, Cambridge, Massachusetts, May 29-31, 1990.

Violette, D.M., C. Lang, and P. Hanser. 1990. "A Framework for Evaluating Environmental Externalities in Resource Planning: A State Regulatory Perspective," presented at National Conference on Environmental Externalities, Jackson Hole, Wyoming, October 1-3, 1990.

6.5 Externalities and Least-Cost Planning: The Threshold of a Revolution

Stephen G. Brick
MSB Energy Associates, Inc.
Middleton, WI 53562/USA

Revolution...2 (b) A fundamental change in political organization...(c) activity designed to effect fundamental change in the socio-economic situation.
Webster's, *Ninth New Collegiate Dictionary*

Introduction

Electric utility planning and regulation stand on the brink of a revolution. This revolution involves the adoption of a new regulatory model based on societal benefits and costs. The issue of whether and how to reflect external costs in the planning calculus has crystallized the issues surrounding this fundamental shift. How it is finally resolved will do much to shape regulation and planning in the future. In the U.S. we have been brought to this point through the advent, evolution and adoption of least-cost planning (LCP). In this paper I briefly describe two fundamental changes in regulatory perspective that have occurred as a result of LCP. I suggest that our present efforts to include external costs into planning have functioned as adjuncts to newly-evolved LCP benefit-cost procedures. I argue that we have approached the question of externalities too narrowly, focusing on a single segment of the benefit-cost procedure and improperly limiting the universe of externalities we consider. I examine techniques and results of some recently-developed methods, and conclude that they rely on uncertain and limited data and may have little actual effect on utility-resource acquisition. I conclude by proposing a more-comprehensive program for incorporating externalities in utility planning and acquisition and discussing the prospects and prerequisites for the success of such an approach.

New Regulatory Paradigms

LCP has led to the examination and reformulation of two fundamental premises that have guided utility regulation and planning for the past half-century. the definition of the services electric utilities are required to provide to their customers is most important among these. Traditionally, this definition was confined simply to electric power. Now, it has been expanded to include energy efficiency, load management - so-called demand-side management options (DSM) - as well as the production of electricity by alternative fuels. The rationale underlying the shift is straightforward: if it is less expensive to conserve a kilowatt-hour, or to generate electricity using alternative means, than to produce one conventionally, this course of action is preferable. It is also consistent with the utilities' historic mandate to minimize costs. LCP had its genesis in a period marked by both increasing cost and by declining public acceptance of conventional coal and nuclear power plants. Substantial electric-rate increases caused by power-plant construction-cost overruns and fuel-price inflation, combined with growing public distaste for the local effects of plant construction, spurred the search for cheaper alternatives. Electric utilities are now being required to provide customers a wide array of *energy services,* rather than just kilowatts and kilowatt-hours. The challenges are manifold: appropriate administrative vehicles must be found for delivering energy-conservation services; staff must be rapidly acquainted with a wide range of demand-side technologies; and utility management must refocus the mission of its company.

New possibilities for satisfying the demand for electricity called forth new techniques for analyzing their benefits and costs. This is a second crucial area of ferment and change. Traditionally, utilities selected from among alternatives the plan that minimized the net present value of revenues per kWh collected from ratepayers, while reliably meeting the expected need. In the case of new power generation resources, all ratepayers - present and future - were assumed to benefit equally from the project. As planners began to consider additional alternatives new factors began to pop into the benefit-cost calculus. Winners and losers appeared where previously there had been none. When a utility installs an energy efficiency measure on a customer's premises all ratepayers benefit by reducing the need to build new generating capacity and by reducing the need to operate existing capacity. But the participating customer receives an additonal benefit in the form of a reduced electricity bill. The effects on customers who do not participate in a utility's DSM programs are suddenly important. The question of whether a customer would have acted without utility intervention - the *free-rider* problem - becomes significant. The revenue requirement equation could not adequately accommodate these new variables and an improved formulation was sought. A new approach was needed because it was recognized that plans built using traditional forms of analysis (seeking to minimize the cost of *electricity*) would be inconsistent with those conforming to the stated goals of LCP (seeking to minimize the cost of *services provided by electricity*). The revenue requirement perspective has gradually given way to a *societal* or *all-resource* perspective. Although there is not unanimous agreement that a societal perspective is the only appropriate standard for analysis, a majority of regulators appears to have embraced the societal test, either as the sole benefit-cost criterion, or as one of a panoply of tests. Regulators view the societal perspective as a means of advancing the goals of LCP, and expect that direct costs to ratepayers will be reduced thereby. Electric rates may increase, but greater end-use efficiency will reduce total consumption to more than offset higher rates.

Against this backdrop comes a resurgence of environmentalism. Two significant environmental threats - acid deposition and global climatic change - are clearly linked to utility combustion of fossil fuels. Regulators have been quick to recognize that LCP, and the societal perspective it embraces, provides a potential vehicle for addressing these issues. There has also been widespread recognition, extending far beyond the confines of the utility regulatory community, that DSM offers a potent means of *preventing* pollution. The U.S. Environmental Protection Agency has, for example, instituted a Pollution Prevention Initiative, in which energy efficiency plays a central role. The Congress is now conferring on revisions to the Clean Air Act that include rewards for states that utilize DSM as an air-pollution control strategy. The shift in emphasis from controlling pollution "after the fact" to preventing it is significant. This provides a touchstone for assessing the effects of our efforts to include external costs in utility planning: planners and policy-makers believe, *a priori*, that successfully including social costs has not been driven by an abstract desire to investigate and total all such costs, ultimately permitting a bloodless, objective model to define the societal optimum. Rather, it is founded on the belief a legitimate one, I think - that effective consideration of externalities will result in more DSM and less pollution. This should inform our work: if our efforts do not lead to more DSM either our working hypothesis is incorrect or our planning models are flawed.

Current Practice

To date, externalities have been treated quantitatively in one of two ways: (1) as adjuncts to recently-developed LPC benefit-cost procedures ans (2) as adjuncts to state-mandated all resource bidding programs. These are different sides of the same coin. In both cases, the benefit-cost procedure - its functional form and its components - are specified in detail.

Recall that LCP seeks to ensure that DSM and other non-traditional alternatives compete fairly with the old guard. This underlies the move to abandon the revenue requirement approach in favor of a societally-based model. Regulators have discovered, however, that utilities continue to favor traditional means of providing service and that without carefully specified benefit-cost procedures, often stack the deck against DSM and other alternatives. As described briefly above, a lengthy episode in the LCP drama has been devoted to developing benefit-cost procedures. Many of the states that have abandoned regulatory responsibility in favor of market forces have, neverthess, looked over their shoulders and built the same, carefully-specified procedures into bidding regulations.

Since we have devoted so much energy to devising these methods, it should not surprise us that when we began our efforts to internalize social costs we turned immediately to our beloved benefit-cost procedures. In almost all cases, we have done so without pausing to consider whether this is appropriate or sufficient. It is a perfectly logical turn of events, after all, to take a hitherto-ignored category of costs and treat it in a manner consitent with other, more familiar costs. As long as the following conditions are true, we should have no concern over continuing in this fashion:

- The benefit-cost formulation must be without flaw in functional form;
- The data used to represent the conventional components of the benefit-cost equation must be reasonably accurate; and
- The data used to characterize social costs must be reasonably accurate and complete.

Of these three areas, the functional expression of the benefit-cost measure is the least troublesome. Without considering the accuracy of the inputs, the arithmetic used to reduce the costs and benefits of each of a diverse set of energy resources to an equivalent metric is comparatively reliable. The data used to represent the conventional components of cost and benefit (performance data, capital costs, operations and maintenance [O&M] costs, avoided-energy benefits, and avoided-capacity benefits) are of variable quality and, in some cases, mere educated guesses. Capital and operating costs for conventional power generation technologies are well-researched and documented; the same cannot be said for many DSM technologies. For a mature DSM technology performance data, capital and O&M costs are typically well described. But information on market potential, existing market penetration levels and consumer attitude are frequently lacking. The estimates used for these figures - which are essential to developing a DSM program - can spell success or failure. For immature technologies the problem is compounded by highly uncertain cost information. For emerging technologies, performance data itself may be highly uncertain. The second condition is not satisfied. The third condition - the subject of most of the presentations at this conference - clearly is not satisfied. Consider the broad range of environmental externalities. The lion's share of our attention has focussed on the social costs imposed by atmospheric pollutants. Where direct costing is possible the results are surrounded by wide margins of error. Where direct costing is not possible the use of shadow prices must suffice, but is fraught with uncertainty. Take, for example, the practice of using afforestation costs as shadow prices for the cost of global climatic change. The validity of these figures depends not only upon the cost of afforestation in developin nations (for which we have very little in the way of actual data) but equally upon the practicability of such actions themselves, which is highly questionable. If we expand our consideration to environmental externalities beyond those caused by atmospheric pollutants, we find that the quality and availability of data rapidly falls off. Finally, we have, for the most part, ignored tha vast array of social costs that are not strictly environmental in nature. While the subject of this conference is environmental externalities, it is worthwhile for utility planners and policy-makers to enquire whether such a limitation is appropriate. I believe it is intellectually disengenuous to lop off a subset of the universe of social costs will prove to be consistent with environmental improvement, and not, as some have argued, work against it.

Having suggested that two of the three conditions I described above are not satisfied, it may appear that I am rejecting completely the methods currently in use. This is not the case. But it is important to examine these methods, determine their potential impact and limitations, and explore other means of achieving policy objectives should these fail.

In jurisdictions empoying LCP, the incorporation of external costs has, for the most part, occurred at the resource screening and ranking phase. This is true whether the figure used is a simple percentage adder, a shadow price based on pollution-control costs or direct valuation of external damages. "Screening and ranking" comprise one step in the benefit-cost procedure, designed to provide a straightforward means of comparing the costs and benefits of all candidate options. Each of these methods serves to lower the required benefit-cost threshold for alternatives having positive attributes. In Wisconsin, for example, alternatives not involving direct combustion of fossil fuels receive a 15 percent credit. These options, therefore, pass the benefit-cost screening with a ratio of 0.85 rather than 1.00. Once DSM options have passed the screening phase they pass into the program design stage.

In Wisconsin the adder was devised to insure that DSM programs narrowly failing to pass resource-screening procedures were not rejected, but were carried forward to the program design stage. The adder grew out of the beliefs that the inputs to the benefit-cost analysis, described above, were surrounded with considerable uncertainty and that it was reasonable to give the benefit of the doubt to options offering obvious but unaccounted-for environmental benefits. As such, Wisconsin's policy was never envisioned as a comprehensive means of treating external costs, but as a remedy to a specific problem.

Utility regulatory commissions have adopted a wide range of figures to represent external costs in the resource screening process, ranging from Wisconsin's 15 percent proxy adder to a comprehensive set of pollutant-specific externality values adopted recently by the Massachusetts Department of Public Utilities. What effect will these procedures have?

I tested the effect of a wide range of adders on the cost effectiveness of a set of commercial and industrial DSM measures that we recently analyzed for implementation in the commercial and industrial sector of a large North American utility. The following table sets forth the results:

Effective Adder	Number of Cost-Effective Measures
0	32
15%	34
25%	34
50%	35
75%	36
100%	36
200%	38

It is clear that even a small adder can turn a marginal failure into a successful measure. It is surprising, however, to discover that an adder of 200 percent did not advance all measures to the winning column. The options analyzed are shown in the next table, along with the energy savings and benefit-cost ratio (calculated using no externality credit). I include this lengthy table to illustrate an important point: a survey of the alternatives listed below indicates that a majority (32 out of 42) pass the cost-effectiveness screening without any added benefit for avoided environmental impacts. I believe this raises a critical question: Is the research and development of DSM technologies informed by the prevailing *economic* value for the conventional generation it replaces? Although the data presented below do not represent a statistically-valid sample of all DSM technologies currently available, they were assembled in the course of a thorough evaluation of potential DSM measures applicable in the commercial and industrial sectors. If the market for new DSM technologies is defined by current electric rates, the fact that over 75 percent of the options reviewed turn out to be cost-effective should not surprise us. If the externality credit is dropped at the scrreening stage , as it is in many jurisdictions, it is likely that the proper signal will fail to reach those exploring new technologies. At the very least, the credit must be carried through to the program design phase, to ensure that technologically cost-effective DSM is not eliminated because of program implementation costs. This should increase the amount of DSM implemented, but will not itself guarantee that the proper price signal reaches the DSM research and development market. Only by reflecting external costs in the actual price charged for electricity will a clear signal be sent.

End-Use	Option	(kWh/1000 square feet)	Benefit -Cost Ratio
Lighting	Dummy replacement C	2449	551.43
Air Conditioning	Heat gain reduction from lighting C	794	530.57
Lighting	Daylighting controls C	1855	34.78
Water Heating	Fiberglass insulation C	874	24.60
Air Conditioning	Chilled water reset C	794	21.92
Refrigeration	Floating-head pressure control C	3931	19.56
Lighting	Low-wattage fluorescent C	1113	12.53
Motors	High-efficiency motors, 200 HP I	9578	10.63
Lighting	HID interior C	3710	9.73
Motors	High-efficiency motors, 75 HP I	5398	9.72
Motors	High-efficiency motors, 15 HP I	3503	8.70
Air Conditioning	Cooling towers C	397	8.05
Motors	High-efficiency motors, 40 HP I	1228	6.72
Air Conditioning	Outside air economizer cycle C	1191	6.58
Air Conditioning	High-efficiency cooling equipment C	1588	6.10
Refrigeration	Glass doors C	19655	6.07
Refrigeration	Walk-in strip curtains C	5897	5.29
Water Heating	Foam insulation C	187	5.27
Air Conditioning	Energy-efficient motors C	72	5.04
Lighting	Occupancy sensors and timers C	2226	4.57
Air Conditioning	Reduction of outside air ventilation C	397	3.59
Lighting	Compact fluorescent C	5565	3.40
Lighting	Reflectance C	2449	2.75
Motors	High-efficiency motors, 150 HP I	6787	2.75
Air Conditioning	Glazing C	794	2.65
Air Conditioning	Window film C	794	2.65
Motors	High-efficiency motors, 5 HP I	11050	2.62
Lighting	Electronic ballasts C	1484	2.27
Lighting	Energy-efficient ballasts C	742	2.20
Air Conditioning	Thermal energy storage C	0	2.08
Refrigeration	High-efficiency compressor C	4717	1.85
Water Heating	Heat-trap addition C	62	1.76

Lighting	Micro-metal halide lamp C	1484	0.99
Lighting	High-pressure sodium C	1484	0.99
Air Conditioning	Interior shade C	794	0.79
Appliances	Clothes washer - High-efficiency motor C	73	0.61
Ventilation	Reduction in fan flow rate C	216	0.50
Water Heating	Heat-pump water heater C	4368	0.42
Appliances	Clothes dryer heat exchanger C	292	0.32
Air Conditioning	VAV system C	360	0.30
Ventilation	Economizer - Fan shutoff C	144	0.22
Air Conditioning	Adjustable-speed drive C	288	0.10

The prospect for added externality costs - even substantial ones - having a marked effect on the development and implementation of DSM in states with all-resource bidding is not necessarily better. Initial results from states with bidding indicate that DSM bidders do not propose programs that are marginally cost-effective, but rather try to maximize their profit by proposing only highly cost-effective technologies. This is consistent with our understanding of the high rates of return required by entrepreneurs and their investors. Although the externability adders used in present bidding mechanisms will undoubtedly increase the profitable DSM resource, it is unlikely that they will have an immediate or pronounced effect of the emergence of new DSM technologies.

The mechanisms now in place have a definite value, and should be retained. As seen above, even a modest adder can help a few more alternatives pass the initial screening and ranking; this is a crucial step on the road to comprehensive incorporation of social costs, however small its effect. Were these mechanisms not present the options would be dropped permanently from consideration and important opportunities to reduce societal costs lost. But our current approach to externalities does not satisfy all our needs. We need to build additional means of dealing with social costs into the LPC process that has evolved and is now in use.

Future Directions

We need above all, to express our policy aims and expectations at the outset of the planning process. There is good reason to wonder whether it is appropriate to bury the results of the compelling research on external costs in a mechanistic benefit-cost process. These findings should be used to formulate appropriate policies, which in turn should govern the development of plans. Findings on the staggering external costs of our energy-consumption habits may have a more direct effect on our action if they operate as constraints rather than as inputs to a planning model. I suggested above that early attempts to include social costs in utility planning were founded on an assumption that the methods would increase the use of DSM and other non-polluting resources. If the political will cannot be mustered to let policy declarations drive planning, at the very least we should check the results of our planning to see if they meet our expectations. Little, if anything, has been done or proposed along these lines. If we fail to do this we are permitting a morass of data to act as a substitute for energy and environmental policy, which is clearly not appropriate.

Most importantly, we must consider allowing external costs to flow directly to consumers as a surcharge on the price of electricity. Such action will have a direct effect on consumers' behavior, and send, as well, appropriate signals to developers of DSM and other desirable technologies. Although considerable thought needs to be given as to how a transition can be accomplished between the present (in which many U.S. electric rates do not even reflect direct marginal cost, much less social marginal cost) and the future (where electricity is twice or three times as expensive) such a solution is a more direct and sure way of accomplishing the internalization of social costs. Under such a scenario, there will be a period during which utilities collect from ratepayers much more than is necessary. This problem has perpetually plagued marginal-cost pricing schemes. In this case, where money collected represents the social damage inflicted by the generation and consumption of electricity, there are a couple of logical uses for the surplus. One suggestion is that the surplus go to an environmental restoration fund. As appealing as this proposition is, I do not think it is good policy. First, it is unlikely that exact determinations can be made as to who the benefactors of the fund should be, given that electric utilities are not the only producers of environmentally harmful agents. Second, I doubt that effective mechanisms can be established to channel the funds to appropriate places, once they are identified. Third, since environmental damages are usually cumulative, the fund would not be sufficient to cover preexisting damages or those that continue after the utility system reaches equilibrium. Finally, and most significantly, the presence of such a fund might send the signal that any damage, once inflicted, can be repaired with enough money. This is manifestly untrue, and a bad message to send at this or any time.

I believe the fund might be more profitably used to hasten the development, demonstration, and implementation of innovative DSM and other non-polluting technologies. Such action would move us closer to the point where pollution prevention replaces retrospective clean-up as our chief environmental policy.

The regulatory and analytical advances needed to effect such a change are small in comparison to the political ones. For the time being, U.S. utility regulators are still expected to keep electric rates low, not to minimize societal costs. Although LCP has adopted a societal perspective, the community at large has not. Without this change, it is unlikely that we will ever fully achieve the benefits expected by comprehensive internalization of societal costs.

6.6 Internalization of "External"Costs: Necessary, but not Sufficient!

Prof. Dr. P. Hennicke
FH Darmstadt
D-6100 Darmstadt

1. Introduction

Exactly speaking, there is no production process which does not cause so-called "external" effects in one form or another. Maximization of private profits or - probably even more - the directives of bureaucratic planning each are actually prototypical for a system of motivation and regulation which offers an enormous incentive for reducing individual costs (for "easier" fulfilling of plans, respectively) to the debit of third parties. In the following, we refer to capitalistic market economies as reference system because we are more familiar with the way they work, not because we estimated that bureaucratic (said to be real socialist) planning economies were able to control "external" effects in a better way.

Today, in many areas of mass production (energy industry, chemical and other basic industries, transport, industrialized agriculture) transferring damages to third parties is no longer a more or less bothersome marginal phenomenon. An increasing number of industrial production processes has to be named more exactly a co-production of "bads" (waste, emissions, consumption of resources) as well as "goods". For example, the production of each kilkowatt-hour of electricity from central power plants (coal) in the Federal Republic of Germany releases one kilogramme of carbon dioxide at the same time.

The only reason why these typical forms of co-production did not rise as a problem to economists, is because their science did not care for the pollution connected with production, and even less for the material destiny of a good after its economic decease through productive or consumptive use.

From the perspective of macroeconomics, the society as a whole, and of future generations, however, in the meantime the sum of externalized damages caused by an increasing number of production areas has become greater than even its private benefit. It is almost certain, that the current model of industrialization dominating worldwide will on the long run produce so many "bads" that it is going to question its own economic functioning as well as the survival of mankind, if the world keeps to it without changes. (1)

It would certainly be naive, if we attempted to capture the global catastrophes possible today, like ongoing stratospheric ozone depletion, the anthropogeneous greenhouse effect, or nuclear devastation after a severe nuclear accident (core melt-down), only by the category of "external effects" - like Pigou did 70 years ago. (2)

Therefore, if we call for "internalization of external costs", we have to set out clearly what we are talking about:

Either, we talk about those phenomena played down as marginal by neoclassics of otherwise intact market economies. In that case, a tax or fee will be sufficient to readjust the economist's world. Apparently, though, the world is not as idyllic as the neoclassical environmental economist imagines in his lecture books. Or, we talk about the real risks to environment and world climate acting globally and virtually time- and borderless. In this case, the so-called internalization of "external" costs through a tax or fee is something like the most beautiful minor detail in the world. Certainly necessary; but by no means sufficient, as we will point out in the following.

2. Why do we concern mainly for the minor detail?

An answer to this question is: because ruling theory teaches to do so!

The concept of "external effects" is based on the theory of efficient allocation under decentralized decisions. This theory assumes basic conditions extremely far from reality concerning preferences, availability of factors, and technology. It states, then, that given these basic conditions, a system of perfect competition will lead to efficient allocation of resources in equilibrium - the so-called Pareto-optimum. Such a pareto-optimal allocation is understood as a state in which it is impossible to improve the situation of one economic subject without causing loss to another economic subject.

Among the mentioned extreme assumptions is the absence of "external" costs. Costs are defined as "external", if they are to be borne by society or other economic subjects, without being considered in the economic calculations of the economic subject that has caused them. Even in Pigou's times, substantial social costs (e.g., damages to health, losses of property value and natural riches) were more the rule than the exception, but environmental damages were a quantité négligeable compared to today's situation. Nonetheless, even then it was a smart idea to define that ruinous exploitation of human and natural resources apparently immanent to the logics of market and profits as simply "external" costs which do not question the internal rationality of market allocation. (3)

In order to stick to the theory of optimal market allocation in spite of accumulating evidence for ecological crises since Pigou and in the future, a timelessly functioning deus-ex-machina is needed to make social and private costs meet again. The Pareto-optimum can be saved if "external" effects can be neutralized regarding prices and costs *through government intervention* (taxes/subsidies and norms/standards) or through private agreements. (4) Besides defining standards and norms for pollutant emission control through legislation, mainly taxes (sometimes subsidies as well) are regarded as appropriate forms of government intervention for taking into account "external" effects.

The neoclassical theory of efficient allocation under decentralized decisions belongs to the studies of an economy student like formal training to the soldier. But, as little as soldiership is only saluting, the neoclassical Nirwana of efficient allocation and Pareto-optimum is sufficient to explain how capitalistic market economies function. Nevertheless, we can imagine neither a soldier without formal training nor, apparently, an economist without neoclassics.

What is the reason for neoclassics being the ruling theory still?

An answer might be: really existing capitalism has proven in history to be so astonishingly dynamic and attractive to just compete the economy of bureaucratic central administration out of the race *in spite of* neoclassical theory. Really existing capitalism, on the other hand, has never been convincing enough intellectually to not need a theory of optimal allocation - namely neoclassics - for justification.

Neoclassics, however, have hardly contributed anything *to the analysis of reasons* for environmental damages and to the foundation of an effective policy to contain environmental destruction. Their main function is to play down environmental destruction as a phenomenon principally external to the rationale of market allocation. Therefore, some praise the "self-control" of the market, aided by suitable government intervention through strictly market-conform instruments (certificates, taxes, fees), as the only effective means against any form of environmental destruction. Thus, the question if the dynamics of markets and profits themselves might be among the most important reasons for global destruction of the environment, could not even arise within traditional economic science.

Within the ruling neoclassical theory, since Pigou's invention for the treatment of "external costs" until today any true progress in research on the causes of environmental destruction has beeen blocked.

It is almost tautological, if a well-known book on environmental economics states today: "The fact that the external costs exist, has to be regarded as one of the fundamental reasons for environmental destruction in market and planned economy systems." (5) Vice versa from this follows, that after internalization of "external" costs only by means of "correct prices" one of the fundamental reasons for environmental destruction has been abolished. This statement, though, can neither be tested empirically, nor gives a practicable concept for avoiding environmental destruction: there will be an argument on "correct" prices as long as there will be economies and the professional guild of economists. Whenever and whereever there is environmental destruction, the reason will be "incorrect" prices.

So, on one hand, the concept of "external effects" and the theoretical efficiency gain through internalizing "external costs" can only be founded exactly in connection with the neoclassical theory of efficient allocation. On the other hand, however, real markets, and especially the markets in energy industry, are hopelessly far from the conditions of perfect competition. Particularly for the electric utility industry, *the exclusion of direct competition* through area-bound monopolies and a high concentrated supply side are essentially constitutional. Two conclusions follow from this:

1. It has to be proven if the internalization of "external" costs actually leads to an efficiency gain on real markets with monopolist suppliers. This cannot just be assumed referring to the theory of efficient allocation.

2. It has to be analyzed if taxes and fees are in reality sufficient instruments for avoiding or containing global environmental damages. The neoclassical theory of "external effects" cannot give a satisfying answer to this.

3. Searching for the "True" Social Costs?

A justified objection to this strict interpretation of the concept of "external" costs is: Even without neoclassics, common sense tells us that the transfer of individually caused damages (costs) to third parties must be stopped by government intervention and that, vice versa, individual actions to prevent damages (and save costs) in favour of third parties must be rewarded. I agree with such a pragmatic understanding of "external" effects and their internalization: we must then search for a general, practicable concept to establish the form and volume of ecological taxes (fees) which would make sense in both market and planned economies, without referring to the theory of optimal allocation.

The recent discussion on internalization of "external" costs, though, has a different goal.

For example, O. Hohmeyer's stimulating book starts with the following statement: "It has become apparent that the market diffusion of systems using new and renewable sources of energy ... is taking place more slowly than expected. This indicates that various barriers to the market introduction of such technologies have been underestimated." (6) One can only agree with that. The more astonishingly, however, O. Hohmeyer picks out only one of the "various barriers" in his further analysis: the lacking internalization of external costs of conventional nuclear and fossil-fueled power production on one side and the lacking consideration of net benefits of renewable energy resources on the other. After a governmental correction of price relations through taxes and subsidies everything will be in order agaih: "Therefore, research efforts to estimate the full costs of energy systems to society are necessary. The knowledge of these full social costs of energy could enable government to take corrective action to help the market mechanism achieve an optimal allocation of resources." (7)

"Optimal allocation of resources"? This unappriopriate reminiscence to neoclassics is misleading, in the best case meaningless in the context of Hohmeyer's sound effort to quantify and monetarize the damages of conventional and the benefits of renewable power production. In the worst case, it may lead to boundless theoretical discussions on the "true" extent of social costs.

The merit of O. Hohmeyer's work is *first*, that it gives a solid foundation why electricity from conventional (fossil as well as nuclear) generation has to be put a tax on, while substantial subsidies to accelerate market penetration of wind and photovoltaic power plants are macroeconomically benign, because damages of conventional power generation can hereby be avoided.

Second, the magnitude of a possible governmental taxation/subsidization has been estimated systematically for the first time, thus precising *one* necessary instrument of energy and environmental politics for accelerating market penetration of renewable sources of energy.

Further research, however, should not concentrate on ever more complex attempts to estimate "external" costs, driven by the illusion that it would be necessary to evalue the "true" social costs in order to get closer to the Nirwana of optimal allocation.

It is essential for the foundation of global instruments that costs and benefits be analyzed and quantified in a way so pragmatic and sound that the proposals on the magnitude of taxes and subsidies are on the safe side. It appears to me that O. Hohmeyer's estimation has achieved that goal.

Taking the highest, though conservatively estimated values for each figure from his analyses, the corresponding tax rates for conventional power generation will anyway be on a level considered impossible to realize by our politicians. (8) There is no danger politicians might confront us with the call for even higher rates. Given this, what benefit lies in proposing even higher and expensively calculated tax rates to politics *for rejection*, even if they claim to internalize the costs of the climate catastrophe as well?

Instead, we should turn our attention towards several unsolved theoretical questions, but first of all to practical solutions to the problem.

4. Some Open Theoretical Questions

4.1 Markets for Energy or for Energy Services?

Most analyses of "external" costs of the energy system concentrate on the supply side, regarding energy markets where the price per unit of energy is the decisive parameter.

On the other hand, exactly those studies concerned with "external" costs and the environment focus on the conceptions of "energy services" (ES) (Energiedienstleistungen - EDL) and "energy service utilities" (ESU) (Energiedienstleistungsunternehmen - EDU).

The main cause for the existing confusion is the fact that we let the market concept particularly for electric and gas industries unchanged though at the same time using the energy service concept.

The neoclassical concept of "external costs" is based on the conception of an "energy market" where sellers offer "kilowatt-hours" (energy and power, e.g., in the forms of electricity, natural gas, heat), and where the "consumption of kilowatt-hours" satisfies the needs of customers. Nobody, however, is consuming energy for its own sake. Energy is a means to fulfill a purpose. What consumers really need are energy services like a warm home, sufficient lighting, specific uses of mechanical power, communication purposes.

From this point of view, energy is nothing but an *intermediate product* helping to "produce" the respective energy services - i.e., the respective end products (useful effects). Technically seen, ES always are a "package" consisting of energy and of a conversion technology. and only their simultaneaous use leads to the useful effect wanted.

The same is true for the costs. The user wants to buy this "package" of energy and conversion technology as cost-optimal as possible. To satisfy this need in today's reality, however, the user has to shop on at least two, often three or more markets: on the "energy market" he gets kilowatt-hours from the utility, whose traditional interest is to sell him as much as possible. On the "market for household appliances" he buys for example a refrigerator at the appliance retailer's, who is interested in selling the appliance giving himself the biggest profit, and not the most energy efficient one.

When the "package" is more complex, e.g. building a house, the consumer has to purchase advice and financial services on the "markets for information and capital", because he does not have any survey e.g. over active and passive solar energy use, thermal insulation, or possibilities of financing.

The number of markets/acting people involved and the complexity of information acquisition increases further, when the concept of ES is applied not only to single objects (like houses), but to utility territories or whole national economies.

From this *systems analysis* point-of-view, there is a radical change in the optimization criterion on all levels of the energy system: it is not energy, but energy services, that have to be provided "as cheap as possible" ("so billig wie möglich"). It is no use for the home-owner if the prices for one kilowatt-hour is cheap, but his energy bill for the completely uninsulated home is high. The other way round: an ESU selling the home-owner in a package energy, advice and a financing aid for his thermal insulation investment, can recover this surplus service from the customer by raising the price for the kilowatt-hour and simultaneaously lowering his energy bill.

The price of one unit of energy is no longer suitable as a measure for judging the economic activities of an ESU and its new range of products - packages for ES which fit exactly to the needs of its customers.

What conclusions for the debate on internalization do we have to draw from this?

4.2 The substitutional competition between energy and capital

The crux of the matter is that for analytical purposes, one cannot forego the above-mentioned new market conception and an innovative marketing for ES, but that the politics of most energy utilities are not of that kind, and real markets are not (or not yet) organized in such a way. For this reason, we have to use the category of "substitutional competition" between energy and capital ("efficiency", technniques of more efficient energy use) as an auxiliary.

This is because utilities operating in the traditional manner and the lack of marketing for ES would not have as large an effect as they do if - in a situation where *direct competition* between energy suppliers does not function - at least the *substitutional competition* between energy and capital functioned.

A. Lovins has a very optimistic view of this. (9) His essential argument states that, because the purchase of NEGAwatts is more cost-effective for the energy user on a microeconomic scale (without internalization of "external" costs) than the purchase of power, NEGAwatts are going to compete MEGAwatts out of the market sooner or later. Utilities may - to their own advantage - accelerate this process through profitable NEGAwatt sales, but neither utilities nor other barriers will stop the triumph of NEGAwatts.

This is no doubt a challenging idea: while some people waving the banner of "deregulation" have been striving for years for cheaper energy supply (i.e., MEGAwatts) through more *direct competition* between energy suppliers, A. Lovins with his smart ideas comes in through the back door of *substitutional competition*. Only if competition with the even cheaper NEGAwatts is included, "deregulation" of energy supply will make any sense. (10)

If we apply A. Lovins' ideas pragmatically to the problem of internalization, this will lead us to the call for introducing a specific fee/tax for all non-renewable energy sources - if possible, specifically according to the respective, different "external" costs. (11) Such a general energy fee/tax is a useful and necessary extension of Hohmeyer's proposal. In that connection it would have to be proven if, and to what extent, the production and use of energy saving technologies causes less "external" costs related to a specific ES than additional energy supply from fossil or nuclear fuels. As already stated above, one should proceed pragmatically in quantifying appropriately specified taxes/fees (for traditional supply techniques) and subsidies (for efficiency technology).

Yet another question is, if A. Lovins' fascinating NEGAwatt concept implies a come-back for the neo-classical competition equilibrium: if, for example, one would say that the "efficiency revolution" will win by means of market and competition alone, under the condition that prices for NEGAwatt and MEGAwatt are set "correctly" in the sense of neoclassical optimal allocation of resources, i.e., including the so-called external costs.

We stay, nevertheless, to our contrary position:
Government interventions influencing substitutional competition in favour of techniques for more efficient use of energy are truely of even greater importance for containing environmental destruction than interventions for controlling direct competition between renewable and non-renewable energy sources. But also in the efficiency area, controlling with only one non-specific instrument would not at all be sufficient facing numerous barriers particularly for substitutional competition.

4.3 Barriers and Market Failure

Whoever comes from the dreamland of competition equilibrium and neoclassical instruments into real utility and energy politics gets a "practice shock". No "energy market" is functioning like in the lecture book, the world is full of barriers. Nonetheless, up to date there is no theoretical analysis of the main barriers to an energy policy for climate stabilization. Also, there is no reliable analysis of the effectiveness of possible instruments for reducing barriers.

An evaluation of national and international programmes for accelerating market penetration of techniques for more efficient energy use and of renewable energy sources (12) shows, however, such a great variety of barriers, that only *a package of instruments including global taxes and fees as well as measures specific to sectors of target groups* promises success in reducing and overcoming barriers.

It is crucial for the debate on internalization to notice that the steering effect of non-specific instruments controlling via prices (certificates, taxes, fees) is limited by certain forms of structural, institutional, and legal barriers. At this point, market failure cannot be corrected by any kind of internalization, be it the most sophisticated. The following sections may highlight this problem:

"Split market"
A functioning substitutional competition, for instance between electricity and caputal (efficiency), would require that *on one hand* suppliers of efficiency technology were

comparable to suppliers of electricity regarded to market position, liquidity, and capital stocks. Exactly speaking, this would only then be the case, if a large number of homogeneous and competing investors were to decide between *investment in MEGAwatt or in NEGAwatt*. So far, this is at best the case for those few energy service companies that have been founded by utilities. In most cases, however, we are dealing with diametrically different types of investors. Big, monopolist electric utilities on one side contrast with e.g. many single departments of mixed company groups offering a variety of efficiency technology on the other.

On the other hand, market survey and decisions of millions of customers without market control (except for large industries) are being systematically distorted. Incorrect energy price structures and (des-)information by utilities generally suggest them mainly the purchase of energy as means to provide ES, and not possibly much cheaper efficiency technology.

The barriers for market introduction of more energy efficiency are therefore extremely numerous, if the single customer is left alone with his need for ES on the monopolized markets for electricity and natural gas. Because there are hardly and markets for ES so far, the single energy user has to evalue his optimum of total costs for cost-effective ES "hand-made" from market parameters on different markets. Often, he is not capable to do so alone (e.g., as private household or a small firm), or not willing without a change in utility regulation (e.g. energy suppliers), because this total optimum is a priori contradictory to his private interest in selling energy.

Energy consulting and systematically organized additional and graduate training (Fort- und Weiterbildung) particularly for decision-makers in municipal authorities, in small- and medium-size industrial and commercial enterprises and in craftmanship are for this reason a conditio sine qua non. They must, however, be accompanied by new methods of "discovery planning" and utility regulation like "least-cost planning" and new types of organization and transaction for efficiency investments (energy agencies [Energieagenturen]; performance contracting).

"asymmetrical market position"

Apparently, the supply of ES generally involves actions of different people, whose economic power, market position, and social interest may vary to a great extent.

- "David-Goliath"-constellations: traditional energy suppliers (utilities) differ from energy users systematically. For instance, the market position of energy suppliers is de facto far superior to the position of energy demand (with the exception of large industrial customers), solely through their possession of natural resources (e.g., brown and hard coal of hydropower), through concentration of technical production factors (power plants, grids), through their enormous financial power, liquidity, market survey, and planning competence.

- "state-guaranteed cost-transmission" ("staatliche Kostenüberwälzungsgarantie)": in addition to their market position, the investment decisions of utilities supplying electricity to the public are also de iure secured close to risklesness. In the Federal Republic of Germany, for example, this is achieved through a web of legal regulations (areas of exception from monopoly prohibition [Ausnahmebereiche] according to § 103 GWB) strenghtening even more their market-dominating position (territorial cartels), as well as through privileged parameters of action (e.g., in their policies of setting prices and buying from non-utility producers [Einspeisevergütung]). Therefore, the electric utility industry in the FRG has been able to produce mismanagement and overcapacities for decades, without facing any economic consequences. Utilities have been able to transfer all the costs they had produced to their customers.

"pay-back gap"
For these reasons, owners of power stations in the FRG can typically operate with extremely long lead times and with pay-back periods of 20 to 25 years, while industry usually requires 3 to 5 years. Residential as well as small commercial customers cannot at all calculate "production costs" or pay-back periods of measures increasing energy efficiency without instruction.

For public investors, systematical barriers for financing even highly cost-effective energy-saving investments result already from legal dispositions (division of budgets into administrative and investive parts [Verwaltungs- und Vermögenshaushalt]), but also from lack of financial liquidity.

For these reasons as well as for other results of market structures, subjective and objective pay-back periods of efficiency investments differ a lot, especially for households, small business, and public administrations ("pay-back gap").

The following example may illustrate this: let the normal (i.e., objective) pay-back for an efficient household appliance be 9 years at an average electricity price of 25 Pf/kWh (around 16 US-cents per kwh). Given a technical life-time of 15 years for the appliance, this is a cost-effective investment. If, however, the household requires as usual a pay-back period of only one year, the price of electricity will have to be around 2,25 DM/kwh for the investment to pay.

This also highlights the exorbitant level required for an energy tax, if all the theoretically cost-effective, but in reality "hindered" potentials ("gehemmte Potential" nach E. Jochem) were to be mobilized by *pretial control instruments alone*. One can be sure that exactly those politicians who prefer the easier-to-impose energy tax to legal requirements and prohibitions as well as to structural reforms would never consider energy taxes of that magnitude politically "feasible".

"disparity of power plant planning" ("stromwirtschaftliche Disparität") (13)

Particularly a forced market penetration of cogeneration plants as well as of district and central heating cannot be achieved through pretial control alone against the existing structural and legal barriers. Although there is an enormous cost-effective potential for industrial and municipal cogeneration plants in the FRG (14) - assumed the calculation of cost-effectiveness is made with the same parameters and methods as used by a big utility (Verbund-EVU) for planning central power plants - it is realized very slowly. An energy tax would not be able to change this substantially. Prominent among other reasons for this is the fact that many municipal or industrial newcomers on the cogeneration market have to compete with full costs (i.e., long-term marginal costs) against the mixed-price calculation, respectively, the short-term marginal costs of the partially depreciated power plant stock of their previous suppliers (often a big, supraregional utility [überregionales Verbund-EVU]). The big utility can thus invest in cost-effective central power plants producing nothing but electricity, simultaneaously preventing the undesired competition from newcomer's cheaper cogeneration plant by offering dumping prices below its own long-term marginal costs ("stromwirtschaftliche Disparität").

Without an appropriate help from energy politics (e.g. legal dispositions for non-utility power producers; obligation to least-cost planning), decentralized power production in cogeneration plants will, in spite of its theoretical economic advantage in most cases, not be able to succeed against the big centrally-generating electricity monopolies.

"Split incentives" ("Investor/Nutzer"-Problematik)

The largest technical potentials for energy savings in the FRG can be found in heating applications, particularly for space heating (15). There is no doubt that the equivalent heat price and thus an appropriate energy tax will play an important role for the cost-effectiveness of, say, an investment in thermal insulation of a single-family home. Nonetheless, barriers are generally high in the field of building efficiency, so that without legal standards (like Swedish low-energy building standards) it will be impossible to even approximately realize the existing potentials.

This holds true especially for multi-family houses and the lodging sector. When an appartment house is sanitated with respect to energy efficiency, the tenant earns the benefit of reduced energy costs, while the owner of the house has to bear the risk and the higher investment, at least in the short run. High and rising energy prices improve the economy of investments in thermal insulation, but according to every experience they are not sufficient to solve the mentioned conflict of interests.

A possible measure is a certificate on specific energy consumption for space and water heating per square meter and year according to the Danish model.

In the existing buiding stock in general, efficiency investments have to be stimulated through consulting and financial support specific to target groups with priority when they are cheapest: in connection with renovation or sanitation measures carried out anyway. Otherwise, there will be "lost opportunities" impossible to correct even by high energy taxes during the long technical life of buildings and heating equipment.

Conclusions from these facts:

- There is little use in limiting the discussion on internalization to the comparison between supply technologies. It surely has been useful to point out by Hohmeyer's study that market introduction of wind and photovoltaic plants is handicapped compared to conventional power generation by non-consideration of external effects. But barriers for market introduction of energy-efficiency technologies are more effective and quantitatively important.

- Rising energy prices (by means of a fee or tax) may give an economic incentive for "jumping over" existing barriers for market introduction of technologies for more efficient use and generation of energy, but they do not remove the barriers themselves. One might say, higher prices are the whip forcing the horse over the hurdle, but the hurdle itself (i.e., a variety of institutional, legal, and structural barriers) is not removed by them.

Particularly the "pay-back gap" leads to the result that, on one hand, uncontrolled market processes continously direct too much capital into expansion of energy supply and not into efficient use of energy on the demand side, and that on the other hand, newcomers on supply side also (like industrial or municipal owners of cogeneration and/or solar facilities) are being "handicapped" systematically compared to owners of conventional power plants.

Therefore, energy policies have to direct themselves towards systematically supporting the substitutional competition between investment in energy savings or generation through, among others, legal requirements and prohibitions as well as concepts for consulting and financing.

Governmental energy policies, in addition, have to take care through appropriate control instruments especially for the supply side - the utilities - to direct their activities from the beginning towards the societally most efficient investment, systematically including energy savings. Otherwise, continued misallocation of capital into expansion of power generation instead of realization of societally more cost-effective NEGAwatts would cause severe losses to society for subsequent corrections through legal requirements and prohibitions or financial support programmes for the demand side. The new corporate policies of an ESU and the methods and instruments of a more intelligent utility regulation (through "least-cost planning") are essential especially for *avoiding* "external" costs *ex ante*.

5. Climate Catastrophe as an "External Effect"?

5.1 Necessary Measures Against the Imminent Climate Catastrophe

Our basis thesis has been: If we proceed with "business-as-usual" energy policies, those consequences of our economic model played down as "external" effects are going to make the earth uninhabitable in the long run. This holds true particularly for the ongoing emissions of climate-affecting gases (especially CO_2) and the global climate changes imminent from them. We want to conclude our contribution with a brief discussion of the implications which these effects have for the debate on internalization.

In the Federal Republic of Germany, the possibility of a climate catastrophe is meanwhile taken seriously by science as well as politics. The inquiries, studies and publications of a commission on climate protection of the German parliament (Enquête-Kommission "Vorsorge zum Schutz der Erdatmosphäre" des Deutschen Bundestages) have also contributed to this.

The federal government (Bundesregierung) has decided in June of 1990 to install a working group from several ministries, "which will work out proposals aiming at 25 % CO_2-reduction in the FRG until 2005 - relative to the emission volume of 1987 - and analyse reduction possibilities for other climate-affecting gases. ... The government will elaborate a concept for realization of the national CO_2 reduction potential. On realization of CO_2 reduction, it will consider international agreements and effects on economic goals, like employment, price stability, economic growth, equilibrium in foreign exchange and safety of energy supply." (17) (own translation)

The FRG has hereby *announced* probably the most comprehensive CO_2 reduction policies of a big industrialized country - contrary to the so far delaying policies of, e.g., the US, the USSR, and Japan. Yet until now, this is not a decision for realization, because the execution of this CO_2 reduction policy is bound to a number of national and international conditions (see above citation).

The recommendations of the parliament commission, however, exceed the government's 25 % decision by far. The Enquête-Kommission has recommended the following CO_2 reduction schedule (base year 1987: emission 715 mio. t CO_2; without non-energetic consumption):

- until 2005 30 % (minus 210 mio. t CO_2)
- until 2020 50 % (minus 350 mio. t CO_2)
- until 2050 80 % (minus 560 mio. t CO_2)

The same percentages are now being demanded for the united Germany.

There is a consensus that these targets will not be reached without radical interventions into the energy and transport systems. But how radical do these interventions have to be? And what is the possible contribution of global instruments? Opinions differ greatly on this question.

5.2 Non-specific instruments are not sufficient for climate stabilization

Recently, a report to the German minister of economy (BMWi) attempted for the first time a forecast of future energy consumption in the FRG under assumptions on basic conditions set by energy policies to support rational enregy use in principle. Besides an energy tax, a number of political measures aimed at reducing presently existing barriers for efficient energy use are assumed. (18)

According to this forecast, primary energy consumption will decrease until the year 2005/2010; capacity for power generation will also slightly go down, while nuclear power generation figures remain approximately constant. For these reasons, and because of accelerated coal substitution through natural gas and increased imports of (nuclear) power, CO_2-emissions also decrease slightly until 2005 ("Status-Quo-Projektion"). In a "sensitivity analysis" ("Sensitivitätsrechnung") a possible further CO_2-reduction until 2005 is forecasted under the assumptions that the real oil price will rise to 35 $/b until 2010 (instead of 25 $/b), and that energy tax in 2010 will be 40 % instead of 20 % in the base case.

This forecast performed by ISI/Prognos can thus be considered as the first quasi official model simulation on the effects of CO_2 reduction policies acting mainly through the market and higher energy prices. Its results, as judged in comparison with the requirements of energy policies for stabliizing climate, should discourage a politician dedicated only to market conform control. The "Status-Quo-Projektion", where energy taxes rise from 5 % (1995) to 20 % (2010), results in a CO_2 emission reduction of 7 % until 2010. In spite of a 60 % higher level of oil prices (real including taxes) until 2010 compared to the "Status-Quo-Projektion", "Sensitivitätsrechnung" can only reduce CO_2 emissions by 10 % more. This is a CO_2 reduction by far insufficient for the Federal Republic of Germany.

6. Conclusions

To date, all federal governments of the FRG have dedicated themselves more or less to "market-conform energy policies". So far, so-called "market-conform" energy policies, however, have only reacted to acute crises, under apparent pressure enforcing action, and with a time horizon of a few years. Continuing this short-breathed pattern of action and reaction would be disastrous for stabilizing climate.

Concerning this, the recently intensified debate on the need for internalization of "external" costs means progress. It has broken up the dogma that confidence in self-control of energy markets and uncorredcted market prices is already the best form of energy policies. *From this point-of-view*, one of the most important decision groups for economic and energy policy in the FRG, the conference of ministers for economy of the states (Ministerkonferenz der Landes-Wirtschaftsminister) has given a remarkably realistic statement. Under the headline "The basic economic problem: steering against world market" they state that today's goal is to inhibit a development of fossil energy consumption corresponding to market (!). And: "The reason why market alone cannot induce the reduction of consumption necessary for solving the global warming problem is because the threat to climate connected to the use of fossil fuels as so-called external effect is not being internalized, i.e. not included in prices and cost calculation. In such a situation, the system of market economies requires to administratively correct market processes in a way that scarcety relationships (here: climate compatibiity) are reflected in market prices." (19) (own translation).

Again, we find the classical form of playing down climate catastrophe as an "external" effect. From this neoclassical diagnosis follows inevitably the illusion that *market and price control alone*, namely internalization of "external" costs, is the means able to make otherwise unchanged energy policies compatible to climate.

To implement successful policies for climate protection in the energy system, the relationship between government and energy monopolies has to be changed radically through a more democratic, independent, and competent regulatory authority, and through an effective planning and regulation process oriented towards optimization for society as a whole. Comprehensive packages of measures and instruments are to be introduced, including structural reforms of property order (decentralization and municipalization), legal requirements and prohibitions, and reorientation of corporate policies of utilities as well as "market conform" systems of fees, financial support, and incentives.

Furthermore, climate protection policies have to be strictly preventive policies or they will fail, because of the long time lag of the climate system. In other words: when we will be able to estimate the costs of global climate change in order to internalize them, it will inevitably be too late to prevent climate catastrophe.

Despite these facts, however, in the public debate on measures for climate stabilization policies in the FRG all political parties still focus on the question in what forms (fee on CO_2-emissions, or general primary or final energy tax) and to what extent "external" costs should be internalized. This discussion might seriously impede the realization of the needed complex packages of measures and instruments of climate stabilization policies, unless several illusions on the omnipotence of market-conform instruments that have been fed by the debate on internalization are destroyed on time. A result of this debate might then be not to occupy oneself with the essential question: namely, how *the necessary reconstruction* ("Umbau") towards a "soft" energy, transport, and industry system compatible to environment and climate can really be achieved.

The above-mentioned theoretical arguments have shown that this reconstruction, and climate compatibility of the energy system, cannot be achieved by just introducing an energy tax. If you have not been convinced by these arguments so far, you may perhaps be so by the global empirical test made possible by the energy price crises of the seventies.

In a phase where the oil price exploded nominally more than sevenfold (from 1973 = 82 DM/t to 1985 = 622 DM/t; in real terms fivefold), and energy policies acted procyclically, primary energy cnsumption in the FRG was "only" held constant at about 380 mio. tce from 1973 into the eighties, and CO_2 emissions were reduced from 784 mio. t (1973) to 715 mio. t (1987) - some 9 %. How, then, are we to reduce CO_2 emissions by 30 % in the next 15 years by pretial control alone?

The goal for the future is not only decoupling energy consumption and economic growth: energy consumption of industrialized countries will have to be reduced dramatically, if we want to prevent global warming. This will only be possible, if *the quality of growth* changes radically as well. The time horizon for this change is extremely short (10 to 20 years). Who does seriously believe that a tax is sufficient for implementing this change in time?

Remarks

(1) see the work of the Enquête-Kommission "Vorsorge zum Schutz der Erdatmosphäre" des Deutschen Bundestages, especially the third report (Dritter Bericht zum Thema Schutz der Erde, Bundestagsdrucksache 11/8030 vom 2.10.1990)

(2) Pigou, A.C., The Economics of Welfare, London, 1920, repr. 1950

(3) see Kapp, W.S., Soziale Kosten der Marktwirtschaft, Frankfurt 1979

(4) comp. Baumol, W.S., On Taxation and the Control of Externalities, in: The American Economic Review, 62/1972, p. 307ff.; Coase, R.H., The Problem of Social Cost, in: Journal of Law and Economics, 3/1960, p. 1ff.; and Buchanan, P./Tullock, G., Polluter's Profits and Political Response: Direct Controls versus Taxes, in: The American Economic Review, 65/1975, p. 139ff.

(5) Wicke, L., Umweltökonomie, München 1989, p. 43

(6) Hohmeyer, O., Social Costs of Energy Consumption, Berlin/Tokyo 1988, Introduction

(7) ibid.

(8) O. Hohmeyer calculates mean gross social costs of electricity generation of about 4 to 9 Pf/kwh for fossil-fueled and about 10 to 21 Pf/kwh for nuclear power plants (at present, 1 US cent = 1,5 Pf).

(9) Lovins, A., NEGAwatts: A Practical Remedy for Megagoofs, Old Snowmass 1985; idem, Saving Gigabucks with NEGAwatts, Old Snowmass 1986

(10) e.g., in the form of "competitive bidding"

(11) We explicitly vote *against a fee on CO_2 emissions*, because

- other climate-affecting gases (CH_4 and O_3/NOx) are neglected
- the one-sided concentration on the goal "climate stabilization" leads to risk shifting, and to discrimination and neglecting of other "external" effects (nuclear risks; consequences of road accidents)
- this implies a decision for nuclear energy "through the back door"
- the control effect would be prohibitive for coal use in electricity generation, but small for automobile gasoline

(12) comp. Enquête-Kommission "Vorsorge zum Schutz der Erdatmosphäre", Dritter Bericht (s. note (1)).

(13) comp. Stumpf, H./Windorfer, E. (WIBERA), Fernwärme in der Bundesrepublik Deutschland. Hindernisse für ihre Förderung, Düsseldorf 1984

(14) comp. Enquête-Kommission "Vorsorge zum Schutz der Erdatmosphäre", Dritter Bericht (s. note (1)); and Fritsche, U./Köhler, S. (ÖKO-Institut), Das CO_2-optimierte GRÜNE Energiewende-Szenario 2010, Freiburg/Darmstadt 1990

(15) The technical energy savings potential is being estimated as 35 to 45 % for the FRG in total and as 70 to 90 % for existing and new buildings (compared to 1987 consumption); see Enquête-Kommission, as cited

(16) comp. Hennicke, P., Wie kann die Aufheizung der Erde verhindert werden? Eckpunkte eines CO_2-Reduktionskonzepts für die Bundesrepublik, in: MEMO-Forum, Nr. 15, Bremen, März 1990; Hennicke, P./Müller, M., Klimaänderungen und Treibhauseffekt, in: WSI-Mitteilungen 7/1990

(17) BMU/UBA, Zielvorstellung für eine erreichbare Reduktion der CO_2-Emissionsn, Vorlage vom 13.6.1990

(18) ISI/Prognos, Die energiewirtschaftliche Entwicklung in der Bundesrepublik Deutschland bis zum Jahr 2010. Kurzfassung, Basel 1989, S. 15

(19) Wirtschaftsministerkonferenz vom 14./15.9.1989, in: Arbeitsunterlage der Enquête-Kommission "Vorsorge zum Schutz der Erdatmosphäre" 11/366 vom 5. Oktober 1989, S. 48ff.

(20) comp. Hennicke, P./Bach, W.. Zusatzvotum zu Abschnitt E "Nationales Vorgehen zur Reduktion energiebedingter klimarelevanter Spurengase", in: Enquête-Kommission "Vorsorge zum Schutz der Erdatmosphäre", Dritter Bericht, as cited

Translation: Dipl.-Phys. S. Thomas

7. SUBJECT AREA 5:

WAYS OF INTERNALIZING EXTERNAL COSTS

PRACTICAL APPROACHES

7.1 Incorporation of Environmental Externalities in the United States of America

Richard L. Ottinger
Pace University Center for
Environmental Legal Studies

1. Introduction[1]

Internalizing the environmental costs imposed on society by polluters is the wave of the future in addressing environmental degradation. By signalling to industry the true societal costs of their operations, inclusion of environmental costs in the price of goods produced gives an economic incentive to industry to reduce pollution. This can be an important supplement to regulation.

Government regulation of pollution has proved to be generally inadequate to address the severe threats to the planet posed by global warming, acid rain, urban, urban smog and toxic contamination of our air, water and food supplies. While some improvements have resulted from regulation in the United States, particularly where contaminants have been totally prohibited as with asbestos, lead in gasoline and DDT, generally pollution increases stay well ahead of mandated regulatory controls. Inevitably economic growth, in both developing and industrialized countries, receives higher government priority than environmental protection and outpaces governmental efforts to control pollution or require switching to less polluting industrial resources.

Governments are just starting to consider supplementing pollution regulation with pollution or fees that will introduce into the marketplace prices that reflect the damages to society inflicted by polluting resources. The OECD has just published a review of pollution levies indicating a total of 85 pollution taxing regimens in six of its principal countries. Thus, Germany, France and Holland have imposed waste water effluent charges ($2, $9 & $39 per capita respectively); Switzerland has imposed extra landing fees on noisy aircraft; Sweden and Norway have required returnable deposits on automobile bodies to prevent their being dumped. West Germany is considering an auto tax based on tailpipe exhausts.[2] And the U.S. House of Representatives' Ways & Means Committee recently held hearings on pollution taxes (March 6, 7 & 14, 1990).

The environmental organizations, which historically have resisted pricing environmental impacts on grounds that it would constitute a licence to pollute, have now embraced the idea. Daniel Dudek of the Environmental Defense Fund has been a leading advocate of marketplace treatment of pollution; he helped draft the Administration's Clean Air Act proposal to create emissions trading rights.

It is remarkable that the conservative electric utility industry and its usually equally cautious regulatory commissions have pioneered in applying marketplace principles to valuation of environmental externalities. In twenty-nine jurisdictions, utility regulatory commissions have started formal consideration of incorporating these externalities into utility planning, bidding and other resource selection procedures.

2. Significance of Utility Consideration of Environmental Externality Costs

United States electric utilities are a major contributor to the damages imposed on the global environment. They account for about two-thirds of United States sulphur dioxide emissions and one-third of United States nitrogen oxide emissions, both of which cause acid deposition. They contribute one-third of the national emissions and eleven percent of the global emissions of carbon dioxide, the principal greenhouse gas.[3] Thus, United States electric power production produces a significant environmental impact on the United states and the world, the cost of which is not included in the price of electric power.

Until recently, most utilities throughout the country have selected their supply and demand side resources on a least cost basis, without regard to the very real environmental costs imposed on society. The utilities and regulatory commissions have proceeded on the basis that full compliance with all applicable environmental laws and regulations is sufficient.

However, government-imposed standards and regulations do not purport to eliminate pollution. Rather, they merely limit emissions to levels established by a combination of legislative compromise and regulatory feasibility.[4] The environmental impacts that remain after regulation are significant. Not all emissions are regulated, and even those emissions that are regulated are not eliminated. To the extent that power producers price and select resources solely on the basis of complying with government standards, the effect is to assign the very substantial residual environmental externality costs a value of zero.

As difficult as it is to fix exact dollar figures on environmental damages, one always has to come back to the basic tenet that "A crude approximation, made as exact as possible and changed over time to reflect new information, would be preferable to the manifestly unjust approximation caused by ignoring these costs,"[5] and thus valuing environmental damages at zero.

Incorporation of the dollar values of environmental externalities in utility resource selection procedures will enable utilities and regulators to take environmental costs into account in making resource acquisiton decisions. In practical terms, the result will be to encourage utilities to invest in less polluting resources.

3. Why Utilities Should Incorporate Environmental Costs

It could be argued that consideration of environmental externalities should be addressed solely by pollution controls, taxes or fees enacted by Congress and state legislatures, rather than by utility actions and regulatory commission orders affecting resource selection. Indeed, currently, environmental costs for other industries such as automobile and chemical manufacturing, smelting, etc. are addressed only through legislated controls.

National pollution fees could internalize environmental externality costs for all polluting sources, thus sending universally improved price signals to the marketplace.[6] Of course, complete regulation, eliminating all pollution that poses a threat to human health and the environment would also internalize environmental externality costs universally. Neither appears to be politically feasible at present, however.

In the absence of legislative internalization, utilities nevertheless will continue selecting resources which, after regulation, will still impose significant environmental costs on society. If utilities and their regulatory commissions continue to assign only traditional direct costs to these resources as a principal basis for their selection, these environmental costs still will be ignored. For a full accounting of costs in this continuing resource selection process, it is important that the environmental costs be counted.

There are two principal reasons why utilities and regulatory commissions should consider environmental externalities in their resource selection processes: (1) utilities are franchised monopolies vested with a duty to serve the public interest, a responsibility that includes environmental protection[7]; and (2) foreseeable international, federal and state environmental laws and regulations are likely to impose more stringent environmental controls over the thirty to forty-year life span of electric power plants, making it imprudent for utilities to invest in resources which will have to be abandoned or which will require expensive environmentally-required retrofits.[8] It is within the traditional role of utility regulatory commissions to oversee utility public interest obligations and to prevent imprudent investments.[9]

It should be noted that when commissions require consideration of environmental externalities in utility planning and resource selection, they are not internalizing these costs in the existing rate structure, but are using them as a means of making prudent selection of new sources.

Accounting for environmental externaltities in utility decision making will not only help to avoid the economic costs of future environmental controls, it will also reduce the level of uncertainty associated with utility resource plans and increase system reliability.

4. Magnitude of Environmental Costs

The following values are presented as "starting point" figures from the Pace 1990 study. "Starting point" values, as used in that study, are those that seem to most reasonably represent the range of values in the studies reviewed, considering the locations at which the studies were performed, their documentation and thoroughness. The term "starting point" is used because the figures do not pretend to be cost estimates - too many relevant costs were excluded from this study[10] and the studies reviewed are often too uneven quality and documentation - but the "starting point" figures do give the order-of-magnitude of the values in the studies reviewed, and they should provide a good starting point for further research.

a. Fossil Fuel-Fired Plants

Emissions from electric power plants are generally available on a plant specific basis. However, it is difficult to estimate a "typical" set of emissions for any resource type. Emissions depend on several factors other than simply the plant type, including: the age and type of the plant; fuel type, grade and sulfur content; installed emissions control technology; plant operations, including heat rate, combustion temperature, and steam or limestone injection. Finally, the estimate of emissions per unit of energy output can be further complicated by the allocation of emissions to joint products, such as steam and electricity. In the examples presented, all of the externalities have been allocated to electricity output.

The externality values are estimated for three coal-fired technologies, two oil-fired technologies with three different sulfur content assumptions, and two natural gas-fired technologies with differing control equipment. These examples are illustrative of the relative externalities of different fossil-fired resource options, both existing and new, but are not meant to represent average externalities of these technologies.

The reports reviewed estimate the following "starting point" values, in dollars per pound of emissions: \$2.03/lb SO_2, influenced primarily by the estimated health effects; \$0.82/lb NO_x, also dominated by health effects; and \$1.19/lb particulates, based primarily on visibility effects (\$0.83/lb.), with a strong contribution from health effects (\$0.36/lb.).[11] The value of reducing SO_2 emissions was estimated to be 2.5 cents/lb carbon (or 0.68 cents/lb CO_2) using a mitigation cost estimate for tree planting from the middle of the range found in the studies. These ranges vary from \$17.08/T of carbon to \$80/T, depending on fixation rates, discount rates and other assumptions.

For coal-fired facilities, dollar per pound figures are converted into a cent per kilowatt hour cost. See Table A. No pollution control equipment is used for any of the example plants for SO_2, NO_x or CO_2. This assumption applies for many older coal-fired boilers in most of the country, and it provides a base-line for comparing plants with control technologies. The absence of control could also apply to newer facilities built in the 1970's because they may be able to meet the standards by controlling emissions through improved operating procedures.[12] For particulates, some pollution control equipment (e.g. bag-houses) is assumed to be installed on each plant. The "starting point" externalities from the older coal-fired boiler burning 1.2 % sulfur coal are on the order of 5.7 cents/kWh generated, or 6.7 cents/kWh delivered. For an Atmospheric Fluidized Bed Combustion (AFBC) plant with the same heat rate and burning 1.1 % sulfur coal, these externalities are on the order of 2.8 cents/kWh (generated). For an Integrated Gas Combined Cycle (IGCC) plant with the same heat rate and sulfur content of 0.45 % sulfur these externalities are on the order of 2.5 cents/kWh (generated).

For oil-fired facilities, the results are shown in, Table B. Again, and for the same reasons as for coal-fired plants, no pollution control equipment is used on any of the example plants for SO_2, NO_x or CO_2. This is typical for many older residual oil-fired boilers, and controls are not generally necessary for a distillate oil-fired combustion turbine to meet NSPS standards. The SO_2 and particulate emissions are strongly dependent on the sulfur content of the oil. Table B shows "starting point" estimates of the externalities from two different oil-fired technologies: boilers burning residual oil with three different sulfur contents and a combustion turbine burning distillate oil. The "starting point" externalities from the residual oil-fired boiler burning 0.5 % sulfur oil are on the order of 2.7 cents/kWh (generated). For the same boiler burning 1 % sulfur oil these externalities are 3.8 cents/kWh (generated), and for the same boiler burning 2.2 % sulfur oil these externalities are 6.7 cents/kWh (generated). The "starting point" externalities of a combustion turbine burning distillate oil, and having lower efficiency than the boilers, are on the order of 2.5 cents/kWh (generated).

For natural gas-fired facilities, the results are shown in Table C. No special NO_x control equipment is assumed to be used on the gas-fired boiler or the combined cycle unit. However, selective catalytic reduction (SCR) and steam water injection (SWI) are applied on the BACT unit, which is a combined cycle unit fitted with best available control technology. SO_2 emissions from natural gas combustion are shown to be negligible. The "starting point" externalities from a natural gas-fired steam plant are on the order of 1.0 cent/kWh (generated). For a combined cycle unit these externalities are estimated to be on the order of 1.0 cent/kWh (generated). For a combined cycle unit fitted with SCR and SWI, these externalities drop to 0.7 cents/kWh (generated).

These results show that the externalities of older coal- and oil-fired boilers are dominated by SO_2 and NO_x emissions. For newer coal and oil-fired technologies, and for gas-fired plants, the externalities are largely dominated by CO_2 emissions. This shift in the emphasis of the externalities is shown by the studies reviewed to be largely due to improved boiler design and efficiency, improved control equipment efficiency, and increasing use of control equipment and emissions-reducing operating procedures in response to ever-tightening emissions regulations.

Finally, these externality estimates leave out so many potential external effects that they are likely to be conservative rather than overstated. To derive more complete cost estimates, other externalities should be incorporated into future environmental cost estimates for fossil fuel-fired power plants, including: air emissions from combustion not included in the studies reviewed, such as the greenhouse gases methane and N_2O; air toxics including heavy metals, ozone precursors, VOCs; and other externalities related to fossil-fuel combustion for which valuation studies could not be found, such as water use, land use, and solid waste disposal; and externalities related to the front end of the fuel cycle, such as extraction, transportation and processing, which were excluded from the Pace 1990 report.

Table A: Externality cost for coal-fired units

(Emissions - lbs/MMBTU)

Externality	$/lb	Existing Boiler (1.2% S)	AFBC (1.1% S)	IGCC (.45% S)	NSPS (N/A)
	[A]	[B]	[C]	[D]	[E]
[1] SO_2	$ 2.03	1.80	0.55	0.48	1.2
[2] NO_x	$ 0.82	0.607	0.3	0.06	0.006
[3] Particulates	$ 1.19	0.15	0.01	0.01	0.03
[4] CO_2	$ 0.0068	209	209	209	209
Totals:					
[5] $/MMBTU Input		$ 5.76	$ 2.80	$ 2.46	$ 3.90
[6] Heat Rate (BTU/kWh)		10,000	10,000	10,000	10,000
[7] $/kWh Generated		$ 0.058	$ 0.028	$ 0.025	$ 0.039
[8] $kWh Delivered		$ 0.068	$ 0.033	$ 0.028	$ 0.045

Notes:

[A]: Unit Values derived in Chapter V.
[B][C][D][E]: Emissions are from PLC (1989); SO_2 and CO_2 emissions have been restated as lbs SO_2 and lbs CO_2. All emissions are expressed as lbs/MMBTU fuel input.
[E]: NSPS regulations require 1.2 lbs/MMBTU and 90% reduction for plants with emissions greater than 0.6 lb/MMBTU; for plants with emissions less than 0.6 lb/MMBTU; NSPS requires 70% reduction in emissions.
[1]: No SO_2 scrubbers are installed on the first three plants.
[2]: NO_x emissions are uncontrolled in each case.
[3]: Particulates emissions vary widely and are extremely dependent on the ash content and sulfur content and sulfur content of the coal. NSPS requires 0.03 lbs/MMBTU and 90% reduction.
[4]: CO_2 emissions are derived in PLC (1989).
[5]: Sum of (value x emissions for each externality) for each plant.
[6]: Assumed heat rates for each plant.
[7]: [5] x [6]/1,000,000.
[8]: Assumes 15% marginal energy losses.

Table B: Externality cost for oil-fired units

(Emissions - lbs/MMBTU)

Externality	\$/lb	Boiler # 6 Oil (.5% S)	Boiler # 6 Oil (1% S)	Boiler # 6 Oil (2.2% S)	Combustion Turbine # 2 Oil (1% S)
	[A]	[B]	[C]	[D]	[E]
[1] SO_2	\$ 2.03	0.54	1.08	2.38	0.16
[2] NO_x	\$ 0.82	0.357	0.287	0.357	0.498
[3] Particulates	\$ 1.19	0.055	0.09	0.174	0.036
[4] CO_2	\$ 0.0068	169	169	169	161
Totals:					
[5] \$/MMBTU Input		\$ 2.60	\$ 3.68	\$ 6.48	\$ 1.87
[6] Heat Rate (BTU/kWh)		10,400	10,400	10,400	13,600
[7] \$/kWh Generated		\$ 0.027	\$ 0.038	\$ 0.067	\$ 0.025
[8] \$kWh Delivered		\$ 0.032	\$ 0.045	\$ 0.079	\$ 0.030

Notes:

[A]: Unit Values derived in Chapter V.
[B][C][D][E]: Emissions are from PLC (1989); SO_2 and CO_2 emissions have been restated as lbs SO_2 and lbs CO_2. All emissions are expressed as lbs/MMBTU fuel input.
[1]: SO_2 emissions are uncontrolled in each case.
[2]: NO_x emissions are uncontrolled in each case.
[3]: Particulates emissions are calculated from EPA Ap-42 using the formula: 0.02 + 0.07 x S, where S ia the sulfur content in percent.
[4]: CO_2 emissions are derived in PLC (1989).
[5]: Sum of (value x emissions for each externality) for each plant.
[6]: Assumed heat rates for each plant.
[7]: [5]*[6]/1,000,000.
[8]: Assumes 15% marginal energy losses.

Table C: Externality cost for natural gas-fired units

(Emissions - lbs/MMBTU)

Externality	$/lb	Existing Steam Plant	Combined Cycle	BACT (SCR, SWI)
	[A]	[B]	[C]	[D]
[1] SO_2	$ 2.03	0	0	0
[2] NO_x	$ 0.82	0.248	0.42	0.042
[3] Particulates	$ 1.19	0.003	0.003	0.0002
[4] CO_2	$ 0.0068	110	110	110
Totals:				
[5] $/MMBTU Input		$ 0.95	$ 1.10	$ 0.78
[6] Heat Rate (BTU/kWh)		10,400	9,000	9,000
[7] $/kWh Generated		$ 0.010	$ 0.010	$ 0.007
[8] $kWh Delivered		$ 0.012	$ 0.011	$ 0.008

Notes:

[A]: Unit Values derived in Chapter V.
[B][C][D]: Emissions are from PLC (1989); SO_2 and CO_2 emissions have been restated as lbs SO_2 and lbs CO_2. All emissions are expressed as lbs/MMBTU fuel input.
[1]: SO_2 emissions are zero from gas combustion.
[2]: NO_x emissions are uncontrolled in the first two cases; For the BACT case, Selective Catalytic Reduction and Steam Water injection are assumed.
[3]: Particulates emissions are estimated for CEC (1989). BACT assumes fabric filter control.
[4]: CO_2 emissions are derived in PLC (1989).
[5]: Sum of (value x emissions for each externality) for each plant.
[6]: Assumed heat rates for each plant.
[7]: [5]*[6]/1,000,000.
[8]: Assumes 15% marginal energy losses.

Tables A-C summarize the environmental costs for fossil fuel plants from the studies reviewed and "starting point" cost for them.

b. Nuclear Plants

The external costs of nuclear power have been divided into a number of categories including: (1) the health, property value and wildlife costs caused by routine radiation emissions; (2) the costs caused by non-routine emissions, for which no studies were found; (3) the health and property damage costs associated with accidental emissions; and (4) decommissioning costs that are not internalized.

The "starting point" environmental costs from nuclear operations are estimated to be 2.91 cents/kWh, which is the sum of: (1) 0.11 cents/kWh for routine and non-routine operations, (2) 2.3 cents/kWh for accidents, and (3) 0.5 cents/kWh for decommissioning costs. This analysis omits all front-end costs from extraction and transportation of uranium, excluded from the Pace 1990 study.

Table D below represents the nuclear plant's "starting point" values.

Table D: Starting point values for nuclear power externalities

	Cents/kWh
Routine operations	0.11
Accidents	2.30
Decommissioning	0.50
TOTAL	2.91

c. Waste-to-Energy Plants

The only study located estimated the externalities of waste-to-energy plant emissions at 11.14 cents/kWh. If half of this value were allocated to the energy production function of the plant, then the environmental externality costs would be approximately 5.5 cents/kWh. The study did not appear to take into account the risks from exposure to toxic substances in the incinerator ash and may have undervalued the risks from toxic emissions.

Since the types of emissions from garbage incinerators are similar to those from coal-fired plants, one costing approach is to apply the dollar-per-pound values for coal plant emissions to incinerators. This provides a lower value of approximately 2.8 cents/kWh solely for SO_2, NO_x, HCL and particulate emissions, excluding a value for human health effects from HCL, for which no social costs were found.

Table E below displays the environmental externality costs of waste-to-energy plants from the studies reviewed.

Table E: **Calculation of conventional waste-to-energy plant emission externalities**

Pollutant	Cost Per Pound	Pounds Per Kilowatt Hour	Externality Cost in Cents/Kilowatt
SO_2	$1.72	.014	2.4
NO_x	.98	.002	.2
HCL	.14	.077	.1
Particulate	$1.18 - 2.36	.00075	.09 - .18
Carbon Monoxide	-	.00085	-
Volatile Organic Hydrocarbons	-	.00015	-
CO2	-	-	-
		TOTAL	2.79 - 2.88

d. Solar, Wind and Biomass

"Starting point" costs are given in Table F below from the environmental externalities of solar, wind, and biomass power generating systems. These costs cannot be directly compared, however, because the figures were developed using different methodologies for each technology.

Table F: Summary of environmental costs for various renewable energy technologies

Technology Type	Cents/kWh
Solar	0 to 0.4
Wind	0 to 0.1
Biomass	0 to 0.7

5. Federal Incorporation

The studies reviewed in Pace 1990 indicate that FERC has taken no action to incorporate environmental externalities in the wholesale and interstate markets it regulates. Indeed, the recent Notices of Proposed Rulemaking it has issued on PURPA bidding guidelines don't even address DSM options, to say nothing of environmental costs. Federal consideration of environmental costs has been primarily through environmental regulation, which by reducing pollutants, internalizes some of their costs.

Present federal subsidies and tax credits, on the other hand, are found to benefit fossil and nuclear electricity which bear high environmental costs much more than less polluting electricity options such as renewables and DSM. All subsidies are found to distort marketplace pricing of energy resources.

6. Incorporation by Utilities and Regulatory Commissions

A surprising number of states, twenty-nine, have required or are actively considering requiring, their utilities to include environmental externality costs in some manner in planning, bidding or other resource acquisition procedures.[13] Nine states plus Bonneville Power Administration and the Northwest Power Planning Council have sought to quantify environmental costs or have quantification measures pending; three other states have such measures under active consideration.[14] Those entities are approaching incorporation in several ways:

Quantitative. Some utility commissions either increase the cost of a proposed new resource by a specific amount/kWh, reflecting the environmental consequences of using the resources (e.g. New York) or allow a credit to reduce the effective kWh cost of what are deemed to be less-polluting resources (e.g. Wisconsin). The adder or credit is based on quantified costs.

Qualitative. Several utility commissions require utilities to consider environmental impacts without specifying how the costs are to be calculated or incorporated (e.g. Arizona).

Rate of return. A number of utility commissions award an increased rate of return on specific non-polluting resource investments (e.g. demand-side management in Washington) or an increased rate of return based on over-all performance in utilizing non-polluting resources (e.g. Idaho).[15]

Avoided cost. Some utility commissions increase avoided costs in order to take into account environmental costs (e.g. Virginia).

Ranking. A few utility commissions permit a utility to be guided in selecting resources by a commission-imposed or commission-approved ranking system that weights resources by their environmental impacts or by other criteria, not their environmental costs (e.g. New England Electric System (NEES) in Massachusetts was allowed to use a ranking based on a customer survey[16]).

Collaborative. Several utility commissions authorize utilities to determine environmental costs and their treatment through a collaborative negotiating process involving regulators, state agencies and public interest intervenors. Outside experts are usually hired to design programs, evaluation procedures and incentives rates to utilities, and to determine how environmental costs will be calculated and incorporated. The process is sanctioned and subject to oversight by the commission.

Several innovative incorporation methodologies have not yet been applied in any state. These include "*Environmental LCUP,*" in which pollution reduction targets are set for particular pollutants and the utilities are required to meet these targets at least cost; *environmental dispatch*, in which utilities or a power pool are required to include enviornmental costs in least cost dispatch of existing resources; *cost-charging*, in which quantified environmental costs in least cost dispatch of existing resources; *cost-charging*, in which quantified environmental costs are charged to the resource-owner and the proceeds used to establish a pollution mitigation fund; and *set-asides*, in which utilities are required to set aside a specified portion of their capacity needs for designated resources considered to be relatively environmentally benign.

Table G below shows graphically the status of state actions incorporating environmental externality costs, and Table H shows the type of action taken by states which have acted to incorporate these costs.[17]

Table G: Status of state actions incorporating environmental externality costs

<u>Legend</u>

O - Incorporation Ordered
P - Incorporation Order Pending
U - Under Consideration
N - No Action

STATE	O	P	U	N		STATE	O	P	U	N	
ALABAMA				X		NEBRASKA				X	
ALASKA	X				1	NEVADA	X				1
ARIZONA	X					NEW HAMPSHIRE				X	
ARKANSAS				X		NEW JERSEY	X				
CALIFORNIA	X				2	NEW MEXICO				X	
COLORADO	X					NEW YORK	X				
CONNECTICUT		X				NORTH CAROLINA			X		
DELAWARE				X		NORTH DAKOTA				X	
DISTRICT OF COLUMBIA			X			OHIO	X				
FLORIDA				X		OKLAHOMA				X	
GEORGIA				X		OREGON	X				
HAWAII			X			PENNSYLVANIA	X				
IDAHO	X					RHODE ISLAND			X		
ILLINOIS				X		SOUTH CAROLINA				X	
INDIANA				X		SOUTH DAKOTA				X	
IOWA		X				TENNESSEE				X	
KANSAS	X					TEXAS	X				
KENTUCKY				X		UTAH			X		
LOUISIANA				X		VERMONT	X				
MAINE			X			VIRGINIA	X				
MARYLAND			X			WASHINGTON				X	
MASSACHUSETTS	X				2	WEST VIRGINIA				X	
MICHIGAN	X				2	WISCONSIN	X				
MINNESOTA	X				2	WYOMING				X	
MISSISSIPPI				X							
MISSOURI				X	3	BPA	X				1
MONTANA			X			NWPPC	X				1

Source: Pace 1990, Chapter X, Section G

(1) Established by legislation.

(2) Order issued to consider externalities; implementation pending.

(3) Commission has stated that it may consider externalities.

Table H: Type of action by states acting to incorporate environmental externality costs

Legend

O - Incorp. Ordered
P - Incorp. Order Pending
U - Under Consideration
QN- Quantitative Consid.
QL- Qualitative Consid.
ROR- Rate Consideration
AC - Avoided Cost Consid.
ED - Environmental Dispatch
C - Collaborative Action
PL - Planning Consideration
B - Bidding Consideration
L - Legislative Action

State	QN	QL	ROR	AC	ED	C	PL	B	Comment
ALSK				O			L		
AZ		O					O		
CA	OP				U	C*	O		*Collab. ended
CO	O*							O*	*by fuel type
CT	P		L*			C	P	P	*5% ROR adder
DC							U		
HW							U		
ID			O	O					
IA				P		C			
KA	P		O						
ME		U					U		
MD	U					C		U	
MA	P*					C	P	P	*DSM Evaluations
MICH	U			O			U		
MN*		O					P		*Law caps SO2
MT		U	L*				U		*2% adder for DSM
NV		L					L		
NH							L*		*Law under consid.
NJ	O			O				O	
NY	O		O	L		C	O	O	
OH		O					O		
OK			O*						*ROR, trash only
OR*	O	O					O		*Law caps CO_2
PA		O*					O*		*Not implemented
RI							U		
UT							U		
TX		L					L		
VT	O*					C	O*	O*	*15% on DSM
VA				O*					*15% DSM adder to AC
WA			L*						*2% ROR law for DSM
WI	O		O		U		O		
BPA	L						L	U	
NWPPC	L						L		

Source: Pace 1990, Chapter X, Section G.

7. Barriers to Incorporation

Despite this heartening process on the part of utilities and regulatory commissions to incorporate environmental externalities in planning and resource selection, a number of barriers exist to more widespread adoption.

First, many commissions have not yet adopted least cost planning procedures, leaving it to their utilities to determine the selection of resources. Since inclusion of externalities is a societal function not reflected in utility profits, it is unlikely that utilities will on their own act to incorporate externality costs in their selection procedures and none have done so.

Second and related to the first barrier, only a very few utility commissions have provided ratemaking revisions that reward utilities for considering environmental externalities in their resource selection determinations. Just nine states give rate of return consideration to environmentally benign DSM measures, citing environmental externality cost rationales, for example.[18] No commission has adopted rate incentives specifically tied to avoidance of environmental costs of rewarding investment in environmentally benign resources other than DSM resources. It is unrealistic to expect utilities to take environmental costs into account unless they are rewarded for doing so (or penalized for failure to do so).

Third, there are a number of rate impact and equity issues which derive from utility consideration of environmental externalities.

Requiring incorporation of environmental costs could have a significant impact on rates and the competitive position of electric utilities. So far, rate impacts have not been a significant problem because the environmental adders (or credits to non-combustion sources) have not been large enough to have a significant effect on rates, and the costs have been applied only to selection of new sources, rather than to more highly polluting existing resources.

If, however, the full environmental damage costs are ascertained and used, and particularly if application is made to existing resources, the impact on rates could be considerable. On the other hand, if consideration of environmental costs causes utilities to invest more in energy efficiency resources, which generally have economic and social costs lower than traditional supply alternatives, then the electric bills of participating customers may be reduced (even though utility rates charged to consumers will increase).

Another equitable issue that must be addressed is whether these rate impacts should be borne by all ratepayers or only the class of customers eligible to participate in the utility DSM investment. All customers will benefit by environmental enhancement, and to the extent that DSM is undertaken, all will benefit from avoidance of more expensive supply resource costs. However, if all ratepayers are required to pay for DSM investments, the ineligible classes of customers will pay for DSM without being able to enjoy any of its dirct benefits. New York has permitted utilities to experiment with both treatments with respect to allocation of DSM rate impacts.

In this respect, making DSM programs available to low income customers has particularly urgent equity considerations. Unless utilities mount aggressive programs to enable low income customers to participate, they will end up paying higher electricity rates for programs from which they cannot benefit in terms of energy and bill savings. The result will be subsidization by low income ratepayers of more affluent customers.

There also may be competitive problems arising from rate impacts. If the impact of consideration of environmental costs is sufficiently great, it may overprice electricity in relationship to resources that are not incorporating environmental costs. Thus, it may promote more direct use of polluting fuels (such as oil or gas heating rather than electric heating, for example), or self-generation of electricity by major customers who do not have to consider environmental costs.

Fuel-switching may take place as a result of utilizing environmental externality costs. This may result in higher rates even though environmental improvements result. If significant environmental costs are incorporated, natural gas, which is cleaner than oil and coal, may be preferred to an extent that will drive up gas prices. The much-debated role of nuclear fuels as an environmentally preferable substitute for fossil fuels will probably not be affected because nuclear plants bear large environmental externality costs in addition to their heavy capital costs and long lead times. Solar electricity, wind generation, biomass and DSM measures, however, should be considerably advantaged by incorporation of externalities, because their environmental costs are small.

These rate impact and equity issues should be taken into account in devising programs for utility incorporation of externalities.

8. Conclusion

Incorporation of environmental externality costs in electric utility planning and resource selection is an important first step towards internalizing these costs. It is remarkable that twenty-nine states have taken some action to require utilities to take these costs into account and eighteen have actually issued orders to do so. The commissions in these states are wrestling innovatively with how to value these costs.

While damage costs are by far the best to use because it is the environmental damages to society that are sought to be avoided, it is apparent from the damage cost literature that much research on damage costs remains to be done. All commissions to date have used control costs as a proxy.[19] Most commissions that have acted have either had their staffs estimate the damages as in New York, used a proxy adder or credit as in Wisconsin, given some kind of rate of return consideration or merely ordered their utilities to consider the costs without specifying how they should do so.

Despite the uncertainties, environmental costing studies demonstrate that the environmental externality costs of most electricity-related environmental impacts can be valued. The studies also show the means of valuing them. But the literature also shows that a great deal of research is still needed to quantify many of the externalities. Electricity is found to have substantial environmental costs, not presently included in most utility resources selection. While many of the cost studies are incomplete and of varying adequacy, they do indicate the order of magnitude of these costs.

If environmental costs are to be internalized universally in the future through pollution taxes or fees, reliable damage figures are as important to their accurate determination as they are to the incorporation of accurate externality costs in utility resource selections.

The barriers to adoption of environmental costs are considerable. Adoption of least cost planning is required and rate consideration for utilities that consider environmental costs. Important rate impacts and equity issues must be resolved.

That nore than half of the utility commissions have started to consider environmental externality costs in encouraging, although only a handful (nine) have actually attempted to quantify those costs and use quantified environmental costs in utility resource selection procedures.

The studies reviewed in Pace 1990 indicate that many utilities are reluctant to incorporate environmental externality costs. Progress has been most pronounced when commissions have required quantification of environmental cost and when rate of return consideration is given for investment in less polluting resources.

Despite the reluctance of many utilities, the electric utility industry (including independent power producers and demand-side energy service companies) is pioneering in consideration of environmental costs in resource selection. The results for improvement of the environment could be of great significance.

References and Footnotes

1 This paper is based on a study performed by the Pace University Center for Environmental Legal Studies (Center) for the New York Energy Research and Development Authority (NYSERDA) and the U.S. Department of Energy (DOE), *Environmental Costs of Electricity*, Oceana Publications, Inc., Dobbs Ferry, N.Y., September, 1990, also distributed by the American Council for an Energy-Efficient Economy, Washington, D.C. [hereafter referred to as "Pace 1990"]. The study was performed by Center staff, Pace University Law School students, and consultants, Shepard C. Buchanan of Bonneville Power Administration, Paul Chernick and Emily Caverhill of PLC, Inc., and Alan Krupnick of Resources for the Future.

2 "Money from Greenery." *The Economist*, October 21, 1989, pp. 16, 17.

3 Electric utilities contributed in 1985 68 % of national SO_2 emissions (16,204,000 tons of 23,699,000 tons nationally), and 33 % of national NO_x emissions (6,989,000 tons of 21,054,000 tons nationally). National Acid Precipitation Assessment Program (NAPAP), Interim Assessment, Vol. I (1987).
U.S. electric utilities accounted for 33 % of U.S. CO_2 emissions in 1988. CO_2 accounted for 66 % of the global warming from 1880 to 1980. U.S. E.P.A., Office of Policy, Planning and Evaluation, *Policy Options for Stabilizing Global Change*, Draft Report to Congress, February, 1989 [EPA Report]. It has been predicted that CO_2 will account for 80 % of future warming because of its long duration in the atmosphere. Lashof and Ahuja, *Relative Contributions of Greenhouse Gas Emissions to Global Warming*, 344 Nature 529, April 5, 1990. The remaining greenhouse gases, CH_4, CFC 11 12, N_2O and others, contributed the balance of global warming. *EPA Report, id.*

[4] The political nature of standards is well demonstrated by the political compromises now being made in the U.S. Congress over the extent of additional air pollution regulation in the Clean Air Act.

[5] Bland, Paul F. 1986. "Problems of Price and Transportation: Two Proposals to Encourage Competition from Alternative Energy Resources." *Harv. L. Rev.* 10:2.

[6] Another way to place pollution values in the marketplace is by setting stricter standards and allocating emission trading rights to facilitate pollution reduction at least cost. This has been proposed legislatively for the first time on a large scale in the Administration's submission for Clean Air Act amendments now pending befor Congress; these provisions have recently been adopted by both the House of Representatives and Senate.

[7] See. for example, the New York Environmental Quality Review Act (SEQRA), New York Environmental Conservation Law, Section 8-0107 and its implementing regulation, 6 NYCRR, Part 617.9(c) (3), which require all state agencies to assess and minimize or avoid environmental impacts "to the maximum extent practicable." As noted above, however, these regulations do not eliminate substantial environmental impacts and costs to society.

[8] Tighter national controls are already in prospect, such as the revision being considered in Congress of the law which governs disposal of solid wastes. Resource Conservation and Recovery A. 42 U.S.C. 6901 et. seq. International negotiations for accords that would affect electric utilities are also in progress such as those before the Intergovernmental Panel on Climate Change, Working Group III, considering a 20 % reduction in carbon dioxide by 2020 and 50 % thereafter. If and when these national laws and treaties are enacted, however, there will still be large residual environmental impacts and damages requiring valuation.

[9] Many utility regulatory commissions still do not require least cost planning or rule in advance on the prudence of utility investments. However, utilities in these jurisdictions still are subject to prudency reviews after they have invested, in their next rate-setting proceedings.

[10] The Pace 1990 study was restricted to the costs of damages to human health and the environment (generally excluding control and mitigation costs) of the operations of electric service facilities through waste disposal (excluding all front end costs such as mining, fuel processing, equipment manufacturing and transportation of fuels, equipment and materials to the generation site.

[11] Sulfates (SO_4), which are very small particulates, are believed to contribute significantly to particulate-related health effects. Sulfates can be emitted from a power plant either as particulates or as SO_2, which is then converted to SO_4 in the atmosphere.

[12] This would not be true for coal-fired power plants built after the most recent New Source Performance Standards, however. These plants are required to have scrubbers regardless of sulphur content of fuels. This may change, however, in the Clean Air Act Amendments of 1990, presently being considered by Congress, which may modify the requirement for scrubbers on some newly constructed plants. A cost figure for a NSPS-controlled plant is provided in Table A.

[13] Orders: AZ, CA, ID, KA, MA, MICH, MN, NJ, NY, OH, OR,
TX, VA, VT, WI
Statutes: ALSK, NV
Pending: CT, IA, MN
Consideration: DC*, HW, ME, MD, MONT, RI, UT
No Consideration: ALA, ARK, DL, FL, GA, IL, INMD, KY, LA, MISS, MO,
NB, NH, NM, MD, NC, OK SC, SD, TN, WA, WVA, WYO

* The District of Columbia (DC) is considered as a state for the purposes of this analysis

Pace 1990, Chapter X discusses these incorporation issues and Chapter X, Section G, summarizes state-by-state, with citations, the treatment of environmental externalities. At the end of this section, Table G shows the status of state actions incorporating environmental externality costs and Table H shows the type of action taken by states that have acted to incorporate environmental externality costs.

[14] Orders: CA, CO, NJ, NY, OR, WI, VT
Pending: CT, MA, CA (limited order; more extensive order pending)
Consideration: DC, MD, MI

[15] The New York PSC was recently upheld in providing utility rate incentives for engaging DSM programs. *Multiple Intervenors v. New York State Public Service Commission, No. 249-90 (N.Y. Sup. Ct.*, June 7, 1990. The Pace Center for Environmental Legal Studies filed a brief in the case in support of the PSC.

[16] *Investigation by the Department of Public Utilities on Its Own Motion Into Proposed Rules to Implement Integrated Resource Management Practices for Electric Companies in the*

Commonwealth, Mass. D.U. No. 89-239, Comments by Massachusetts Electric Co. (a NEES subsidiary) at pp. 17-21 and Attachment 2, February 23, 1990.

[17] These tables are based largely on a 50-state survey of state regulatory commissions in November, 1989, recently updated, on state treatment of environmental externalities. The survey was done for the National Association of Regulatory Utility Commissioners (NARUC) by Lawrence Berkeley Laboratory (LBL), to be published shortly. Cohen, S.D., J.H. Eto, J. Beldock, and G. Crandall. *A Survey of State Regulatory Commission Activities to Incorporate Environmental Externalities into Electric Utility Planning and Regulation*, draft of May, 1990. For the other sources, *see* Pace 1990, the State-by-State Survey Sources at the end of Chapter X, Section G.

[18] Commission Orders: ID, KA, NJ, NY, OK, WI
Statutes: CT, MONT, WA

The Oklahoma commission gives an incentive rate of return on environmental grounds solely to resource recovery plants.

[19] The Oregon Commission, however, has ordered its utilities to seek to quantify damage costs in evaluating resource selections, *Re Least Cost Planning*, UM 180, OR PSC Order No. 89-507 (April 20, 1989), and the New York Commission has ordered a pooled study by its utilities of their environmental externality damage costs, NY PSC Case 28223, *Electric Utility Conservation Programs*, Opinion and Order 89-15 (May 23, 1989).

7.2 A Comparison Of Methods Used By Regulators To Account For External Costs

Stephen H. Burrington
Conservation Law Foundation
3 Joy Street
Boston, Massachusetts 02108 - 1497, U.S.A.

Introduction

Regulators who have embraced the goal of accounting for environmental externalities in utility resource planning have found themselves confronted with a task of daunting complexity. A multitude of environmental impacts require evaluation. Questions about causation abound. Many impacts cannot be considered on a generic basis, and others may be truly unique to a particular generating facility.
Assigning values to non-market environmental goods presents perhaps the most difficult task to all, and one that some regulators may fear to be controversial as well as difficult.

The difficulty of accounting for externalities initially prompted regulators and advocates to look for simple solutions and to grasp at precedents. Crude regulatory tools were fashioned, and were then sometimes proposed by others for uses for which they were not intended. Utilities, showing a distinct lack of enthusiasm for the task of quantifying externalities, generally declined to come forward with methodological solutions or new data. Until the end of the 1980s, a well conceived initiative seemed at risk of becoming mired in the difficulties of implementation.

As the following discussion shows, however, dramatic progress has been made. This paper first reviews the tree most important initial efforts to ensure that externalities would be reflected in electric utility resource planning[1] -- efforts that have often been misunderstood

[1] The regulatory actions which this paper does not examine consist principally of requirements -- such as those imposed in Arizona and Minnesota -- that utilities give "qualitative" consideration to externalities, and requirements -- such as that of Oregon -- that they quantify externalities in some unspecified manner (Cohen et al. 1990; Foley and Lee 1990; Ottinger et al. 1990). Experience to date strongly suggests that such requirements will have little effect, if any, on utility resource planning. Pending proceedings

and misused as precedents. The paper then discusses a regulation of the Massachusetts Department of Public Utilities that has opened a promising new chapter in the transition to social cost-based electricity resource planning.

The Northwest Power Planning Council: Quantification of Externalities Plus Ten Percent for Efficiency

The Northwest Power Planning Council -- a federal energy planning agency for four states, and the first entity to incorporate externalities into electricity resource planning in the United States -- has received considerable attention for the 10 percent adjustment to generation costs that it uses to favor conservation resources. The 10 percent "adder" is thought by many to reflect an assessment of the environmental impacts of electricity generation that do not flow from conservation. Yet that interpretation of the adjustment contains no more than a grain of truth.

The Pacific Northwest Electric Power Planning and Conservation Act, the federal statute establishing the Council and governing its activities, defines the "system Cost" to be used in cost-effectiveness calculations to include "quantifiable environmental costs and benefits.... directly attributable to [a] measure or resource" (U.S. 1982, sec. 3 (4) (B)). The Act then goes on to provide for the 10 percent adjustment by stating that the "incremental system cost" of a conservation measure or resource "shall not be treated as greater than that of" any nonconservation measure or resource unless the former is 110 percent of the latter (*id.*, sec. 3 (4) (D)). Congress thus did not intend the 10 percent adjustment to account for externalities. Rather, Congress required the Council to include the value of quantifiable externalities along with price in the "system cost" of each resource, and then to go on and give conservation a further 10 percent advantage.

In affording conservation this additional advantage, Congress appears to have had two purposes. First, it intended the 10 percent adjustment to "address [] the traditional

in which regulators are considering how to account for externalities, such as those underway in Vermont and California (*id.*), also are not discussed herein. While an interim order requiring use of an "adder" for externalities has been issued in Vermont (Vermont Public Service Board 1990) and has received attention from commentators (Cohen et al. 1990; Foley and Lee 1990; Ottinger et al. 1990), it is indeed just an interim order and is based only on a very limited record. A more extensive inquiry into externalities has recently been initiated in Vermont (Conservation Law Foundation 1990).

problems associated with conventional resources, specifically cost overruns which raise their costs well above original estimates, and the use of fuel which is subject to drastic price increases" (U.S. Congress 1980). Second, Congress apparently intended to provide some reflection of the environmental benefits of conservation that elude quantification (id.). The Council cited a similar mix of planning and environmental considerations when it recommended the continuation of the 10 percent adjustment in 1987, listing the following specific reasons: (1) non-power benefits of conservation -- reduced wood use, uncreased comfort, and environmental benefits; (2) load stability and predictability; (3) benefits to local economies in the Northwest; (4) flexibility to adapt to changing circumstances; (5) costs of conservation not borne by the system; (6) and potential increased protection of fish and wildlife (Northwest Power Planning Council 1987).

Quantification of the environmental costs of each resource option -- not a crude 10 percent "adder" -- has been the primary means of accounting for externalities under the Act. Quantification studies carried out pursuant to the Act have indeed provided an important data base for those in the United States who have sought to tally up the environmental costs of power. But in the context of pressure to adopt a simple, conservative adjustment and to defer quantification, the misconception that the 10 percent adjustment represents a value placed on overall environmental costs has become widespread.

Wisconsin: Externalities Analysis Plus a Non-combustion Credit

Another of the most widely-cited regulatory initiatives, the 15 percent "non-combustion credit" adopted by the Wisconsin Public Service Commission ("PSC") in 1989, relies on the 10 percent adjustment used by the Northwest Power Planning Council. The PSC requires the cost of all non-combustion options to be reduced by 15 percent to account for avoided environmental impacts, principally ones resulting from air emissions, that are associated with combustion (Wisconsin Public Service Commission 1989). In adopting the non-combustion credit, the PSC wrote:

> The 15 percent level for the non-combustion credit was based on a review of the Northwest Power Planning Council (NWPPC) policy, in which conservation options receive a ten percent credit. Additional factors not considered by the NWPPC, particularly concerns over global climate change and human health effects of fine particulate emissions, suggest that the level should be increased. The commission finds that fixing the credit at 15 percent, and applying the credit to the cost of any option that does not involve combustion is appropriate.

(*Id.*, p. 12.) Unlike the 10 percent adjustment of the Northwest Power Planning Council, which favors only conservation resources, the 15 percent non-combustion credit favors all resources not requiring combustion -- including, presumably, nuclear power and hydropower. But like the 10 percent adjustment, the 15 percent non-combustion credit was intended to supplement, not substitute for, an evaluation of environmental externalities. Applying the credit, wrote the PSC, "does not in any way reflect *all* of the [external costs] of options, and does not supplant a full analysis of" those costs (*id.*).[2]

New York: First Quantification of the Costs of Power by a State Public Utility Commission

In 1989, the New York Public Service Commission took the important step of specifying, on the basis of data for the major externalities of coal-fired generation, how externalities of the same type should be valued in relation to one another and in relation to the price of generation. Relying primarily on the costs of pollution control as measures of the value of environmental damage, the PSC set the following values for externalities of coal-fired generation:

Externality	c/kWh
Air emissions	
Sulfur oxides	0.250
Nitrogen oxides	0.550
Carbon dioxide	0.100
Particulates	0.005
Water discharges	0.100
Land use impacts	0.400
Total	1.405

(New York State Public Service Commission 1989a and 1989 b). For at least two utilities, the PSC staff found that the total value for externalities amounted to 24 percent of expected generation costs (*id.*; New York State Department of Public Service 1989). The

[2] Wisconsin utilities also carry out resource planning subject to a statutory mandate to achieve specific reductions in emissions of acid rain precursors (Wisconsin 1986) and subject to a regulatory mandate to anticipate specific required reductions in carbon dioxide emissions (Brick 1989, p. 3; S. D.. Cohen et al. 1990, p. 29).

PSC adopted the above value as maximum ones, and subsequently established scales to be used in calculating values for specific resource options with lower pollutant levels than the coal-fired plant whose pollutant levels had been used to set the maximum values (Putta 1990).

The PSC's initiative represented a dramatic step forward. It recognized that environmental impacts within the same state or region should be valued consistently across utilities. Since data about impacts in a particular area virtually never exist, the risk that the PSC has prevented utilities from accounting for local variations in impacts is almost nonexistent and certainly far less significant than the risk that utilities in areas with similar geographic and demographic characteristics would, if left to their own devices, inappropriately assign different values to identical impacts. Also, setting values through a regulatory process that is open to agencies and organizations with relevant expertise has proven to be by far the most successful course of action. In both New York and Massachusetts, a public exchange of information and ideas has produced good or even exemplary records on how externalities should be valued, while utilities in Wisconsin and Massachusetts that had been ordered to consider or quantify externalities either failed to comply or made it impossible for the public or regulators to review what they had done. Also, given utilities' limited experience in this area and the lack of a clear consensus on methodologies, the setting of values by the PSC has the advantage of substantially reducing the uncertainty utilities face about how their methods and calculations, and resource choices made on the basis of them, will fare in PSC proceedings.

The PSC plans to modify its approach as its staff and the utilities learn from experience. Certain key issues deserve the attention of the PSC and of other regulatory agencies who follow in its footsteps. First the PSC seems not to have protected the relation between price and external costs from being distorted. The PSC has authorized utilities evaluating bids to use total externality values in either of two ways: by adjusting bid prices, or by translating the externality values into point scores to be added to scores for price and other factors (Putta 1990). In the latter case, it is unclear how the PSC ensures that utilities preserve a balanced relationship between externality values and price. The real value of the same externality score could vary considerably depending upon what scale a utility used to assign points in calculating the price score. For example, one dollar value would implicitly be assigned to the externality score, and the externality score would have a certain effect on decisions among resources, if a utility assigned points for price using a scale ranging from 0 to 25 cents per kilowatt-hour. A different value would be implicitly assigned to the "same" externality score, and the score would have a different effect on the utility's decisions, if the utility magnified the differences between price scores by assigning the same number of points for price along a scale from 5 to 15 cents per kilowatt-hour.

A similar problem lies in the use of a scale for translating externalities into scores. Unless the upper bound for each externality scale has been set appropriately, and the rest of the scale calibrated accurately, the PSC's original quantification will not have the effect of imparting realistic values to the impacts of scored resource options. In sum, a system in which scores can be calculated separately for externalities and for price is inevitably fraught with the potential for imputing different values to what should be equivalent internal and external costs. Real-life costs borne by third parties could often turn out to be worth less than costs borne by the utility itself.

Second, the values the PSC initially adopted need to be reviewed, as the PSC no doubt intends they will be, in light of new and better data that have become available. Analyses completed for the Division of Energy Resources (Massachusetts 1990a and 1990b) and Boston Gas Company (Chernick and Caverhill 1989, 1990a and 1990b) in Massachusetts, for the Conservation Law Foundation in Vermont (1990), and for the Pace University Center for Environmental Legal Studies in New York (Ottinger et al. 1990), all draw on much more extensive data bases and adopt considerably higher values than the PSC did. For example, the Pace study provides a value of 4.5 cents per kilowatt-hour for emissions of sulfur dioxide, nitrogen oxides, carbon dioxide and particulate matter from a coal-fired plant meeting new source performance standards (Ottinger et al. 1990, p. 21). That value -- which approximates the values yielded by the other analyses -- is almost exactly five times as great as the value adopted by the PSC.

A methodological difference between the initial PSC analysis and all of the studies just mentioned may underlie the differences in resulting values. The PSC used average costs of control -- costs for a combination of advanced and low-cost control technology -- as measures of the value society attaches to avoiding environmental impacts (Putta 1990, p. 44). All of the other studies relied on marginal costs of control as measures of the same value. If average rather than marginal costs are used, resources that would enable the more expensive, marginal-cost measure to be avoided -- such as conservation or low-pollutant generation -- could be improperly not selected (Chernick 1990). Although a discussion of the difference between the two approaches lies beyond the scope of this paper, the more general use of marginal costs -- which seems to be the better-justified approach -- certainly warrants a re-examination by the PSC of its initial approach.

Massachusetts: A Watershed in the Regulatory Treatment of Externalities

In August 1990, the Massachusetts Department of Public Utilities ("DPU") issued a regulation requiring electric utilities to calculate monetary values for the externalities of resource options, and to add those values to the costs of the options to determine the total social costs of the options (Massachusetts Department of Public Utilities 1990, pp. 56-60, 72-75, 83). The DPU ordered the utilities to use the following values, in 1989 constant dollars, for the eight most significant air emissions:

Externality	$/ton	$/lb.
Nitrogen Oxides (NO_x)	6,500	3.25
Sulfur Oxides (SO_x)	1,500	0.75
Volatile Organic Compounds	5,300	2.65
Total Suspended Particulates	4,000	2.00
Carbon Monoxide (CO)		
Ambient Air Quality	820	0.41
Greenhouse	50	0.02
Total	870	0.43
Carbon Dioxide	22	0.011
Methane (CH_4)	220	0.11
Nitrous Oxide (N_20)	3,960	1.98

(*Id.*, pp 70-72, 83, 85.) These values are to be used "unless it can be demonstrated in subsequent proceedings that other values... are more reasonable" (*id.*, p. 72) -- something that is unlikely to happen, except to a very limited extent, given the extensive body of data on which the values are based (Massachusetts Division of Energy Resources 1990a and 1990b; Chernik and Caverhill 1989, 1990a and 1990b) and the broad support they enjoy among the parties who commented on them. The DPU directed the utilities to propose externality values for nuclear power, hydropower, waste-to-energy facilities and load management (*id.*, p. 59), and "to use [marginal] control-cost estimates as a proxy for environmental damages in the absence of comprehensive damage cost estimates" (*id.*, p. 70).

The DPU found that, as of the time of its order, the inclusion of externalities associated with fuel extraction, transportation and preparation would be too "burdensome and complicated" to be required:

> [I] deally, all environmental externalities associated with energy production and use should be incorporated in the resource evaluation process. It would be difficult, however,

> at least at the present, for the Department to estimate externalities associated with early stages of the fuel cycle... To complicate the problem, externalities associated with early stages of the fuel cycle for each fuel type have varying degrees of site-specific attributes....
>
> ....A priority should be placed on estimating environmental externalities that are the direct result of power-plant operation including all downstream effects (*e.g.*, solid-waste disposal, waste-water disposal). The Department directs electric companies to consider... all impacts resulting from plant operation including [externalities associated with risk of accidents,] air, water, solid waste and spent fuel disposal impacts, and resource use.

(*Id.*, pp. 78-79) The DPU also declined to consider site-specific externalities (*id.*, p. 77). It found insufficient information to require inclusion of economic and social externalities such as government subsidies in the form of research and development, military support and clean-up support, but permitted in inclusion of genuine economic and social externalities such as these -- as opposed to transfer payments -- whenever their existence and level could be determined (*id.*, p. 80).

The DPU considered in some detail the arguments for and against expressing externalities as monetary values and using marginal costs of control as measures of those values. The alternative to using monetary values that had been presented to the DPU consisted of a "weighting and ranking" scheme developed for Massachusetts Electric Company, the state's largest electric utility, and similar to schemes used by some utilities in New York (Putta 1990, pp. 46-47). Noting the objections to that scheme -- that, for example, by assigning externalities a fixed weight, the scheme imposed a cap on the influence externalities could have in relation to price -- the DPU indicated that weighting and ranking schemes could be developed that would overcome at least some of those objections (*id.*, pp. 57-58). The DPU concluded, however, that an effective weighting and ranking approach would require quantification of impacts and their values. In that event, "forming weights becomes unnecessary because the quantified externality values could be monetized and added directly to project costs" to determine which resources minimized the total cost to society (*id.*, pp. 58-59).[3]

If anything, this view of weighting and ranking underestimates its potential to introduce inaccuracy and to result in the over-weighting or under-weighting of externalities in relation

---

[3] Acknowledging that "monetizing externality values does not constitute the elimination of subjective judgments in the [ir] evaluation," the DPU stated its expectation that utilities and other parties proposing values would "reveal all assumptions and judgments so that their merit can be discussed" in a public forum (*id.*, p. 60).

to one another and in relation to price. On the other hand, reducing externalities and price to a common denominator, the dollar, ensures that different external costs, and external costs and internal costs, will be given balanced treatment.

While ensuring that disproportionately high or low values are not assigned to some externalities, or to externalities in general, is a critical consideration favoring the use of dollar values, it is not the only one. To begin with, of course, if "total social cost" is the relevant "cost" in cost-effectiveness calculations, external as well as internal costs should be expressed so that a single "total social cost" figure can be derived. Many studies about health effects, resource damage and so on express results as dollars, which provide the only such widely-used, interdisciplinary common denominator. Assigning a dollar value should facilitate the integration of new data as they become available and the reduction of uncertainty surrounding the initial value.

In addition, valuing externalities in dollars ensures that changes in a project that will lessen its environmental impact will have an effect on the project's score that is certain and can be readily apprehended by the utility and project developer. Such changes will be reflected in a project's competitive position in just the same way that an equivalent reduction in the price of the project would be reflected. Similarly, the calculation of dollar values will facilitate better-focused, more intelligent public review of utility decisions.[4] The calculation of the dollar cost per kilowatt-hour of the effects of certain emissions can be described, and a criticism of a particular step in the calculation made, more quickly and straightforwardly than the attribution of a certain number of points to the effects of the same emissions can be explained, justified and criticized. And a critic could in any event determine the dollar value implicitly assigned to the emissions through a weighting and ranking scheme.[5]

---

[4] This proved to be the case in proceedings leading up to the issuance of the DPU regulation, in which dollar values presented by the Massachusetts Division of Energy Resources (1990a and 1990b) and the Boston Gas Company (Chernik and Caverhill 1989, 1990a and 1990b), and the calculations underlying those values, were thoughtfully criticized and compared by several parties. The weighting and ranking scheme presented by Massachusetts Electric did not lend itself to a similarly thorough review, although the criticism it attracted on that and other grounds indicated that its use would prove to be controversial.

[5] In that event, any of our consequences could ensue. First, the utility, taken by surprise, could conclude that the weight was indeed inappropriate and make an adjustment. Second, already knowing the dollar value implicit in the weight, and concurring in the critic's derivation of the value, the utility and the critic could proceed with the same discussion of its accuracy that would have occurred if the weight had been expressed as a dollar value in the first place. Third, the utility could dispute the critic's calculation of the implicit dollar

In concluding that the marginal cost of control constitutes "the best available proxy at this time" for the value society attributes to uncontrolled impacts (*id.*, p. 70), the DPU rejected an argument that marginal control costs overstate that value (*id.*, pp. 63-70). Society, went the argument, "has already mandated restrictions on emissions such that the point has been achieved where the marginal cost of controlling emissions equals the marginal value of potential environmental benefits", and "the marginal cost of abating any additional pollution beyond that required by government regulation must be greater than the marginal benefit society receives from such abatement" (*id.*, pp. 63, 65). This argument, responded the DPU, misses the point.

> [The] argument implies that pollution emitted by resources meeting government-mandated emission limits cause no loss to net societal values and, therefore, need not be taken into account in the comparison between alternative energy resources. Clearly, this is not true... [I]f society were presented with the option to purchase either a relatively dirty or a clean resource (assuming that both resources comply with government-mandated emission limits), and these resources were alike in all other respects, society would prefer the cleaner resource to avoid costly incremental environmental damages. If we were to accept [the] argument to its logical conclusion, we would be unable to compare the differing levels of environmental externalities caused by different energy resources that comply with government emission standards.

(*Id.*, pp. 65-66.) The DPU in any event found it unlikely that the cost of mandated pollution controls actually reflects the full extent of society's current willingness to pay for avoiding environmental damage (*id.*, pp. 66-67).[6] The DPU concluded that while control costs

---

value, and the debate could focus on the question of what dollar value the weight actually implied. Finally, the utility could respond by saying that the choice af any weight or value involved the use of discretion or "judgment", and contend, in effect, that one weight or value was as valid as another -- thus declining to present a reasoned explanation of the value assigned to the externality in question. The DPU fortunately avoided all of these possibilities and set the stage for more effective review of utilities' resource choices.

[6] Control costs indeed do not represent expenditures at the full level the public is willing to reach to avoid environmental damage. First, the political process tends to depress the level of costs that are imposed (*id.*, p. 67). Environmental controls generally shift costs from a diffuse public experiencing them as externalities to a smaller, better-defined group of actors who find it in their interest to organize and defeat, delay or minimize the imposition of costs. Second, controls do not even purport to avoid all forms of environmental damage. For example, air quality standards have in the past generally been set at levels deemed necessary to protect human health but not crops, building materials or ecosystems. Third, the marginal cost of control represents only the level that the public has indicated it is willing to pay -- the level of expenditure that some sources have been required to spend on

cannot be known with certainty either to underestimate or to overestimate the value of externalities, they probably underestimate that value -- particularly "[g]iven the ongoing movement to make environmental standards more stringent rather than less stringent, and... the increasing sensitivity of the public to environmental issues" (*id.*, p. 70).

The Massachusetts order leaves utilities and other parties to DPU proceedings with some difficult valuation tasks ahead. At the very least, accident risks, water discharges, solid waste and spent fuel disposal impacts, and resource use associated with plant operation all require evaluation (Massachusetts Department of Public utilities 1990). Yet the proceedings in Massachusetts during 1989 and 1990 give cause for optimism: parties to the proceedings introduced a considerable volume of well-organized data, and that data provided a basis for a surprisingly extensive consensus. The setting of values for the eight air emissions enables utilities to rely on those values and to focus on valuing other externalities -- a task made vastly more manageable by the DPU's acceptance of marginal cost-of-control data. Finally, the DPU's mandate to express externalities in dollars means that valuable time and energy will not be wasted on determining whether some dollars have become worth more than others through the design of a weighting and ranking scheme.

Conclusion

As others have noted (Ottinger et al. 1990, pp. 15-16), the likelihood that federal, state and local governments will impose increasingly more stringent environmental controls, but in a manner that will remain difficult to predict, makes accounting for externalities an element of sound utility planning. Keeping a step ahead of future regulatory developments should reduce the level of uncertainty that surrounds utility resource plans, providing protection against radical adjustments of the sort that some midwestern utilities could find to be necessary as a result of the 1990 amendments to the clean Air Act. Accounting for externalities should also prepare the way for a consensus about the resource utilities should acquire, and reduce the frequency and bitterness of the battles that are fought over new resources.
If that is the case, 1990 -- a year in which important efforts to compile and analyze the data were completed and a path-breaking regulation applying the results of those efforts was

---

control, not the level that all sources have been required to spend on control, not the level that all sources have reached. Also, observed costs of control are the costs only of available technology and not of all technologies that may be required under existing laws.

issued -- should be remembered as a year in which the work of utility planners and regulators started to become a bit easier.

References

Brick, S. 1989. Strategies and Methods for Incorporating Environmental Concerns into Utility Planning and Utility Regulatory Decision-Making. Middleton, Wisconsin.

Chernik, P. 1990. A Brief Introduction to the Estimation of the Value of Environmental Externalities. Paper distributed at a technical Public Utilities, Boston, Massachusetts.

Chernik, P., and Caverhill, E. 1989. The Valuation of Externalities from Energy Production, Delivery, and Use: Fall, 1989 Update. Report filed by Boston Gas Company on January 5, 1990, with the Massachusetts Department of Public Utilities in Docket no. 89-239.

---. 1990a. Report to the Boston Gas Company on Including Externalities in 89-239. Report filed by Boston Gas Company on April 12, 1990, with the Massachusetts Department of Public Utilities in Docket no. 89-239.

---. 1990b. Comparison of Total Costs of Supply Options. Memorandum to Greg Tomlinson filed by Boston Gas Company on May 7, 1990, with the Massachusetts Department of Public Utilities in Docket no. 89-239.

Cohen, S.; Eto, J.; Goldman, C.; Beldock, J.; and Crandall, G. 1990. What States Are Doing About Externalities. Electricity Journal. July 1990, pp. 24-35.

Conservation Law Foundation. 1990. Proposed Rulemaking: Additional Adders for Environmental Externalities Arising from Energy Services Provided by Regulated Electric and Natural Gas Utilities. Petition submitted on August 9, 1990, to the Vermont Public Service Board.

Foley, L.O., and Lee, A.D. 1990. Scratching the Surface of the New Planning: A Selective Look. Electricity Journal. July 1990, pp. 48-55.

Massachusetts Department of Public Utilities. 1990. Investigation by the Department of Public Utilities on Its Own Motion into Proposed Rules to Implement Integrated Resource Management Practise for Electric Companies in the Commonwealth, order dated August 31, 1990, and accompanying Rules Governing the Procedure by Which Additional Resources Are Planned, Solicited, and Procured by Investor-Owned Electric Companies, 220 Code of Massachusetts Regulations 10.00.

Massachusetts Division of Energy Resources. 1990a. Comments of the Division of Energy Resources on Environmental Externalities. Submission filed on March 2, 1990, with the Massachusetts Department of Public Utilities in Docket no. 89-239.

---. 1990b. Final Comments of the Division of Energy Resources. Submission filed on May 4, 1990, with the Massachusetts Department of Public Utilities in Docket no. 89-239.

New York State Department of Public Service. 1989. Consideration of Environmental Externalities in Competitive Bidding Program to Acquire Future Capacity Needs of Niagara Mohawk Power Corporation.

New York State Public Service Commission. 1989a. Order Issuing a Final Environmental Impact Statement and Adopting Staff's Response to Agency Comments, No. 88-E-241 (Mar. 24, 1989).

---. 1989b. Opinion and Order Establishing Guidelines for Bidding Programs, No. 88-E-241 (Apr. 13, 1989), pp. 45-50.

Northwest Power Planning Council. 1987. A Review of Conservation Costs and Benefits.

Ottinger, R. L.; Wooley, D. R.; Robinson, N. A.; Hodas, D. R.; and Babb, S. E. 1990 Environmental Costs of Electricity. Dobbs Ferry: Oceana Press.

Putta, S. 1990. Valuing Externalities in Bidding in New York. Electricity Journal. July 1990, pp. 42-47.

U.S. Congress, House Committee on Interstate and Foreign Commerce, Pacific Northwest Electric Power Planning and Conservation Act: Report to Accompany S. 885, 96th Cong., 2d sess., 1980, H. Rept. 976, p. 70, reprinted in 1980 U.S. Code Congressional and Administrative News, p. 6065 (statement of Rep. Williams).

U.S. 1982. Pacific Northwest Electric Power Planning and Conservation Act, sec. 3, U.S. Code title 16, sec. 839a.

Vermont Public Service Board. 1990. Board Decision Adopting (AS Modified) Hearing Officer's Report and Proposal for Decision in Investigation into Least-Cost Investments, Energy Efficiency, Conservation, and Management of Demand for Energy, Docket no. 5270.

Wisconsin, 1986. Wisconsin Statutes Annotated, secs. 15.347, 16.02, 144.385-144.387.

Wisconsin Public Service Commission. 1989. Findings of Fact, Conclusions of Law and Order, Docket No. 05-EP-5 (Apr. 6, 1989).

7.3 State Regulatory Responses To The Threat Of Global Warming

R. C. Cavanagh
Natural Resources Defense Council
90 New Montgomery
San Francisco, CA 94105

In this 1988 Presidential campaign, George Bush promised to counter the greenhouse effect with a "White House effect" initiated by his new Administration. Nothing discernable as such has emerged in the ensuing two years, however; those concerned about greenhouse gas emissions have had much greater success in eliciting a "Statehouse Effect".

On both symbolic and practical levels, such actions are urgently needed. Treated as nations, the fifty U. S. states account for ten of the world's top thirty emitters of carbon dioxide. When the focus shifts to emissions per capita, thirty-seven of the top fifty entries - and four of the top five - are U. S. states and territories.

States also are amply endowed with legislative and regulatory leverage for improving this record. Slightly over half of U. S. fossil-fuel-derived carbon dioxide emissions reflect the consumption of electricity and natural gas, most of which is sold by state-regulated utilities. Of the remaining emissions, more than three fifths originate in a transportations sector whose energy-efficiency is now receiving heightened statehouse scrutiny.

U. S. leadership at *some* level is urgently needed. As of 1990, no energy technology's costs reflected any of the potential damage associated with carbon-dioxide-induced global climate change. The United States tolerates no comparably flagrant and potentially devastating failure to internalize environmental costs. Despite two years of sustained public and media attention, national political leaders have been unable to fashion a response.

At the root of the problem is an apparent inability, at the highest levels of government, to associate lower fossil fuel consumption with global economic or environmental advantages. The President has repeatedly equated such reductions with economic hardship, and not even a new Middle East crisis has convinced the Congress to approve more stringent fuel economy standards for a transportation sector that consumes more than 60 percent of U. S. oil.

There is no want of support for the rudimentary point that the U. S. retains large untapped opportunities to reduce both energy use and carbon dioxide emissions, while at the same time yielding substantial net economic benefits. In principle, it should not matter that U. S. officials cling to a "no regrets" climate strategy, which rejects all protective measures lacking independent economic or environmental justification; the store of "no regrets" opportunities is ample. However, most of the relevant federal policy options have been languishing for at least a decade. There is little institutional memory of federal success in reducing energy needs, and even less surviving institutional capability to deliver such reductions.

Under the circumstances, it is hardly surprising that the Administration's understanding of energy efficiency options sometimes seems to begin and end with punitive taxation. Asked to evaluate the implications of a mere 20 % cut in carbon dioxide emissions, White House computer models assume that the economy already has exhausted its cost-effective efficiency options in response to today's energy price signals; what inevitably emerges from the calculation are large doses of simulated economic anguish.

Many of the states know better. California affords an instructive case study; from 1973-1985, the amount of energy required to produce a dollar of California goods and services dropped by 35 percent. This rate of improvement substantially outstripped that for the nation as a whole and saved Californians $ 23 *billion*. In contrast with many other states, California has also succeeded in reducing its electrical intensity; per capita electricity consumption dropped by 0.3 percent between 1978 and 1988, while the rest of the U. S. recorded annual per capita *increases* averaging 1.6 percent. Important contributors to the California trends included utility-financed conservation incentives and statewide efficiency standards for appliances and buildings.

Californians recognize, however, that their achievements have only begun to tap the enormous potential for savings that make economic sense even at relatively low energy prices. For example, by majorities of 61-11 and 31-2, respectively, the two houses of the state legislature recently enacted a "Drive-Plus" initiative, which finances rebates for relatively efficient and non-polluting motor vehicles by taxing the poorer performers in the same vehicle classes. This sliding scale fee-rebate system addresses emissions of hydrocarbons, oxides of nitrogen, carbon monoxide, and particulates, in addition to carbon dioxide.

To the regret and mystification of its supporters, California's governor closed out his term in office by vetoing Drive-Plus, but the bill will be back in still stronger form early in 1991. Its market-based approach should avoid federal statutory limits on states' power to set fuel economy standards for automobiles and light trucks.

A related measure that did become law in 1990 commits California's utilities to "improve the environment and to encourage the diversity of energy sources through improvements in energy efficiency and development of renewable energy resources, such as wind, solar, and geothermal energy". Specifically, utilities should "seek to exploit all practicable and cost-effective [energy] conservation and improvements in the efficiency of energy use and distribution" that can match the reliability of energy production alternatives. Moreover, "[i] n calculating the cost effectiveness of energy resources, including conservation and load management options, [state regulators] shall include a value for any costs and benefits to the environment".

These exhortations were backed recently with concrete reforms in the regulation of electric and gas utilities. In August 1990, California's Public Utilities Commission approved a 100 % increase in conservation budgets for investor-owned gas and electric utilities, while at the same time enacting major regulatory reforms that tie the utilities' annual profits in part to their success in delivering energy savings.

If the utilities do well, they stand to increase their profits for 1991 by as much as $ 60 million; poor performance could yield penalties exceeding $ 30 million. The state's largest utility, Pacific Gas & Electric, has responded by becoming the world's largest conservation provider, with a 1991 end-use efficiency budget of $ 150 million (the combined efficiency budgets for all California utilities will exceed $ 350 million in 1991). PG&E's ten year target is at least 2500 MW of efficiency improvements (out of a total of 3300 MW in projected capacity needs); the utility has pledged to avoid any additions of fossil-fueled generating capacity over that period.

In an addendum to the California Commission's order, its President acknowledged the broader significance of these actions:

We are formally reasserting today, at the timeliest possible moment, our determination to regain California's international leadership in energy efficiency. We take that step because energy efficiency remains our largest, cheapest and cleanest untapped energy resource - and this state and this nation have never needed that resource more .... [T]hese programs will enable utilities and consumers to begin to confront the environmental challenges which lie ahead.

Similar measures have proliferated across the United States over the past decade. At least seventeen states have adopted statewide "least-cost" conservation mandates comparable to California's; at least six have created analogous conservation-incentive regimes for their utilities, and many more are considering both measures.

In November of 1990, California's voters decided against taking the additional step of imposing specific goals for reducing current emissions of carbon dioxide (see Appendix). Such commitments are hardly new, of course; fifteen nations have already agreed to freeze or reduce emissions, and the roster includes Germany, Denmark, Italy, Japan, the Netherlands, and Sweden.

The California goal, which reached the ballot through petitions bearing close to one million signatures, was as or more ambitious than recent European enactments: it anticipated the population-adjusted equivalent of a 40 % reduction in total U. S. emissions by the year 2010, with a 20 % cut to be achieved by the year 2000. Although its proponents failed in their immediate objective, they produced technical and policy analyses that ultimately should help catalyze a shift in U. S. policy on greenhouse gas emissions.

That capacity reflects two key themes in the "statehouse effect" campaign, both of which should remain compelling despite the ballot measure's fate. Both are principally pocketbook issues rather than environmental appeals - and necessarily so, since no state acting alone can claim an appreciable impact on the composition of the global atmosphere.

Backers of the California goals argued first that well targeted carbon dioxide reductions offer enormous net economic benefits. One recent study, commissioned by proponents of the ballot measure, identifies a menu of options for improving energy efficiencies that would both achieve the reductions over twenty years and earn $ 27 billion in net present value *after* accounting for the cost of the measures. The analysis demonstrates the remarkably high technical capacity to reduce CO_2 emissions in even a relatively efficient state economy; on a per capita basis, California ranks among the five best U. S. states, yet the authors were able to find in existing technologies a set of reduction opportunities that would cut emissions for the year 2010 to less than half of current levels (despite projected population growth or more than 30 percent).

A crucial effort to transform the debate about climate defense is underway here; the hope is to recast CO_2 emissions as a sign of wasted funds and missed opportunities rather than a hallmark of economic progress. From the perspective of a state or nation's political and commercial leadership, relatively high carbon dioxide emissions should carry the same stigma - and perceived costs - as relatively high rates of illiteracy or infant mortality.

Increasingly competitive international markets will be increasingly intolerant of inefficient energy use.

A second argument for state action on CO_2 is more subtle but perhaps no less compelling in its appeal to local self-interest. It proceeds from the premise that no prudent planner counts on indefinite extensions of licences to pollute at will, particularly when an international outcry is mounting against the pollutant at issue. Pressures will mount steadily in the U. S. Congress for carbon dioxide fees or emissions limits, or for some combination. The first serious federal legislative efforts came in 1990, and more are certain to follow.

States already are beginning to hedge these risks by seeking to avoid energy-resource additions with a high level of exposure. New York State was the first to assign a cost to carbon dioxide emissions in its guidelines for evaluating utilities' resource options, and Massachusetts followed (in August of 1990) with a $ 22 per ton assessment. That represents roughly a twenty-fold increase over the New York value. California regulators are now considering staff proposals to raise this figure further to $ 26 per ton (by contrast, the cost of sequestering carbon through rural U. S. reforestation projects is estimated at roughly $ 15 per ton).

These are not fees actually charged to emitters; rather, they are imputed to prospective resources for purposes of cost-effectiveness evaluations. As a result, economic comparisons of alternatives incorporate an explicit allowance for relative carbon dioxide impacts, increasing the attractiveness of options that minimize or avoid these emissions. The Massachusetts and California adjustments certainly are large enough to affect decisions; for example, a typical new 500 Megawatt coal-fired plant emits about 4.3 million tons of carbon dioxide annually, which yields a $ 95 million/year imputed penalty at the Massachusetts rate.

These actions are designed to protect ratepayers against future legislation that will make the projected costs all too real. We may not get a carbon tax in 1990 or 1991, but even near-term inaction is far from certain, and a growing scientific consensus on the threat of global warming virtually ensures that these costs will be internalized long before new energy facilities approach retirement. These considerations led one major U. S. environmental organization, the Natural Resources Defense Council, to call recently for a 50-state moratorium on investment in new coal-fired power plants. At minimum, investors in such facilities can no longer ignore significant climate-linked financial risks associated with the technology.

Although state action against greenhouse gases can and will be justified solely in terms of narrow self-interest, most who act obviously are responding to a much richer array of motivations. The list of vulnerable resources that cry out for a "statehouse effect" extends far beyond energy users' pocketbooks. Advocates of state action understand this very well, as do many who are responding.

Until a national climate defense campaign is well underway, however, state action must serve economic and environmental masters simultaneously. From the perspective of national and international climate policy, nothing is gained now by equating greenhouse leadership at the sub-national level with deferred economic development or immediate political pain. States that lead must be perceived widely as states that are worth following. California and a number of other U. S. jurisdictions have met that test throughout the 1980s, and they are poised to accelerate their rates of progress. We may dare to hope that the White House effect will not lag far behind.

Appendix: Selected Provisions Of California's
Proposition 128 ("Big Green")
Part. 7 Greenhouse Gas Reduction Plan
44390. By January, 1, 1993, the Energy Resources

Conservation and Development Commission shall adopt and implement a plan to reduce annual emissions of any gases which may contribute, directly or indirectly, to global warming. The plan shall provide for the maximum feasible net effective reduction in the global warming potential of these gases. The plan shall also require a net reduction in carbon dioxide emissions of twenty percent (20 %) by January 1, 2000, measured from 1988 levels, and forty percent (40 %) by January 1, 2010. These percentages shall be adjusted, if necessary, by a correction factor which reflects any difference between the projected rate of population growth in California, and the projected rate for the United States.

For purposes of this Part, "net effective reduction in global warming potential" means a reduction, based on the best evidence available, of those air contaminants which contribute directly or indirectly to atmospheric warming, including but not limited to carbon dioxide, chlorofluorocarbons, halons, nitrous oxide and methane, weighted to reflect their respective contributions to global warming. Carbon dioxide emitted from the generation of electricity which is consumed within California shall be considered a part of California's emissions, regardless of where the electricity is generated.

44391. (a) The provisions of the plan specified by Section 44390 related to energy conservation and development and electrical power generation shall be prepared and implemented by the Commission. In preparing and implementing those provisions, the Commission shall consult with affected Districts to ensure that those provisions will not interfere with the attainment or maintenance of state or federal ambient air quality standards.

(b) The provisions of the plan related to emissions from vehicular sources and motor vehicle fuels shall be prepared and implemented by the Air Resources Board. That Board shall also prepare and implement provisions relating to any greenhouse gas emissions not specified by subdivision (a) or (c).

(c) The provisions of the plan related to stationary sources, indirect sources, area wide sources and transportation system use in a particular district may be prepared by the air pollution control district or air quality management district, at the district's option. If prepared by a district, the provisions shall be submitted to the Board which shall, after public hearing, approve or revise the submission. Districts shall adopt regulations to implement the measures in the plan relating to emissions from stationary sources, indirect sources, area wide sources and transportation system use, provided that districts may adopt and enforce regulations more effective in reducing emissions than measures contained in the plan.

(d) The plan provisions prepared by the Board and any district shall be submitted to the Commission by June 30, 1992.

44392. All state and local agencies shall adopt any necessary regulations to implement the plan prepared and adopted pursuant to Section 44390, and shall not take any action inconsistent with that Plan.

Selected References

1. D. Lashof & E. Washburn, *The Statehouse Effect: State Policies to Cool the Greenhouse* (Natural Resources Defence Council, Washington, DC: July (1990) (includes extensive tables on carbon dioxide emissions by states and nations, both aggregate and per capita).

2. C. Calwell *et al.*, *Clearing the Air: The Dollars and Sense of Proposition 128's Atmospheric Protection Provisions* (Natural Resources Defense Council, San Francisco,

CA: September 1990) (analysis of options for reducing California's net carbon dioxide emissions by 20-40 %).

3. California Senate Bill No. 1905, introduced February 6, 1990 ("Drive Plus" legislation establishing revenue-neutral taxes and rebates keyed to emissions of hydrocarbons, oxides of nitrogen, carbon monoxide, particulates and carbon dioxide from automobiles, light-duty trucks, and medium-duty trucks).

4. California Assembly Bill No. 3995, introduced March 2, 1990 (legislation establishing statewide goals for utility- financed conservation and requiring incorporation of environmental costs in cost-effectiveness evaluations of utilities' energy-resource options, including conservation).

5. California Public Utilities Commission, Decision No. 90-08-068 (August 29, 1990) (order approving expansion of California's utility-financed conservation programs for electricity and natural gas, and linking utilities' profits in part to the success of those programs).

6. Massachusetts Department of Public Utilities, Order No. D. P. U. 89-239 (August 31, 1990) (Massachusetts regulatory decision that utilities must, in evaluating the emissions of energy-resource options, assign a cost of $ 22 per ton of carbon dioxide).

7. California Energy Commission, Staff Issue Paper #3R, Docket No. 88-ER-8 (November 21, 1989) (assessment of sequestration costs for carbon dioxide).

8. California Energy Commission, Conservation Report (October 1988) (documentation of energy efficiency trends and cost savings in California).

9. D. Goldstein, *Conservation Policies in California: Past Trends and New Directions (Natural Resources Defense Council, San Francisco, CA: July 1990) (Electrification Trends for California and U. S.)*.

7.4 Regulator's Perspective On Environmental Externalities

Mary Lou Munts, Commissioner
Public Service Commission of Wisconsin
Madison, WI 53707

History of Wisconsin Planning Process

My comments come from the perspective of a regulator in a state where environmental concerns strike a deep chord among our citizens and where these concerns have been repeatedly reflected in legislative action and agency policy.

Wisconsin passed its own Wisconsin Environmental Policy Act in 1971, paralleling the National Environmental Policy Act, which established procedures for the agencies regarding environmental impact assessments and statements. Then the Legislature in 1975 addressed environmental concerns related to utilities more comprehensively in a Power Plant Siting Act.

This step initiated a two-step approval process for major electric facilities. It requires the major utilities to file long-range plans (called Advance Plans) with the Commission every two years. Each plan contains a forecast of need for electricity for the next 20 years, predictions of power plants where construction would start within the next 15 years and of transmission lines where construction would start within the next 10 years. Utility plans must be found to be in the public interest when considering health, safety and environmental issues before they can be approved. In order to receive a Certificate of Public Convenience and Necessity, power plants (over 12 MW) and transmission lines (over 100 kV and more than one mile in length) must have been approved in a prior Advance Plan and must be found to be in the public interest considering engineering, economic, environmental and other factors, and must be found not to have adverse impact on environmental values such as: ecological balance, public health and welfare, historic sites, aesthetics of land and water and recreational use. In addition, the proposed facility must not unreasonably interfere with the orderly land use and development plans for the affected area. (Wis. Stats. 196.491)

The Commission has deliberated over five Advance Plans since 1975. These planning proceedings have provided a useful forum for dealing with the changing issues in the electric industry. Forecasting and nuclear power versus coal power were the big issues of Advance Plan 1 in 1978, while externalities and the alternatives of coal, natural gas and demand-side and renewables promise to be the big issues of Advance Plans 6 in 1992. Many issues that were hotly contested in earlier plans, such as forecasting and transmission, have disappeared as the technical issues replace the policy issues.

Environmental issues have been aired in all these proceedings but only in the last plan did the Commission begin a more formal consideration of externalities in conjunction with the initiation of integrated resource planning. Nonetheless, the broad discussion of environmental externalities in earlier plans contributed to the Commission decision to pursue integrated resource planning and to the very rapid Commission-encouraged progress in demand-side planning in recent years.

Advance Plan 1 (1978) was notable for its intense debate on coal versus nuclear plants and on forecasting. Questions of fuel price and availability, and health and environmental effects of both fuel cycles were explored in exhibits and testimony. The Commission found that there is no environmentally benign way to generate electricity. In the only administrative ban by any state, the Commission ordered that nuclear power plants could not be part of future plans unless utilities could show reasonable progress on the issues of nuclear waste disposal, the cost of decommissioning nuclear power plants, and the cost and availability of uranium.

Although the Commission framed its decision in economic terms, there were underlying environmental concerns. Subsequent events proved the wisdom of this decision as Wisconsin avoided the misery many states experienced with nuclear albatrosses. This first plan showed the value of allowing a state-wide perspective in the examination of aggregate environmental as well as economic costs often overlooked when a single power plant is considered. The debate about forecasting methods, which was so intense at first, became noncontroversial later as the base of technical information expanded and unterstanding developed. Without the planning forum, these issues would have been fought repeatedly on a case-by-case basis or in prudence proceedings.

Advance Plan 2 (1980) concentrated on the costs and effects of coal-fired plants (which were a given as a result of the Commission's nuclear ban) and, for the first time, the Commission reviewed renewable technologies in depth. The Commission recognized in its findings that costs of air pollutants are borne by the public and are ignored in standard cost-benefit analysis, but it did not seriously address or implement a combustion minimization strategy.

The Commission did order the major utilities to provide $ 500,000 for a study of the extent and effects of acid rain in Wisconsin. This joint effort of the Commission, the utilities and the state's environmental agency eventually bore fruit in 1986 in the form of state acid rain legislation which will cut in half the state's utility SO_2 emissions by 1993. After much study and legislative effort, sophisticated system modelling helped break the logjam. The Commission acid rain cost study played a significant role in this success because it helped convince two of our major utilities that the additional costs were manageable and worth spending to put the controversy to rest.

The Commission's Second Advance Plan order has not led to significant increases in the use of renewable resources in the state. The Commission ordered more aggressive exploration of renewable energy by utilities (such as new coal fired power technologies capable of burning refuse derived fuel, and testing of wind power technology). However, 10 years later there has been only a modest increase, amounting to 55 megawatts produced with RDF and wood wastes. These technologies have their own substantial environmental pollution problems which the Commission is carefully monitoring. Renewable technologies have been repeatedly reviewed in subsequent plans but they have not been cost effective on a broad enough scale to significantly affect Advance Plans.

In the third plan in 1983, the Commission staff developed an alternative plan designed to permit the Commission to consider a significantly different energy future. The plan (the Alternate Electric Power Supply Study) found that existing power plants, renewable energy sources, and conservation could meet the need for electricity for the period from 1981 to 2000 without construction of additional power plants. This study was hotly contested by the utilities and the Commission compromised by ordering the utilities to develop alternate "parallel" plans along with their preferred plans. The Commission, in the next plan, determined that parallel planning was also unsatisfactory, leading it to order single-path integrated resource planning.

In the third Advance Plan order, the Commission also rejected the western Wisconsin utilities' transmission plans, citing economic and environmental concerns. This led to the development of a comprehensive, integrated approach to transmission planning, where environmental concerns are a major factor. This approach has substantially eliminated controversies about the need for facilities. This allows the commission to focus on addressing environmental controversies related to transmission routing that arise in siting cases. Based on information from environmental assessments, the Commission has chosen sites and line configurations providing environmental mitigation where it has concluded that the particular environmental benefits are worth additional expenditures.

Results of Early Plans

What was the result of review of environmental consideration in these early plans? The nuclear ban was the most significant change from the utility supply plans. Although the Commission rested its decision on cost grounds, the motivating force for intervenors and nuclear opponents was environmental. There was no significant consideration of environmental effects from the ensuing coal plants except for the efforts which susequently resulted in state acid rain legislation. Transmission planning was dramatically recast as a result of environmental challenges to the need for particular facilities.

On the demand-side, repeated efforts to quantify and incorporate greater conservation and renewable resources in the plans were largely unsuccessful in a direct and measurable sense. However, the frustrations caused by these efforts did lead to the conclusion by the Commission in Advance Plan 4 that there was a reasonable basis to believe that there was enough conservation potential for the utilities to double their previous efforts. Subsequent orders in annual rate case decisions began that acceleration which increased conservation spending from $ 20 to $ 135 million, about four percent of utility revenues.

Fortunately, the planning process is like a big lantern that sheds light in many directions and informs other aspects of Commission activity such as the ongoing development of demand-side programs in rate and construction cases. It is here that the Commission has expanded demand-side programs. The biggest leap forward was when our largest utility was ordered to undertake a $ 73 million demand-side program so as to postpone an expensive atmospheric fluidized bed life extension project. This decision was a rough justice response to a project that was not adequately considered in the planning process, indicating that it is often easier to make a bold ad hoc decision where there has been a utility default in planning.

Similarly, expansion of demand-side and load management programs has been further accelerated by an unforeseen rash of applications for combustion turbines. Some utilities brought in expanded programs with these applications while others will be required to expand their programs as a condition of approval of the combustion turbine filings.

It is the vision of the planning process that has assisted the implementation of conservation in rate and construction cases rather than the implementation of explicit plans. That is why the unsuccessful efforts to develop better explicit plans through the alternate scenarios of either the staff alternative plan or utility parallel plans led the Commission to order integrated resource planning for Advance Plan 5. In theory, the use of this method provides a more systematic way to use externalities to weigh demand-side and renewable resources versus traditional supply-side measures. In practice, this will take time. It is still in the controversial stage of earlier plan issues like forecasting and transmission planning when technical and policy issues had not yet been disentangled.

Formal Consideration of Externalities

The plans submitted for Advance Plan 5 were required to include an analysis of externalities (given the acronym of NEEDS - not easily expressible in dollars). These issues, following our planning statute, include but are not limited to: health, safety, reliability, environmental effects, use of nonrenewable resources, social issues (jobs, equity, land use), local and regional economic issues, political concerns, effects on planning, risk, and technical innovation. The NEEDS process received largely negative attention because the information presented by many of the utilities was not found particularly useful, nor did it have any apparent effect on utility plans.

There is still a great deal of refinement needed in reflecting the specific externalities that the Commission wants to be addressed. For example, our utilities argued strongly for economic development as a factor in our last proceeding. The Commission found that specific job impacts of proposed projects were a consideration but that broad economic development was rather an end result of an effective planning process.

For Advance Plan 6, utilities have been required to develop externalities information for each option and plan. Guidance to utilities on "how to do this" was intentionally vague to give them the opportunity to be creative in developing individual methods and approaches "to let a thousand flowers bloom". However, how consideration of externalities has affected decision-making must be clearly stated. A more refined statewide NEEDS process will probably evolve out of Advance Plan 6 based on evaluation of what methods prove must be useful to the Commission and on a review of how techniques have evolved elsewhere since mid-1988.

As commissioners we face a difficult problem as we carry out our regulatory responsibility. We must choose between various futures relying on partial information. We know that the various options available to us differ greatly in impact. How to balance these impacts is the basic unanswered question. Choosing our future based solely on the consideration of engineering costs is not adequate. Too much is ignored. We also know that quantifying every effect of each option is probably an impossible task. Somewhere, somehow, we have to devise an approach which is manageable and useful for the utilities, for the Commission and unterstandable by the public.

This will not be easy and from the standpoint of the regulator may have certain perils. Requiring the inclusion of NEEDS in the analysis by utility planners is an attempt to get utilities to consider these factors. This process will be reviewed by Commission staff. It is less clear whether it will ever be a useful tool for the Commission, itself. As a decision-maker, I am concerned about trade-offs between plans on a broader basis. I want these trade-offs to be made as explicit as possible, but I do not want one person's judgment, at an earlier stage, to substitute for my judgment on their relative social value.

It is important to note that the societal perspective of our statute may require looking at trade-offs among environmental and other factors. This is one reason for my skepticism about the use of composite numbers which weight the monetary value of externalities. Such numbers may mask decision trade-offs which should be the province of the Commission rather than the analyst.

Poor decisions may result from pretending to know in a rigorous way what you do not really know. It is also mind-boggling from a process standpoint. Our planning process is a contested case proceeding, relying on expert witnesses. The battle of witnesses for different interest groups on quantification of a wide variety of NEEDS factors, running from reliability to jobs to environmental effects, is not one I care to contemplate either in terms of the resources expended or the usefulness of the result.

Non-combustion Credit and Alternative Scenarios

The dissatisfaction the Commission felt about the usefulness of the NEEDS information in the Advance Plan 5 filings prompted us to look for a brute-force way of insuring that some measure of environmental externalities was reflected in the planning process. This was the origin of our selection of a 15 percent credit for noncombustion resources. This number was based on very broad estimates in the hearing record which placed noncombustion costs not already included in the planning process at 10 to 25 percent. What I was responding to as a regulator was the importance of giving adequate credit to demand-side measures and some renewable energy forms in the option selection process based on their benign qualities. Our choice of the point between 10 and 25 percent was essentially a political compromise as to a reasonable level of the otherwise ignored cost of combustion.

In our next Advance Plan, the 15 percent number will be reexamined. I would prefer to see our debates focussed on such credits, adders or other broad tools of this sort with recognition a number cannot nor should not be established with rigor. The Commission will need information on the range of costs of environmental damage, the costs of control measures and any other data you experts at this conference think will be useful for decision-makers. An analysis of the actual effect that the noncombustion credit has had on choices made in the planning process will be instructive as to what range of numbers make a real difference in results, since that was the intent.

Perhaps the most important kinds of help needed by commissioners are tools at the decision level which permit comparison of the proposed plans with plans which rely on significantly different strategies. Such alternate scenarios must provide unterstandable choices if they are not to be the useless exercise of the "parallel" plans earlier ordered by the Commission.

For our next plan we have asked for a sensitivity analysis which estimates the cost to utilities of a 20 percent reduction in carbon dioxide emissions by the year 2000. It was the cost study for acid rain that changed fear to fact and persuaded our two largest utilities to support the state acid rain legislation which now stands us in such good stead for implementation of the pending federal legislation. Similar cost studies on CO_2 give promise of helping make a difference in how we weigh to costs of change versus the risks of the status qou.

Political and Institutional Barriers

The integrated resource planning process gives promise of providing a better vehicle for looking at alternative scenarios and meaningful modifications to plans than the earlier failed efforts of the Alternative Electric Power Supply Study and "parallel" plans. Whether this hope is too optimistic will await Advance Plan 6. It would be unrealistic, however to pretend that planning happens in a vacuum. There are political and instituional constraints on commissions no matter how good the technical process. Would, for example, a commission implement a 50 percent noncombustion credit even if the record would support such a choice? Or, asked differently, would a commission modify long-range plans to substitute renewable for combustion resources if it knew that it might lead to higher electricity costs of 10 to 20 percent while combustion resources would raise costs 5 to 10 percent?

These questions illustrate the climate within which commissions and utilities have historically operated where concern about rates has been the driving force. Serious reordering of preferred utility choices based on a societal perspective needs considerable courage by commissioners and often needs outside political reenforcement. That is why evidence that suggests an international global warming agreement to reduce CO_2 is imminent will do more to drive serious consideration of noncombustion scenarios than the best of data.

In Wisconsin, the nuclear ban by the Commission was supported by highly effective political clamor. The Commission needed the legislature and environmental groups to take the leadership on acid rain controls, with the Commission's role being one of giving useful technical and policy assistance, because it did not have the power to mandate emission limits.

The statutory authority of the Wisconsin Commission is broader than most states, but its environmental powers primarily derive from planning authority which often produces different results from the specific standards of the state environmental enforcement agency.

Even with our broad statutory power, our last order is being challenged in court because the Commission ordered that planning for life extension of existing power plant be based on meeting New Source Performance Standards for nitrogen oxide and particulates. Our utilities argue that air emission regulation is the responsibility of the state's environmental agency while we argue, and expect to prevail, that our planning process both permits and requires us to anticipate controls that are likely to be required at the time a facility is constructed or soon thereafter.

There has historically been a healthy tension between our agency and the environmental agency because it is limited to enforcement of specific limits or standards while our planning power uses a long-range societal perspective. This led to such serious controversy at one point that the environmental agency succeeded in getting legislation passed which restricts the Commission from denying a certificate for air quality reasons if the environmental agency determines that the plant meets its air standards. Cooperative work on state acid rain legislation has helped ameliorate this rift. Both our agencies viewed this issue from a longer range perspective because the environmental agency accepted that current SO_2 controls were inadequate to address acid rain. The new area of important environmental concern is now CO_2 and there is every reason to believe that there will be agency and public support for Commission recognition of environmental costs of this unregulated pollutant.

Prognosis for the Future

The jury is till out on whether the methods being developed by our utilities will make a real difference to what actually happens in the selection of program options by utility decision-makers. It will also be instructive to understand what difference the 15 percent noncombustion credit made in option selection and what difference the NEEDS process made. This may tell us whether the NEEDS effort is worth the candle as well as whether the 15 percent adder had a significant effect.

If the evidence in the record indicates that the 15 percent credit has made little difference, the Commission will certainly want to consider a higher adder. I would hope, however, that other broad alternative approaches (not mutually exclusive) would be proposed and evaluated such as a required percentage or quota reduction in CO_2 or setasides of renewable generation. A change in system operation, such as the total cost dispatch approach of Stephen Bernow, is also an option to examine to achieve a desired level of pollution reduction.

Because new base load coal plants have been proposed in utility plans for the first time in ten years there will undoubtedly be much debate about the evaluation methods for the variety of coal and gas technologies which may be considered. Our Commission will need the best scientific and technical help it can get to weigh the competing supply side options for intermediate and base load supply side capacity, and their economic and environmental costs and other externalities. Evidence from other states will be helpful, such as the recent Massachusetts order which set 4.6 cents per kWh as the environmental costs of a coal plant meeting new source performance standards and 1.7 cents per kWh for a combined cycle gas turbine.

One question that will affect these decisions is whether the noncombustion credit should treat natural gas differently because ist produces less NO_x and about one-half the CO_2 emissions of coal and no SO_2 or particulants. A countervailing argument will be aired on whether a gas depletion adder should be considered. Such an adder was proposed in the recent Commission proceeding on proposed combustion turbines on the grounds that gas prices do not appropriately reflect the intergenerational equity issues associated with depletable resources. The Commission will consider this issue in Advance Plan 6.

The gas versus coal issue on the supply side may provide debate similar to the coal versus nuclear controversy in the first Advance Plan if concerns about global warming continue to escalate. This debate may be three-cornered, however, with proponents for demand-side and renewable resources arguing that neither gas nor coal resources need to be included in the active 15-year planning horizon.

The question of how much demand-side potential is actually achievable and cost effective has been continuously debated, and the staff review of plan adequacy will determine whether this will again be a major debate. The utilities have also been asked again to look seriously at renewable resources. Will recognition of appropriate environmental externalities, in conjunction with advancing technologies and increasing conventional costs, result in a much larger role for demand-side and renewable measures than in earlier plans?

The 15 percent noncombustion credit will already have been factored into utility plans presented and NEEDS information will be available in some form. It is too late at this date for decision-makers to change the amount of the credit or formality of the NEEDS information in the analysis presented. However, the information provided by the examination of the effect on plans of the 15 percent adder and that provided by the 20 percent CO_2 reduction scenario, plus additional information from the hearing record should at least give the Commission a good basis for reflecting newer data or methods so as to modify the utility proposed plans.

What this points out is how slowly a multi-year planning cycle adapts. In the real world there is a tension between the time it takes to refine and agree on a process, and the immediacy of the issues facing us. We need tools to help us make decisions today.

My comments are not intended to descourage participants in this conference who are devoting such serious professional efforts to work on environmental externalities. My intent is rather to encourage you to focus on the big questions which will help Commission decision-makers use a societal perspective in which environmental externalities receive ther appropriate weight. We need your help to look at whether we get the results intended, and to help us develop other broad options to assist our immediate decision-making and to develop an improved externalities process over time.

Based on our experience with controversial issues in earlier proceedings, it is most important to keep in mind our big picture goals and objectives. Second, better technical information must be pursued aggressively but with the view always to clarify and facilitate progress on policy questions. Lastly, it is important that policy-makers act even with incomplete information, recognizing that policy is set as much by inaction as action.

The difference between environmental externalities and many other controversial issues we have addressed is that it tests more severely the commitment and the will of commissioners to use a societal perspective in which environmental externalities receive their appropriate weight. This perspective requires a willingness to choose options in the face of risk and uncertainty which may cost more ratepayer dollars but which benefit society broadly. The role of experts is to clarify the level of risk and the nature of uncertainty and to provide understandable ways to weigh costs and benefits.

Commissions will make these decisions in a world where the debate about potential risks of global warming is growing rapidly. This debate has many parallels to the debate in our state about the costs and risks of nuclear power which the Wisconsin Commission handled so wisely and which so many commissions without planning processes handled so poorly. Fortunately, more states have joined our ranks and many of us will be struggling together to make critical decisions about how fast and at what cost utilities should reduce carbon emissions.

As regulators, we would do well to heed the caution of a previous chair of our Commission, Charles Cichetti, who said that "It is better to be roughly right than precisely wrong". That is why I ask your help in ways to help us be roughly right.

I would like to thank Stephen Bernow, Stephen Brick and Commission staff members Michael Arny, David Iliff, Michael John Jaeger, Barbara James, Sarah Jenkins, Paul Newman, Anita Sprenger, and Steven Schur for their critical review of this paper and their many suggestions for improvement.

7.5 California's Experience with Including Environmental Values in Electricity Planning

John A. Wilson
California Energy Commission
1516 Ninth Street
Sacramento, CA 95814-5512 USA
Telephone: 916-324-3308

Introduction

California has faced a number of environmental challenges over the years, but none has had as dramatic an affect as the concern over air quality and the drastic actions that will be required to attain healthy air. California, especially the South Coast region around Los Angeles, experiences some of the highest levels of air pollution in the U.S. Many areas of California are currently violating national and California ambient air quality standards, and increasing energy use will exacerbate the problem.

Electricity is both part of the problem and part of the solution. The South Cost Air Basin is the test bed for a starting plan to attempt to comply with air quality standards. Parts of the plan call for drastic reductions in electricity and gas use in homes and businesses, while at the same increasing energy use by electrifying large portions of the transportation system and industrial processes in the basin. At the same time, utility and independent energy producers will be required to reduce the emissions from their power plants. All this creates a complicated planning problem, that the Commission and other interested parties are still trying to sort out. This paper discusses how the Commission has incorporated environmental values in its electricity process, both past and present.

The Role of the California Energy Commission in Electricity Planning

Since 1975 the California Energy Commission has been responsible for long-range electricity planning in the state. The Commission was created in response to the perceived need for about 20 nuclear plants that utilities planned to build on the California coast or coal plants in the California desert because demand was growing at about 7 percent per year. Utilities wanted a "one-stop" siting agency that would issue permits to build the plants. Environmental advocates wanted to insure that alternatives to nuclear power were fully considered before the plants were built. What has happened over the last 15 years is that high energy prices, conservation programs, and alternative energy sources have eliminated the need for any additional nuclear power plants beyond the plants that were already under construction in 1975.

To carry out the Commission responsibilities it has a biennial process to adopt 20-year demand forecasts for each utility as well as assess to need for new generation resources. The assessment is based on several factors: maintaining economic growth and environmental quality, conserving resources, and maintaining public health and safety. The Commission is the permitting agency for thermal power plants over 50 MW. To complement these activities the Commission was also given the ability to affect the need for power plants by setting building and appliance efficiency standards, and affect the selection of future resources by encouraging research and development of new generating technologies.

The Commission has permitted 39 power plants totalling 4,115 MW. About 50 percent are gas-fired cogeneration, and about 30 percent are geothermal. The remainder are several solar-thermal facilities, two coal gasification demonstration projects, and one traditional oil-fired power plant. When the Commission conducts its environmental review of these projects, environmental impacts, or whether the impacts have been sufficiently mitigated that they are not a reason not to build the project and that there are no other alternatives. The Commission maintains a large staff of scientists and engineers to perform this analysis.

While plant-specific analyses have been performed in great detail, environmental analysis was not truly integrated in the long-range planning for energy use. Electricity supply planning focused more on issues of cost and fuel diversity, since they are more quantifiable measures that are easier to evaluate. Environmental values were treated subjectively in terms of "preferences" for conservation measures and alternative generation technologies. However, events of the last year regarding air quality have heightened the concern for including environmental effects explicitly in planning.

The Impact of Air Quality Regulations on Energy Use and Emissions

Air quality in Southern California is the worst in the U.S. About 90 percent of the state's population live in areas that do not meet federal or state ambient air quality standards at least some of the time, although not as often or as severely, as the Los Angeles air basin. Many believe that Northern California is only a few years away from facing the same kind of serious air quality problems as Southern California.

The Federal Clean Air Act mandated attainment of the national ambient air quality standards by 1988, which most California air districts did not meet. In 1988 California adopted its own California Clean Air Act to address California's unique air quality problems and to establish new procedures to attain ambient air quality standards. Beginning in 1988, the California Clean Air Act requires that local districts reduce emission of nonattainment pollutants by 5 percent per year in each air basin. Local air districts are required to develop new air quality attainment plans to meet the standards by December 1991. To date, South Coast Air Quality Management District (SCAQMD) is only district that has adopted standards that affect utility power plants, but others are expected to follow.

The SCAQMD adopted an Air Quality Management Plan (AQMP) as required by federal law to try to meet the federal requirements by the year 2010 (SCAQMD 1989). This plan has gotten a great deal of attention because of the dramatic effect it will have on people and businesses within the region.

Controlling power plant is an important part of the solution to California's air quality problems. NO_x (nitrogen oxides), SO_x (oxides of sulfur), and ROG (reactive organic gases) emissions are the major pollutants from fossil-fired plants, and they are precursors to the formation ozone. Ozone, NO_x, SO_x and ROG emissions also contribute to PM10 (particulate matter less than 10 microns) formation in the atmosphere. NO_x and SO_x contribute to acid deposition, and carbon emissions from power plants can also lead to global warming.

The scope of the plan and its impact on energy use was a bit of a surprise to the Energy Commission. There are three major elements to the plan that affect energy use:

High levels of demand-side management savings. The plan calls for a 30 percent reduction in electricity and natural gas use compared to currently forecasted levels by 2009 in the residential, commercial, and local government sectors (industry and transportation sectors were excluded).

Very extensive industrial electrification. The plan calls for major fuel switching from natural gas to electricity and substantial combustion efficiency improvements. The Energy Commission assumes this means industrial electricity use is 50 percent higher than current forecasts, and natural gas use 35 percent lower.

High penetration of electric vehicles. The plan calls for prohibiting new gasoline engines in the basin and replacing them with electric vehicles. The Energy Commission assumes this means a 20 percent penetration of electric vehicles by 2000, and 70 percent by 2010.

Retrofitting power plants in the basin. Utility power plants in the L.A. basin emit about 4.5 percent of the nitrogen oxides (NO_x) in the basin. The plan calls for reducing total emissions from power plants in the basin by 80 percent.

The Effect of Air Quality Regulations on Energy Demand And Emissions

In the last year Commission has begun a major effort to understand what these kinds of changes in energy use would mean to energy use and energy-related emissions. Table 1 summarizes the results for the scenarios that correspond to the major elements of the plan described above. The utilities included in this analysis that are affected by the SCAQMD plan are: Southern California Edison Co., and the municipal utilities of Los Angeles, Burbank, Glendale, and Pasadena. By 2009 the Commission estimates that combined impact of high DSM, industrial electrification, and electric vehicles, would result in increasing electricity use by about 10 percent, and decreasing natural gas use by 7 percent. In this analysis (Jaske 1990, and McAullife 1990), it was assumed that additional electricity would be provided by new utility-owned gas-fired combined cycle plants, and that existing power plants in the basin were retrofit with pollution control devices as required by the AQMP. actual resource additions would not all be combined cycle power plants, but some mix of utility and independent energy projects, including renewable plants.

The Effect of Valuing Emissions on Electricity Resource Additions

The next step of our analysis was to try to value pollution in our planning models to see what changes occurred in the operation of the existing electric system and in future resource choices. Emission reductions can be achieved by retrofitting existing facilities, by replacing existing facilities with cleaner more efficient gas-fired generation technologies, or by substituting with non-emitting source such as demand-side management or renewables. The Commission uses what is called the "iterative cost-effectiveness methodology" (ICEM). As the name implies, the method uses a production cost model (ELFIN) to iteratively add resources as see if the costs of serving electric load are lower or higher with new resources. This is a very time and computer intensive procedure, but tells us a lot about what are the best resources to add.

The ICEM air quality analysis focused what was the least cost way to meet air quality requirements. The Commission estimated the costs of the different control measures to retrofit power plants or build new more efficient plants. In addition, a value was estimated for residual emissions that would remain even after the standards were met. The commission evaluated two cases:

> Private cost case, which included the cost of retrofitting power plants with control measures, and the cost of obtaining offsets.
>
> Social cost case, which included a value for residual emissions of regulated pollutants that remained after applying control technologies (in addition to private costs), and a value for carbon emissions.

In-basin power plants which emit pollutants are gas-fired utility and independent "QF" (Qualifying Facility) units. Most of these power plants (about 7,000 MW) are older plants that tend to be less efficient and result in greater levels of emissions than newer gas-fired facilities. They have also reached or are near their nominal economic design life. Since many of them are less efficient they are used for peaking operations, which tend to correspond with air pollution episodes. These plants can either be retired and replaced with new, more efficient plants either in or out of the basin, or retrofit with pollution control equipment. QFs would like to replace this with renewable generation Table 1: Summary of Energy Demand Results for the SCAQMD Region

	Forecast Year	
	2001	2009
Electric Energy (Gwh)		
Baseline Forecast	128,260	147,962
High DSM Impact	-15,661	-31,824
Indust Elec Impact	+ 8,580	+16,277
High EV Impact	+18,812	+29,438
Combined Impact	+11,731	+13,891
Comb/EV#2 Impact	+11,731	+13,891
Comb/EV#2/CNG Impact	+3,361	+ 793
Peak Demand (MW)		
Baseline Forecast	29,521	34,682
High DSM Impact	- 4,099	- 8,435
Indust Elec Impact	+ 1,443	+ 2,766
High EV Impact	+ 243	+ 380
Combined Impact	- 2,413	- 5,288
Comb/EV#2 Impact	- 1,646	- 4,086
Comb/EV#2/CNG Impact	- 2,017	- 4,790
Natural Gas (mill. therm)		
Baseline Forecast	13,183	14,149
High DSM Impact	-1,653	-3,176
Indust Elec Impact	+ 452	+ 764
High EV Impact	+1,261	+1,862
Combined Impact	+ 35	- 930
Comb/EV#2 Impact	+ 54	- 907
Comb/EV#2/CNG Impact	+ 759	- 110
Liquid Fuels (mill. gallons)		
Baseline Forecast	6,135	5,422
High DSM Impact	0	0
Indust Elec Impact	0	0
High EV Impact	-1,226	-2,645
Combined Impact	-1,226	-2,645
Comb/EV#2 Impact	-1,226	-2,645
Comb/EV#2/CNG Impact	-2,097	-2,726
No_x Emissions (tons/day)		
Baseline Forecast	735.9	710.6
High DSM Impact	-11.0	-19.4
Indust Elec Impact	- 5.4	- 11.1
High EV Impact	- 61.3	-137.7
Combined Impact	- 77.7	-168.8
Comb/EV#2 Impact	- 77.5	-169.0
Comb/EV#2/CNG Impact	- 89.0	-126.6

	Forecast Year	
	2001	2009
ROG Emissions (tons/day)		
Baseline Forecast	339.5	309.8
High DSM Impact	- 0.9	- 2.7
Indust Elec Impact	+ 0.2	+ 0.5
High EV Impact	- 50.7	-105.7
Combined Impact	- 51.9	-108.5
Comb/EV#2 Impact	- 51.9	-108.4
Comb/EV#2/CNG Impact	- 66.2	- 83.3
CO_2 Emissions (tons/day)		
Baseline Forecast	374,421	371,126
High DSM Impact	-23,250.	-47,441
Indust Elec Impact	+ 7,122	+11,846
High EV Impact	-15,608	-40,613
Combined Impact	-35,368	-84,498
Comb/EV#2 Impact	-34,960	-92,855
Comb/EV#2/CNG Impact	-44,674	-72,369

Note: Several different regions are used in this analysis. Electricity demand is forecast at the level of planning areas for the three major utility groupings in Southern California -- SCE, LADWP, and BGP. Electric generation resources are analyzed for SCE and the combination of LADWP/BGP. Natural gas demand is forecast for the Southern California Gas planning area. Liquid fuels, NO_x and ROG are projected for the SCAB. CO_2 is projected from natural gas demand at the SCG planning area and from liquid fuels at SCAB.

which would be outside of the air basin, since that is where the geothermal resources are, and enough land for solar-thermal plants. However, Southern California utilities claim they have to maintain a significant portion of their generation in the basin to maintain system stability, since much of the generation comes long distances from other portions of the Southwestern U.S.

To evaluate the impact of air quality requirements of the Commission identified the emissions from each power plant and did a plant-by-plant evaluation of the options available to retrofit or retire each plant based on the cost of the retrofit, and resulting emissions and efficiency. The analysis would retrofit or retire enough plants so that the emissions from the utility system would meet the requirements of the SCAQMD plan.

To implement its plan SCAQMD has adopted rules that apply to existing gas turbines (both utility and independently-owned) and utility boilers. SCAQMD has suggested that urea injection be used for older plants and the newer plants be retrofitted with selective

catalytic reduction (SCR), which is both more expensive and cleaner. Utilities can also retire plants, but new plants must use "best available control technology" (BACT) and acquire "offsets" for emissions that are created by the new plants that might increase capacity beyond the original system capacity in the basin.
The Table 2 shows the types of control technologies and their estimated costs (all costs are in 1987 dollars).

Table 2: Cost of NO_x Control Technologies

Type of control	NO_x reduction	Cost (1987$/kW)
New plants:		
SCR	90 %	$45
Existing plants:		
Urea injection	35-75	3-21
SCR	90	72
Low NO_x burners	30-60	3-45

The Commission also had to estimate the cost of offsets for new power plants. Historically offsets have been obtained from the shutdown of existing sources owned or controlled by the project proponent, although some third-party offsets have been approved. Offsets are going to become more expensive as their availability declines. For this analysis the Commission assumed offsets were the same as the cost of applying NO_x emission controls to existing utility boilers, or S45/kW (1987 dollars), or $4,300 per ton per Year ($43,600 per ton one-time cost). This value was escalated at a rate greater than inflation, which resulted in costs of $57,000 in 1995 and $107,000 in 2000 (per ton one-time costs).

One of the paradoxes worth noting of the offset requirement is that adding gas-fired generation *results in less emissions than if renewable generation were added.* The reason for this is that a new source must provide offsets at a ratio of 1.2 to 1, which means that a new source must not only reduce emissions by as much as they are increasing emissions, but 20 percent more.

The value of residual emissions for social cost case was based on the "revealed preference" concept which assumes that the cost that air quality regulators impose is a reasonable proxy for the value to society. The Commission developed estimates of in-state and out-of-state emissions separately based on the notion that while California imports a significant amount of electricity from both the Northwest and Southwest U.S., the air pollution situation in those places is not nearly as severe as in California. In addition, carbon was valued as a "global" pollutant, even though there are no existing emissions requirements. Table 3 shows the values used in the analysis.

Table 3: Values for Residual Emissions (1987 $/ton)

Pollutant	In-state	Out-of-state	Global
NO_x	$11,600	$2,700	n/a
SO_x	11,500	1,000	n/a
ROG	3,300	300	n/a
PM10	7,800	800	n/a
Carbon	n/a	n/a	$26

The results of the ICEM analysis for the *private cost case* shows that the cost-effective resource additions correspond well with the requirements of the SCAQMD regulations. That is, for Southern California Edison Co. about 1,800 MW of existing capacity in the Los Angeles air basin were cost-effective to retrofit with emission control technologies in the late 1990s. Similar results occurred for the Los Angeles, Burbank, Glendale, and Pasadena utilities. The additions both met the emission requirements, and did so at a lower cost than using renewable technologies.

In the *social cost case*, the additions were cost-effective about two years earlier than in the private cost case. However, even with the additional values for residual emission renewable technologies were not cost-effective. Proponents of renewable technologies, primarily geothermal and solar-thermal companies, have argued that the assumed cost of controls, offsets, and residual values are all too low, and the assumed costs of renewable plants are too high; if different assumptions were made then renewable technologies would be cost-effective. In addition, they point out that California is still overly dependent on gas-fired generation, and that the state would not only be better off from an environmental point of view but would also be more diversified in its energy sources if non-gas generation were added.

Conclusion

This analysis showed the importance of understanding the implications of air quality regulations on the electric system. We found that the SCAQMD staff did not know much about how the electric power system operated, and about energy uses and conservation.

We also found that we were equally unaware of the massive changes in total energy use and fuel switching that could be required by air regulations.

A number of important lessons have been learned. We need more communication between agencies that regulate energy and environmental issues to prevent the kind of "blindsiding" that occurred with the SCAQMD regulations. The goal of that communication should be to have more debate and information on the policy options that can achieve a "best solution" from both an energy and environmental perspective. The California Legislature has created a Joint Committee on Energy Regulation and the Environment that is considering changes in the statutory mandate and organization of agencies such as the Energy Commission, Air Resources Board, local air districts, and the Public Utilities Commission to ensure that communication and coordinated decisionmaking occurs.

The Legislature should also consider giving regulators direction on how far agencies should go to achieve ambient air quality standards. The ambient air quality standards were set years ago without reference to what kind of measures would be required to achieve the standards, which puts the air pollution agencies in a potentially untenable position of having to proscribe rules that may be economically and politically unacceptable, but are required by law. Unrealistic mandates do not provide the environmental quality and public health that we are seeking to achieve.

In order to know if the plans will be economically and politically acceptable, we need more research on the feasibility of major elements of the air quality plans, such as high levels of DSM and vehicle of the air basin, and whether it is correct to implicitly assume that the environmental impacts of those resources are preferable to those in the basin.

References

Arikawa, B. (California Energy Commission). 1990. *Staff Methodology for Applying Recommended NO_x Offset Costs in the ER 90 Final Resource Planning ICEM Analysis. Docket* No. 88-ER-8. Sacramento, California.

California Energy Commission. 1990. *Draft Final Electricity Report.* CEC Report No. P 106-90-002. Docket No. 88-ER-8. Sacramento, California.

Jaske, M.R. (California Energy Commission). 1990. *Air Pollution Projection Methodologies: Integrating Emission Projections with Energy Forecasts,* in the Proceedings of the ACEEE 1990 Summer Study on Energy Efficiency in Buildings, Asilomar, California.

McAuliffe, P.K. (California Energy Commission). 1990. *Initial Policy Phase Analyses: Air Quality Scenarios, Electricity Generation and Emissions in the South Coast*, Docket No. 88-ER-8. Sacramento, California.

South Coast Air Quality Management District. 1989. *1989 Air Quality Management Plan*. El Monte, California.

Therkelson, R.L. (California Energy Commission). 1990. *Valuing Emission Reduction for Electricity Report 90. Docket* No. 88-ER-8. Sacramento, California.

7.6 New York State's Environmental Costing Research Program in Context

G. E. Walmet and D. R. Wolcott
New York State Energy Research
and Development Authority
Albany, New York 12223 USA

In this paper we present the research program in environmental externality costing of the New York State Energy Research and Development Authority (Energy Authority). Details about the projects that the Energy Authority has sponsored over the last two years are discussed along with details about a major new project just getting under way. We then present some of our thoughts and preliminary conclusions on this topic drawn from the results of the research program to date. The opinions expressed herein are our own and do not necessarily reflect the views of the Energy Authority.

The Energy Authority's Research Program

The Energy Authority initiated its research program in environmental costing in 1988 through a contract with Pace University Center for Environmental Legal Studies (Pace) that resulted in the publication in September 1990 of *Environmental Costs of Electricity* (Oceana Publications, New York). When the program started, New York utilities were selecting their supply and demand side electric resources without taking into account the environmental costs imposed on society of pollution that was generated even after compliance with all applicable environmental laws and regulations. We recognized that it would be difficult to quantify such "externalized" environmental costs. However, we also knew that a decision not to consider external costs in itself quantifies them by setting their value at zero. This seemed unreasonable given the evidence of massive, transboundary environmental problems (e. g., acid rain, atmospheric ozone depletion, and global warming) that appear to be occurring despite the current level of environmental regulation.

Another motivation for the Energy Authority's research interest was the introduction of integrated resource planning (also known as "least cost" planning) by the New York Public Service Commission (PSC). That concept embodies the notion that it is in New York's interest to minimize all costs borne by its citizens associated with the use of electricity, rather than just minimizing rate payments. A least cost resource would be the one having the lowest total costs, where the total costs include external environmental costs. This theory would gain importance if we were to find significant differences in externalized costs between the various resource options available to utilities (e. g., central station power plants, independent power facilities, and demand side management). We comment on the Pace findings in this regard later in this paper.

To address this issue, the Energy Authority called on Pace to undertake the following tasks:

1) Evaluation of theoretical and methodological issues concerning the estimation of externality costs and their incorporation into utility resource planning;

2) Identification, through an extensive worldwide literature review, of other efforts to estimate externality costs by emissions (e. g., sulfur dioxide) and by electricity resources (e. g., coal plant operations);

3) Assessment of existing analytical methodologies and computer models for estimating environmental costs and development of an "ideal" approach; and

4) Summary of the alternative approaches that could be taken in New York to incorporate externality costs in utility resource planning, given a review of efforts to do so in other regulatory jurisdictions.

There are two important caveats about the study. First, because of budget and scope limitations, the entire fuel cycle of each resource was not addressed. The analyses focused on the environmental costs of operations, decommissioning, and waste disposal. There were no analyses of costs associated with construction or fuel acquisition. Secondly, studies of pollution control technology costs were not reviewed. We considered control costs a fairly mature technical field as compared to the valuation of environmental damages which was the focus of this study.

The study had important linkages with technical and regulatory institutions that provided a real world context. For example, the project was cosponsored by the Integrated Resource Planning Program at the U.S. Department of Energy which provided the federal perspective. The Energy Authority also established a management advisory committee consisting of representatives of the government agencies in New York that might be responsible for implementing the results. These included staff from the PSC, the State Energy Office (SEO), and the Department of Environmental Conservation (DEC). We were extremely fortunate to gain the interest and active participation of German researchers (Olav Hohmeyer and Uwe Fritsche), who had done much of the original trail blazing work in this field. When the draft report was produced, the Energy Authority convened a workshop that was attended by over one hundred people representing all the obvious stakeholder interests for a general discussion of the topic and presentation of comments on Pace's work. The draft report was then subjected to a rigorous peer review process involving 38 people, again representing all relevant interests, including most notably an extensive review by the New York utilities.

While the Pace study was underway, the Energy Authority sponsored another research project that focused on calculating the environmental benefits of demand side management (DSM) through the development of a method to account for externality costs in an existing computer program. Utilities and regulators have increasingly found DSM (e. g., conservation and load management by utility customers) to be an important component of utility integrated resource plans. In this regard, New York utilities are now implementing a wide array of DSM programs pursuant to orders of the PSC.

These efforts have signaled the need for analytical tools that could be used to evaluate various elements of DSM programs (e. g., technology options, end uses, market segments, load shapes, customer acceptance, and utility characteristics). Analyses also need to be undertaken of the costs, benefits, and impacts of such programs on the utilities, their ratepayers and shareholders, and society in general. To meet this need, Synergic Resources Corporation (SRC) had developed its Comprehensive Market Planning and Analysis System (COMPASS), a personal computer-based decision support "screening" model that analyzes DSM and utility strategic marketing options. In New York, a number of utilities, as well as the PSC and SEO, are all now using COMPASS for DSM program planning and analysis purposes.

To support these users, the Energy Authority called on SRC to enhance COMPASS to value changes in environmental emissions at both the end-use and the utility power plant from the implementation of DSM programs. The following tasks were undertaken:

1) Modification of DSM technology characteristics files to include end-use emissions for relevant pollutants;

2) Modification of utility characteristics files to include marginal power plant emissions by season and daily time period;

3) Development of the logic for calculating changes in end-use and power plant emissions by pollutant from the displacement of utility generation by a DSM resource; and

4) Development of the logic for calculating the monetary value of damages to the environment that result from changes in utility operations from the implementation of DSM programs.

As of this writing, the externality costing module of COMPASS has been completed and is undergoing testing and validation. A word of caution regarding this analytical technique is that it is a completely deterministic accounting tool. DSM technology characteristics, utility power plant characteristics, and environmental damage values are all exogenous inputs to the model. It then simply performs the arithmetic to calculate changes in externality costs from the implementation of DSM programs. A nice feature is that the user can specify default values for each input, and in the case of the damage values, the default data presented are taken from the Pace study. The model is also very user friendly (menu driven) and powerful so that relative cost/benefit impacts of various complex DSM options can be calculated quickly and easily. This capability is especially important now that the "societal cost/benefit" test (including the calculation of externality costs) is required for analysis of DSM programs in New York.

The Next New York Externality Research Program

New York has taken the lead in the United States in a number of dimensions of identifying and incorporating environmental externality costs in electric utility resource acquisition decisions. With the completion of the Pace and SRC projects, the Energy Authority has provided a complete review of the topic and the demonstration of an analytical technique to account for environmental costs in resource decisions. The PSC and SEO have also developed a method to explicitly incorporate externality costs in utility ranking and selection of proposals submitted in the integrated resource competitive bidding programs.

The next logical phase of research has now been initiated with the full participation of the electric utilities that constitute the New York Power Pool (Power Pool). This next program was defined when the PSC ordered "the utilities (to) fund and participate in a pooled statewide effort, with staff, SEO, DEC, (the Energy Authority), and outside experts, to develop estimates of external environmental costs for all electric supply and demand options" (New York PSC Opinion No. 89-15, Opinion and Order Concerning Conclusions on Long Range Demand Side Management Plans, May 23, 1989, p. 45).

The two year, $1.6 million program was initiated in October 1990 when a five member program management board convened composed of representatives of the Energy Authority, PSC, Power Pool, and the Electric Power Research Institute (EPRI) which has agreed to cosponsor the program. A representative of Resources for the Future was invited to be the fifth management board member, defined to reflect an independent point of view. A Number of other organizations, including SEO, DEC, and Pace (as a technical adviser to the Energy Authority) have been invited to observe the development and implementation of the program.

The purpose of the program is to develop a comprehensive computer software model to estimate environmental externality costs for the full range of utility resource options in New York including utility generation, independent power production, and DSM. The model will attempt to calculate the value of externalities for specific New York resources (e. g., power plants) based on emissions, control technology, population characteristics (human and biotic), geography, and meteorology. The model will be personal computer-based and may be a useful research and policy analysis tool for the evaluation of utility integrated resource plans and the selection of proposals through the integrated bidding programs. If the program is successful, the model may be the first step in defining what an operational software tool might look like that could be used by utilities in the real time application of "environmental dispatch" of the Power Pool's generation resources that are truly least cost (including social costs).

The program, to be contracted cooperatively by the Energy Authority and the Empire State Electric Energy Research Corporation (ESEERCO), will entail the following tasks:

1) Conduct of a comprehensive literature review that focuses on aspects of the topic that were excluded by budget limitations from the Pace study (e. g., pollution control technology costs, damage values from "front end" technology impacts including fuel acquisition);

2) Conduct of a "critical review" of the combined first task literature review and the Pace study by a prestigious, independent organization (e. g., the National Academy of Sciences);

3) As a result of the critical review, establishment of agreed upon definitions of all terms and data required for specific externality cost analyses including resource options, emissions, pollution transport and fate phenomena, dose/response relationships, and damage functions;

4) Development of an analytical methodology and computer software algorithms to perform externality cost analyses for specific utility resource options;

5) Acquisition of all data required to conduct such analyses for New York utility resource options; and

6) Testing and validation of the model and data through a series of case studies applied to specific New York resource options.

We recognize that this is an extremely ambitious program that could fail in a number of ways both large and small. At best, the result might be a first step in the development and "buying in to" of an agreed-upon analytical framework through the cooperative participation of all of the relevant stakeholders to the process of identifying and incorporating environmental externality costs.

The Use of Externalities in Energy Choices

As our state, nation and world struggle with the difficult issues of balancing energy, environmental and economic needs, we have developed a new tool -- costing of environmental externalities -- to help us make good decisions. The decisions should allow us the numerous benefits of energy use without excessive damage to our health or surroundings. At the same time we need to give proper attention to allocation of scarce economic resources. The concept of externality costing places many of these parameters in a common denominator, money, and hence allows for easier comparison of options and decisions. In this section we will discuss some energy choices which externalities are clearly useful in evaluating, and others in which externalities are less useful, and suggest the direction of future applications. Our views are drawn from our experience in managing the Energy Authority's research program and reflect some of the results to date.

There are many energy choices to be made in the context of integrated resource planning, even more if the concept is applied to all energy use rather than just electricity. Although the work in New York has focused on electric options to date, it is clear that this is an artificial constraint. For reasons explained more fully later, choices have to be made in the total social energy needs context, not just viewing electricity as an isolated domain.

The potential energy choices, to which externalities can be applied are numerous. They include the following:

- Electric generation
  - coal, oil, or gas
  - nuclear
  - hydro, other renewables, wastes
- Electrification versus site-burning of fuels
- New facilities versus old
- Different classes of environmental problems
- Environmental and other externalities
- Supply side versus efficiency

We will comment on each of these choices in turn, examining how externalities can or cannot help make decisions and the need for more work. The Energy Authority's interest in these matters is of two kinds. First, we hope that our support and participation in studying externalities will help all the decision-makers (especially those in New York) who have to make energy choices. Secondly, we conduct a program of research, development and demonstration ourselves. This program is meant to expand the available choices of energy options in the future. Since research resources are severely limited, it is imperative that our own choices for research direction be as well informed as possible. Externalities provide a useful new tool in research planning which is of special interest to us.

Electric Generation

The bulk of the Pace study dealt directly with the topic of the externality costs of electric generation options. Some use of the report has already been made by a group in New York opposing the siting of a 200 MW fluidized bed, coal-fired, cogeneration plant. New York utilities now use externalities in evaluating bids to supply new increments of power or DSM services. Therefore, utilities would penalize a relatively dirty option, although compliant with regulation, when bid against cleaner options. A clear value of the Pace externality work is to help define better numbers, based on actual damages, to eventually update the currently used default values based on pollution control technology costs.
However, a number of problems have been recognized in applying externalities to electric generation. The first problem is simply misuse of the numbers. The value of an externality is dependent on fuel, technology, as well as some other factors such as location and pollutant dispersion. Average, general numbers for a given fuel should not be used if more accurate, specific ones, are available. This error was made by some of the fluidized bed plant opponents in New York. It is not clear if coal can be properly viewed as a "dirty" option. Although coal certainly starts out dirty, it is technically possible to achieve arbitrarily low sulfur, nitrogen oxide and particulate emissions. Gasification/combined cycle power production has already demonstrated close to zero emissions of sulfur and particulates, with moderate amounts of nitrogen oxides, comparable to natural gas-based combustion turbines. Selective catalyst reduction, along with ammonia or urea injection, or catalytic combustion can drive the nitrogen oxide emissions to extremely low values. Alternatively, other clean coal technologies such as ultra-beneficiated coal water slurries achieve emissions comparable to at least good quality residual oil. There are also many other clean coal approaches with emissions comparable to other fossil fuels. There certainly remain legitimate issues about the residue from coal cleaning. Trace metals, sulfur and ash have to go somewhere, but they can be placed back in the mine or some other suitable location with minimal net environmental effect. Therefore, all these conventional pollutants can be technically handled and the issue comes down to economics. Given enough (costly) pollution control technology, coal can be as clean as any fuel -- maybe.

Carbon dioxide (CO_2) can be a problem. Coal has a higher ratio of carbon to hydrogen than oil and gas. This inherently results in more CO_2 per unit energy output than other fuels by a factor of 1.2-1.8 times. In addition, efforts to clean up the other pollutants tend to make the overall conversion less fuel efficient driving CO_2 emissions even higher. In the extreme case of making methanol liquid fuel from coal, the CO_2 emissions may be 3-4 times higher overall than burning natural gas on a net energy output basis. Theoretically, even CO_2 can be abated, at some cost, although the technology is unproven in real use. It would, for example, be possible to lightly gasify coal, driving off primarily light, hydrogen-rich fuel gas, leaving most of the carbon in solid char form to be placed back in the mine with the ash and captured sulfur. Also CO_2 scrubbing of either coal gas or stack gas is technically achievable. Whether, and at what cost, the captured CO_2, could be sequestered in old gas wells or deep oceans, is unproven. How long it would remain trapped is also a valid question. Coal comes to us as a dirty fuel and if burned without controls can result in high externality costs. It can also be cleaned up by control technology to the point of low externality costs, except possibly for CO_2. Therefore, to a degree, the issue of the future of coal may come down to evaluating the externality cost of CO_2 as realistically as possible. We will address this point in a later section on different classes of environmental externalities.

Nuclear energy is another thorny social issue. Clearly, the absence of gaseous combustion emissions helps the externality cost of nuclear power. Other effects on health such as exposure to radiation from mining ore and tailings can be properly and credibly monetized. Like coal emissions, uranium ore problems are amenable to treatment using technology and hence cannot be characterized with generalized externality numbers. However, the most difficult problems with nuclear power are the risk of accident, waste disposal/treatment, and nuclear weapons proliferation. Whether attempts to quantify small probability events of catastrophic consequence turn out to be credible or not to the public and decision-makers is not clear. Hence, although externality work in this area may be useful, for the time being, society will probably make its choices on nuclear power based on other forms of collective wisdom and decision-making.

Externality costs are important to consider in evaluating waste-fired options. However wastes cannot be properly thought of as just another fuel. If a fuel remains in the ground unburned it does no harm. The same is not true for municipal solid wastes or industrial wastes. Something must be done about them anyway. They can best not be produced through waste reduction or avoidance. But if they are produced, they must be recycled, treated or somehow disposed of. All these options, including recycling, result in emissions and other negative effects in the locations where wastes are handled. Therefore, to be equitable, the externalities of all the options also have to be counted. If municipal wastes wind up landfilled, the leachate has to be costed as well as the methane, CO_2 and other trace toxic gases which are released into the atmosphere. At a minimum, methane and CO_2 contribute to global warming. Likewise recycling, although clearly preferable in most cases, always results in disposal residues as well as local air emissions and waste-water effluents. These cannot be ignored nor can movement of a problem from one location to another be assumed to be a solution.

Electrification Versus Site-Burning of Fuels

The decision of wether it is better to burn a fuel to provide a certain energy end-use or to use electricity has been traditionally an economic one. Electricity is such a versatile form of energy that it can do essentially every energy service and could technically replace all end-use fuel burning.

Examining externalities can help decide whether electrification is truly valid. However, partial application of externalities is a potential problem. It is obviously unfair to burden the cost of electricity with an externality cost while not considering externalities for the direct fuel burning option. Perhaps the most important current example of this is transportation. There are serious externality costs from pollution emissions, times lost to traffic congestion, and traffic accident injury and death attributable to fuel burning cars, trucks and, somewhat differently, airplanes. Electric vehicles, especially electric mass transit, can improve all three parameters. Other electrification end-uses such as industrial electrotechnologies and residential space and water heating also provide benefits including avoiding low-level pollution released near populations. Therefore a rational social decision must be based on all compared options being fully burdened with all of their environmental (and other) externalities. The outcome of this decision may well be in favor of electrification in many cases.

Another factor which may prove important in the future is global warming. If avoidance of CO_2 emissions does someday become a world-wide imperative, it is hard to believe that widespread electrification will not be a responsive strategy. There are technical options, for example, the use of CO_2 scrubbers at power plants and the conversion of wind energy, not likely attractive at the end-use. Again the global warming consequences of end-use fuel-burning compared to an electric option have to be analyzed and monetized.

Although it is clear that electric power production causes environmental damage, it is also clear that direct fuel burning alternatives are often damaging. One practical problem in properly evaluating these options is the regulatory segmentation of our energy systems. For example, the PSC can, and has, instituted externality costs in electric options but has no control over the burning of non-utility fuels by end-users. These emissions are the domain of the DEC. Emissions fees or action by an agency like the SEO with energy policy responsibility can conceptually place all these choices on a equitable footing. But care must be taken that vigorous regulation in one sector may bias the decision in favor of an over-all suboptimal result.

New Facilities Versus Old

A profound, generic problem regarding how energy decisions are made by the public and policymakers is that continued operation of old facilities is favored over new ones. This is partly the result of desire to minimize short-term costs of energy by getting as much use out of existing capital stock as possible. Probably even more importantly, the public has become extremely hostile to new facilities located near them. This is at least partially the result of frustrations over existing levels of pollution. The pollution in fact comes from existing plants such as the old, uncontrolled (except for particulates), high sulfur coal plants prevalent in much of the United States. All new plants which are proposed are better than these; many are much better. Yet the only decision point in which the public and regulators can intervene is over the siting of a new plant. Conceptually, it would be much better to have a periodic mechanism to pass judgment on the continued operation of existing facilities vs. their replacement with new technology. The total (including externality) cost of all options, including upgrading the old facility with retrofit pollution control should be compared. Unfortunately the application of externality costs only to new facilities makes retaining old ones even more attractive. But the overall levels of emission are clearly higher, and the decision is likely to be suboptimal.

It is not clear how a process which places old, dirty operating power plants into a periodic decision process could be implemented. Emission taxes, environmental dispatch and some form of periodic relicensing are all possibilities but they must be done on a broad basis so as not to unduly disadvantage an environmentally progressive state or area vs. a less progressive one. Even a fairly large state like New York receives only a modest fraction of its sulfur deposition from in-state sources. With a gas such as CO_2, clearly the consequences of any given increment of emission are insignificant locally but important in global aggregate terms. This is the "problem of the commons" in its worst possible embodiment. Environmental dispatch or any other of these solutions must be practiced on a broad enough scale to be equitable and effective in reducing pollution. Still, this problem is extremely important and has to be solved if we are no to be exposed to the emissions of old plants for the indefinite future. Simply helping find reasons not to build new, cleaner plants is not a reasonable social response.

Different Classes of Environmental Problems

One of the conceptual virtues of the externality paradigm is that it can compare environmental effects of various types using the common denominator of monetary cost. However, a problem arising with the approach is the vastly different nature of the relevant effects, and the highly variable state of knowledge about them. The dominant environmental externality in the Pace study turned out to be sulfur oxides. There has been a vast amount of work done in the last two decades studying the effects of sulfur oxides. Some damage effects such as those to historical buildings are essentially incontrovertible. Others such as effects on trees are in various states of dispute. Human health effects are perhaps the most uncertain. Still, the extensive literature allows rational monetization for sulfur gases. Potentially, the consequence of global warming to mankind greatly exceeds the effect of sulfur gases. However the state of knowledge and scientific consensus is much less developed, hence we have limited ability to credibly monetize CO_2 emissions.

Other environmental externalities are of a totally different type -- for example the devastating effect that extensive hydro development in northern Quebec could have on the relatively small number of Native Americans and other flora and fauna who live there. A great number of tons of CO_2 as well as other air emissions could be avoided by the James Bay hydro development, with enormous externality benefits relative to fossil fuel generation. However, what is the value of a lost culture or species? Even offering the native people environmental compensation (large numbers could easily be justified) is not satisfactory since the consequences of the action are irreversible and subject to error by the wisest of tribal elders. On the other hand, global warming will affect everyone, including the natives of norther Quebec, and may be of paramount importance above all other considerations.

Externalities are most helpful in comparing an environmentally damaging option to more benign ones. However, if the choice comes to different environmental risks the concept is less useful. Clearly fossil fuel burning will tend to look worse than efficiency measures if externalities are taken into account. Externalities are very helpful in making that decision. However, if we consider a trade-off between emissions (say CO_2 vs. sulfur or nitrogen oxides as occurs when sulfur or nitrogen oxide mitigation is practiced) the concept is less useful. We really need good numbers for all the pollutants at issue to make even qualitatively correct decisions. It is true that any estimate of environmental effects is better than a default to zero when comparing an environmental effect to its absence. But if the comparison is between two effects, good numbers are necessary. It could even be possible to make a wrong choice if one well-characterized but less important effect is traded against

an important, under-valued effect. Other complicating environmental effects are toxic air emissions and toxics released to water and the land. In general these are all too poorly understood at this time to allow credible monetization of their externalities. Some of these toxics may become important parameters in certain future energy systems such as photovoltaics.
When two important environmental effects come into direct conflict such as hydro development vs. coal burning, the externality concept is most difficult to apply. Two important but very difficult to monetize effects have to be compared. In less dramatic fashion, the same thing occurs whenever a generally beneficial new facility such as a clean energy plant or a recycling facility needs to be sited. The local opposition can be thought of as based on an implicit externality valuation of the facility which comes out negative for the neighbors. Ways of more credibly quantifying these difficult effects and then providing appropriate compensation and persuading the public and decision-makers is an important challenge to the externality concept.

Environmental and Other Externalities

In the context of this paper "externality" is shorthand for costing environmental externalities. Economists use the term externality for any parameter which is not included in market costs, but could be. Energy can be thought to have other externalities besides the environmental ones. These include various hidden subsidies such as the below "market" value of uranium fuel, the cost of military forces dedicated to maintaining secure supplies of oil from the Mid-East, and somewhat more abstractly, the value of energy security. It is apparent that there is a real social value to energy security. This is evidenced by stock market disruptions in the face of oil supply uncertainty, recessions linked to rapid increases in energy costs, and loss to local energy producing regional economies as a result of rapidly collapsing oil prices.

Economies seem to thrive with stable energy prices and react unfavorably to lack of certainty. Also lack of predictable prices inhibits proper capital investment decisions in efficiency measures. On another level, areas with electric supply problems, even anticipated problems, are cited by businesses as undesirable for location of new enterprises.

The rational way of making choices is to attempt to evaluate all relevant externalities to arrive at total real cost for comparison purposes. Again it is important, conceptually, to do this with as realistic numbers as possible especially if important externalities favor different options. The problem is much easier if the externalities are all the same. For example, if we compare the effects of burning foreign oil in a power plant or an automobile with efficiency measures which avoid burning the fuel, the environmental, military cost, security and reliability issues all favor the efficiency options regardless of their precise values. However, domestic coal, burned with a minimally compliant technology, compared to imported oil is a more difficult problem where the environmental externality favors oil but the security cost favors domestic coal. In this case accurate and credible values are important to achieve good comparative choices. There are many other interfuel choices which also can be said to have externalities that are either similar or dissimilar requiring more difficult decision-making.

Supply Side Versus Efficiency

The final category of energy choices discussed here is the one related to supply side options vs. energy efficiency options of many different types. The issue is conceptually much the same if other, non-electric energy uses such as space heating, industrial processes, or transportation are considered. In general, a capital investment in energy efficiency can reduce use of the energy form. The efficiency equipment, measure, or system should be evaluated for its own environmental externalities. These generally are non-existent or minor compared to the externality cost of the energy supply option involving the combustion of fuel. There have been a few contrary examples such as house air infiltration reductions (a heating efficiency measure) degrading indoor air quality. These effects could be costed and compared to fuel burning externalities or preferably measures such as good design and possibly mechanically controlled ventilation with heat recovery which could eliminate the environmental concern (or internalize the cost).

Another example of environmental externalities with an efficiency measure is relighting a building with efficient fluorescent lights replacing incandescents or less efficient fluorescents. It is possible that the retrofit will stir up some asbestos from an old ceiling or require disposal of PCB laden electric components. Also, fluorescent lights contain small quantities of mercury which present potential disposal problems. Some of these effects, like the remediation of existing environmental problems, are really independent of the efficient device. They would be required, and would be as troublesome, at some time in the future. The mercury should be costed as an externality or internalized by proper disposal or recycling. It has also been observed that coal burning liberates significant mercury and that fluorescent bulbs, even poorly disposed of, liberate less total mercury than saved by reduced coal burning. There are issues of population exposure and concentration which have to be properly evaluated in examining the question of mercury.

At any rate, these examples are fairly minor and there are rare examples of negative effects from efficiency measures. The overwhelming conclusion is that evaluation of externalities (of all sorts even beyond the environmental ones) strongly favors efficiency over supply options. That is the most clear and compelling message of the entire externality concept and the Pace study. It is less critical to know precise externality values for supply options since the result will nearly always be the same. Accurate values will define more correct capital investment levels. It is clear that investment decisions based on just dollar costs of fuels result in underinvestment in efficiency -- just how much underinvestment depends on the level, accuracy and credibility of the externality evaluation.

Conclusion

The preceding discussion of the tradeoffs involved in externality costing analyses highlights the complex and uncertain nature of this field of inquiry. Indeed, we feel like the work over the last two years has raised more questions than provided answers, has identified more issues to be addressed than resolved them. At least we now have a source in the Pace study which has brought together in one place many of the important strands of thought and vital citations of data that heretofore existing in many different reports and locations. And with the forthcoming New York externality modeling program, we have the opportunity to systematically develop a comprehensive analytical methodology which will draw on the important modeling work that has already been done by others. We find that the role that we play is a humbling one: to recognize that so much good work has been done by others in this field and yet to see the substantial amount of work that remains to be accomplished before environmental externality costing will see credible, operational and widespread use in utility resource acquisitions.

7.7 American Public Opinion Regarding Energy and the Environment, 1977-1989

Philip Sparks
Communications Consortium
Washington, DC 20005/USA

Background

This analysis is based on an extensive compilation of nationwide polling data by two dozen survey research organizations on energy issues and national energy policies. The attachment cites specific data from the following sources: Louis Harris & Associates; the Roper Organization; Audits & Surveys; Opinion Research Corporation; Yankelovich, Skelly & White for Time magazine; NBC News/Wall Street Journal polling; ABC News; the Gallup Organization; the Cambridge Reports; The Analysis Group; and RSM, Inc.

Overview

Suffering from the energy crisis and the gasoline price shocks of the mid-1970s, the American public reached an extraordinary consensus in the late 1970s that the nation's energy policies needed to be defined in bold, new directions by the incoming Carter Administration. The Louis Harris & Associates polling in early 1977 indicates that the public was ready to accept tough energy conservation programs, including a gasoline tax to curb energy consumption.

However, by the early 1980s, with the failure of the Carter Administration's energy initiatives and the advent of the Reagan Administration's free-enterprise policies, public perception of the need for a restructuring of a national energy policy bordered on the apathetic. For example, several polls taken in 1983 and 1984 indicated that a majority of the American people thought that the possibility of another 1970s-type energy crisis was not likely. The Roper Organization poll of March, 1983, found that 57 % of those surveyed thought that a new energy crisis was "somewhat unlikely" of "very unlikely." In this light, Americans responded negatively in the surveys cited to any energy efficiency initiatives.

The Mid-1980s, Shifts in Public Opinion Regarding Energy

Even as energy prices remained relatively low in the 1985-1986 period, public perceptions of a need to change energy policy stressing energy efficiency began to shift.

Opinion Research survey research in February, 1985, found that a plurality of Americans felt that there was a "very serious" need to save energy. An NBC News/Wall Street Journal poll in April, 1986, found 89 % of all Americans stating that it was "still important" to conserve energy.

A Reassessment of Energy Policies

A series of polls in late 1987 and beyond indicate that the American public once again saw the threat of an energy crisis. The August, 1987, ABC News poll found that 65 % of the American public thought that a 1970s-style gasoline crisis was "very likely" or "Somewhat likely" and a Gallup Organization poll in December of 1987 found 63 % of those surveyed expressing concern that the United States would once again experience major energy shortages similar to those of the 1970s.

Energy Policy for a Safer Environment

Through 1989, environmental concerns have shaped peoples' attitudes toward energy. Americans want clean energy sources. They emphasize conservation rather than new energy production as the primary means of thinking about energy policies (50 % to 38 %).

The public supports a wide array of policies to achieve these ends. Americans strongly favor an increase in fuel economy standards to build more fuel-efficient cars (61 %);

requiring installation of insulation and more efficient heating and electrical systems (60 %); and a reduction in the use of fossil fuels like oil and coal in favor of renewable energy sources like solar, wind, and water (61 %). They oppose policies that threaten the environment, especially building more nuclear power plants (59 %).

Summary

Forecasting the Energy Crisis

* On the whole, people continue to harbor latent discomfort about energy shortages in the future. Only a third of the population believes that the "energy crisis" is truly over, whereas fear of its return ranges from 49 % to 64 %, depending on external circumstances such as the war in the Persian Gulf or environmental disasters.
* People are concerned about the dependency of the U.S. economy on oil produced in the Middle East. This was a more salient issue in 1980, when 79 % of the public was very concerned. Still, a majority of Americans (52 %) expressed concern over our current level of oil dependency.
* While this latent uneasiness over future energy shortages and dependency on foreign oil producers has people concerned about energy issues, few people (27 %) expect to return to a long lines at the gas station -- no matter what time horizon you ask them to consider -- a few years, five years, or ten years.

Social Cost

* When energy issues are framed in terms of national security, many people are willing to support higher energy prices in exchange for independence. A 1987 Cambridge Reports poll found that 61 % say a secure supply of domestic oil is very important. A 1988 poll found 48 % of Americans were willing to pay higher prices for gasoline, natural gas, or home heating oil if it would promote energy independence, while 34 % were not willing to sacrifice low costs.
* Environmental concerns, particularly global warming, have forced many Americans to examine energy use and production. Americans want clean energy sources, with an emphasis on energy efficiency, rather than new energy production (50 % to 38 %).
* There is a skepticism about the current low price of oil. In 1986 Louis Harris found that most people (59 %) believed that cheap oil now will lead to more expensive oil later on because U.S. oil production will decline, leaving the U.S. vulnerable.

* People are not willing to pay higher oil prices in order to reduce the budget deficit. A 1987 Opinion Research study found that most Americans believe that raising the federal gas tax would not reduce the use of gasoline to the point where it would help reduce the inflow of imported oil (78 %).

Energy Management

* Americans continue to worry about the country's future energy needs. A 1987 Gallup poll found most people emphasized conservation rather than new energy production as the primary means of thinking about energy policies.
* When asked about some possible ways of reducing U.S. dependence on foreign oil, most people would be willing to pay an added tax rather than pursue environmentally dangerous options. A 1987 Cambridge Reports study found most people opposed to increasing the use of coal if it damages the environment (58 %), increasing the use of nuclear energy to generate electricity (56 %), or developing publicly owned wilderness lands (61 %) as a means of reducing our dependence on foreign oil. By contrast, most people (55 %) were willing to pay an oil import fee to reduce foreign oil imports.
* A 1986 NBC/Wall Street Journal poll found that the public believes that it is still important to conserve energy (89 %), but only 41 % claim they make a serious attempt to do so.

Government action

* People wanted a great deal of government intervention during the energy crisis. A February, 1977, Harris poll found that most people expected strong action. They wanted the president to alert the nation through televised speeches (80 %) and to impose tough new conservation programs on both consumers and on industry.
* The public supports a wide array of policies to achieve efficient energy use. Americans believe the best way to do this is to increase fuel economy standards to build more fuel efficient cars (61 %) and are willing to pay an extra $500 to get it (83 %).
* Recent studies have found that while people are not willing to relay environmental standards in order to produce more energy (54 %), they are also opposed to federal intervention to keep up the price of oil in order to keep U.S. production from declining (60 %).

Conclusion

There have been dramatic shifts in public opinion regarding energy policies, the public perception of an impending energy crisis, and specific steps to remedy our energy problems. current public perceptions indicate that through 1989, very much like early 1977, the public sensed the need for a new direction in energy policy.

Americans have become much more environmentally conscious and energy cautions in recent years. Looking for a broad-based public policy that will protect our natural resources, ensure their health and safety, and address issues of global survival, they strongly support increased energy efficiency and the use of renewable resources in order to meet these ends.

People expect things to get worse over the next decade and are looking for national leadership that will act quickly and decisively to protect our environment and promote energy efficiency. The public has little faith in business and industry to set these priorities and is looking toward government to mandate stricter standards for fuel efficiency more energy efficient heating and electrical systems, and a shift away from fossil fuels.

Americans are also prepared to do their share. Many already make serious efforts toward efficiency, through installing insulation and energy efficient heating or air conditioning systems, and buying more fuel efficient cars. They are willing to pay higher prices to promote environmental protection and efficient use of energy. Most people fear there is a greater price to be paid in the long run if we do not change our ways.

Forcasting the Energy Crisis

Q.: "Some people say the "energy crisis" like the United States experienced in the 1970s -- things like gasoline shortages, sharply higher prices, and oil supply disruptions -- will not happen again. Other people say we will once again have periods of energy crisis, just as we did 1970s.

Which view is closer to your own?"

	Crisis Over	Crisis Will Return	DK
* September, 1988	26	67	7
1988	33	49	18
1987	24	64	12
1985	37	49	13
1981	25	60	15

Source: Cambridge Reports; *Analysis Group.

Q.: "How concerned are you about the dependency of the U.S. economy on oil produced in the Middle East: Are you extremely concerned, very concerned, somewhat concerned, not very concerned, or not concerned at all?"

	Q4/87	Q4/80
Extremely concerned	21 %	49 %
Very concerned	31 %	30 %
Somewhat concerned	31 %	14 %
Not very concerned	10 %	4 %
Not concerned at all	5 %	1 %
Don't know	2 %	2 %

Source: Cambridge Reports; March, 1988

Q.:"Within *the next few years*, how likely is it that the United States will experience another energy crisis like the ones we had in the 1970s where there were oil shortages and log gas lines?"

Very likely	28 %
Somewhat likely	37 %
Not very likely	22 %
Not at all likely	8 %
Don't know	6 %

Source: ABC News; August, 1987

Q.:"Please tell me if you strongly agree, somewhat agree, somewhat disagree, or strongly disagree with each statement. ...The United States is so dependent on foreign oil that our national security could be endangered unless we quickly proceed with alternatives to imported oil."

	Strongly Agree	Somewhat Agree	Strongly Disagree	Somewhat Disagree	DK
1988	30	37	16	8	8
1987	34	37	15	8	6
1983	33	35	19	9	4

Source: Cambridge Reports

Q.:"Within *the next five years*, how likely do you think it is that there will be another shortage of gasoline like we had in the early 1970s when there were long lines at gas stations?"

Very likely	27 %
Somewhat likely	38 %
Not very likely	18 %
Not at all likely	13 %
Don't know	3 %

Source: Opinion Research; May, 1987

Q.: "Over *the next ten years or so*, how likely is it that the United States will once again experience the major energy shortages like those which occurred in the 1970s?"

Very likely	26 %
Somewhat likely	37 %
Not very likely	22 %
Not at all likely	11 %
Don't know	4 %

Source: Gallup Organization; December, 1987

Q.: " A few years ago there was an energy crisis in this country, with gasoline, oil and electricity in short supply. What do you think the chances are in the next year this country will have another severe energy crisis like the one a few years ago?"

	3/83	3/84	3/85
Very likely	12	16	12
Somewhat likely	27	31	27
Somewhat likely	27	23	28
Very likely	30	23	28
Don't know	4	7	6

Source: Roper Organization

Q.: "From what you have heard or read, how serious would you say the need is to save energy?"

	3/84	2/85
Very serious	47 %	47 %
Somewhat serious	41 %	41 %
Not serious	12 %	12 %

Source: Opinion Research Corporation

Q.: "People all have different concerns. Will you tell me for each of the following whether right now this is something that worries you?"

An energy shortage in this country?

A lot	26 %
A little	42 %
Not at all	32 %
Not sure	1 %

Source: Yankelovich, Skelly, & White; December, 1984

Q.: "In the next few years, how likely do you think it will be that there will be a shortage of oil in this country?"

Very likely	14 %
Somewhat likely	21 %
Somewhat unlikely	25 %
Very unlikely	33 %
Not sure	7 %

Source: Louis Harris & Associates; February, 1983

SOCIAL COSTS

Q.: "Would you be willing to pay higher prices for gasoline, natural gas, or home heating oil if these higher prices meant the oil industry would be able to develop a more secure supply of domestic oil?"

	Yes	DK	No
Q4/87	48 %	18 %	34 %
Q2/86	45 %	20 %	34 %

Source: Cambridge Reports; February, 1988

Q.: " The United States has a large trade deficit, which means more money is spent on imports, including oil, than is spent by other countries buying raw materials, productions and services. Do you think raising the federal gasoline tax would lower the use of gasoline enough to help reduce our trade deficit, or don't you think that the use of gasoline would go down enough to matter?"

Would lower use	12 %
Use would not go down enough	78 %
No opinion	10 %

Source: Opinion Research Corporation; May, 1987

Q.: "How dependent do you think the United States is on foreign oil: Do you think he United States is very dependent, somewhat dependent, not too dependent, or not dependent at all on foreign oil?

Very dependent	38 %
Somewhat dependent	47 %
Not too dependent	9 %
Not dependent at all	3 %
Don't know	3 %

Source: Cambridge Reports; July, 1987

Q.: "When developing new energy resources should the nation's primary consideration be protecting the environment, or should it be making sure we have sufficient supplies of energy for the future even if this caused some environmental damage?"

	Protect Environment	(Both)	Ensure Supplies	DK
1988	44	36	17	4
1987	45	33	18	4
1986	38	38	20	4
1985	35	39	21	5

Source: Cambridge Report

Q.: "In order to help provide for the country's overall energy for the future, which do you think should be emphasized most: finding ways to produce more energy, or finding ways to use less energy?"

	Produce More	Use Less	Both	DK
1988	38	50	11	1
1987	39	50	9	1

Source: analysis Group (1988); Gallup (1987).

Q.: "Do you think a secure supply of domestic oil is very important, important, not too important, or not important at all to our national security?"

Very important	60 %
Important	30 %
Not too important	5 %
Not important at all	2 %
Don't know	3 %

Source: Cambridge Reports; January, 1987

Q.: "Some people argue that cheap oil now will lead to more expensive oil later on, because U.S. oil production will decline. Do you think they are right or not?

Right	59 %
Not right	34 %
Not sure	7 %

Source: Louis Harris & Associates; April, 1986

ENERGY MANAGEMENT

Q.: "In order to help provide for the country's overall need for energy in the future, which do you think should be emphasized most?

Finding ways to produce more energy	39 %
Finding ways to use less energy	50 %
Both	9 %
Don't know/not sure	2 %

Source: Gallup Organization; December, 1987

Q.: "Here are several things the United States could do to reduce our dependence on foreign oil. Please tell me whether you strongly favor, favor, oppose, or strongly oppose each one as a method of reducing our dependence on foreign oil.

	Strongly Favor	Favor	Strongly Oppose	Oppose	DK
Increasing use of coal even if it damages the environment	11	25	41	16	8
Increasing the use of nuclear energy to generate electricity	12	24	28	28	8
Imposing a tax on all crude oil imported into the U.S.	20	35	22	11	12
Developing oil and natural gas reserves in publicly owned wilderness areas, even if it damages the environment	11	20	34	27	8

Source: Cambridge Reports; July, 1987

Q.: "Do you think that it is still important for Americans to conserve energy as it was a few years ago?"

Yes, still important	89 %
No, not important	10 %
Not sure	1 %

Source: NBC News/Wall Street Journal; April, 1986

Q.: "Today, would you say that you personally try to conserve energy?"

A lot	41 %
Some	44 %
Just a little	11 %
Not at all	3 %
Not sure	1 %

Source: NBC News/Wall Street Journal; April, 1986

Q.: "What would be the most effective way of cutting back gasoline consumption in this country (from a list)?"

Raising the price of gasoline by 50 cents a gallon.

Very effective	74 %
Not very effective	21 %
Not sure	5 %

Source: Louis Harris & Associates; February, 1977

Q.: "Let me read you some statements others have made regarding saving energy. Please tell me whether you agree or disagree with each one. Would that be strongly or just somewhat?"

	Strongly Agree	Somewhat Agree	Strongly Disagree	Somewhat Disagree
An important way for utilities to reduce their use of fossil fuels like oil and coal, is for utilities to help their customers use energy more efficiently.	55	36	4	3
An important way to save energy is to make it easier for people to use mass transportation and carpooling	56	31	6	5
An important way for utilities to reduce their use of fossil fuels like oil and coal is to use renewable energy sources like solar, wind, and water	61	27	4	4
An important way for utilities to reduce their use of fossil fuels, like oil and coal, is to use more nuclear power	23	26	17	27

Source: RSM

Q.: "Let me read you a series of proposals. Please tell me for each one whether you strongly favor, somewhat favor, somewhat oppose, or strongly oppose is."

	Strongly Favor	Somewhat Favor	Strongly Oppose	Somewhat Oppose
An increase in the fuel economy standards of automobiles requiring auto companies to build more fuel efficient cars.	61	29	5	3
New standards for home and office construction that would require more installation of insulation and more energy efficiency heating and electric systems	60	30	4	3
New regulations and incentives for utility companies requiring that they help consumers conserve energy and that the utilities are rewarded for being energy efficient.	48	37	6	6
An expanded federal program that would pay the costs of weatherization in the homes of low income people to reduce their monthly energy bills.	43	37	9	9
A 25 cent per gallon import fee on imported oil to reduce our dependence on foreign oil.	26	30	20	17
A proposal to build more nuclear power plants in the US.	14	22	18	41
A proposal to increase the gasoline tax by 30 cents a gallon, phased over 3 years, to encourage more efficient use of energy and lower energy consumption.	7	17	21	51

Source: Analysis Group

Q.:" If it came down to a simple choice between developing new energy resources and preserving publicly owned wilderness areas, which do you think the nation should choose?"

	Use Energy	DK	Preserve Wilderness
1988	42	11	47
1987	50	13	39
1986	62	11	28
1978	67	14	19

Source: Cambridge Reports

ROLE OF GOVERNMENT

Q: "Because of declining energy prices over the last few years, oil and natural gas companies face serious economic difficulties. While government regulation of oil and natural gas industries has been reduced in recent years, these industries are still governed by many regulations put in place when energy prices were rising. Do you favor or oppose additional reduction in government regulations of the oil and natural gas industries to help them through their current economic difficulties?

	Favor	DK	Oppose
Q1/88	35	28	36
Q3/87	36	23	42
Q3/86	34	25	41

Source: Cambridge Reports

Q: "How strongly do you favor or oppose relaxing environmental standards in order to produce more energy?"

Strongly favor	11 %
Favor somewhat	28 %
Oppose somewhat	19 %
Strongly oppose	35 %

Source: Gallup Organization; December, 1987

Q: "So you think the federal government should intervene to keep up the price of oil in order to keep U.S. production from declining?"

Should intervene	33 %
Should not intervene	60 %
Not sure	7 %

Source: Louis Harris & Associates; April, 1986

Q: "If you had to choose between a fuel that made your car run better and cost less and a fuel that was better for the environment and cost more, which fuel would you choose?"

	Run Better/ Cost Less	Better for Environment/ Cost More	DK
1989: August	39	45	16

Source: Cambridge Reports

Q: "Would you favor or oppose the proposal that cars, on average, get 45 miles to a gallon, if you knew that a new car would cost the buyer $ 500 more but the added cost would be recovered by the gas saved in the first four years alone?"

	Favor	Oppose	DK
1989: October	83	13	5

Source: RSM

Q: "Let me read you some statements others have made regarding saving energy. Please tell me whether you agree or disagree with each one. Would that be strongly or just somewhat? ....US automobile makers have the technology to make cars averaging 45 miles to the gallon well before the year 2000.

	Strongly Agree	Somewhat Agree	Strongly Disagree	Somewhat Disagree	DK
1989: October	54	29	6	4	6

Source: RSM

Q: "Would you like to see new safety rules and emergency cleanup teams in place at all major oil facilities, even if the expense would raise the price of oil and gasoline?"

	Yes	No	DK
1989: May	81	13	6

Source: Media General/Associated Press

Q: "Do you think that the federal government should gradually increase the tax on each barrel of crude oil to encourage continued energy conservation, even though it might mean higher prices to the consumer for oil and gasoline?"

Yes	28 %
No	6 %
No opinion	4 %

Source: Opinion Research Corporation; August, 1983

Q: "Should the President alert the nation to the seriousness of the energy problem with a series of televised speeches?

Yes	80 %
No	13 %
Not sure	7 %

Source: Louis Harris and Associates; February, 1977

Q: "Steps President Carter Might Take on Energy. Recommend a bill that would impose tough new energy conservation programs on both consumers and on industry."

Favor	70 %
Oppose	18 %
Not sure	12 %

Source: Louis Harris & Associates; February, 1977

7.8 The Energy Policy of the Saarland: A Pragmatic Energy Policy in a Global Context Confronted with Uncertainty Concerning the External Costs of Energy

Dr. Frithjof Spreer
Head of the Department for Energy and Technology
Ministry for Economic Affairs of the Saarland
Saarbrücken, Germany

1. The Saarland's Energy Policy

With its 1.1 million inhabitants, the Saarland is one of Germany's smallest states. It is located on the border to France and Luxemburg, midway between Frankfurt and Paris.

The Saarland's present energy policy dates back five years. I am of the opinion that it is Germany's most consistently innovative energy policy. This is particularly due to the people behind the policy: I am backed by Hajo Hoffmann, the Saarland's Minister for Economic Affairs, and Oskar Lafontaine, the Saarland's Minister President. Without the backing of these individuals, I might engineer the world's best programs, but would not be able to implement them.

A state's energy policy is not autarkic. It is integrated in federal energy legislation, we just interpret it within the context of the Saarland.

2. Energy Industry Law and External Factors

The Energy Industry Law of the Federal Republic of Germany dates back to 1935. Its main objectives are providing an economical as well as secure energy supply. The law poses a direct obstacle to the internalization of external costs. There is no obligation to reduce the social costs of energy systems or to aim energy policy towards long-term strategic or global objectives. Thus, what we do in the Saarland is voluntary and verges on this side of the border between the legal and illegal.

Despite this we now have a far-reaching consensus with our slightly fewer than 20 utility companies in the Saarland. By applying the tactic of the carrot and the stick we have transformed them into utility service companies of sorts which broadly adhere to the principle of least-cost planning. This is also the result of personnel policy: Innovative, younger individuals have now gained the upper hand in our utility companies for the most part. They derive great personal pleasure from working for a company whose strategic orientation is recognized by society. In the latest analysis, personnel policy appears to be more significant than making progress with respect to external factors.

3. Justiciability and Incidence

There's nothing to be said against making progress with respect to external factors. Microeconomic views are short-sighted, particularly with respect to energy policy and are generally misguided. The problem of energy is of a global and interregional nature and involves present generations and generations to come.

These realizations do not necessarily result in the right consequences with respect to energy policy. To be sure, applied energy policy has to be justiciable or able to hold up in a court of law. "Unjust" incident effects (i. e. unfair footing of the bill) have to be prevented. Up until now, research into the external factors or social costs has not resulted in justiciable instruments of energy policy. Conversely, we will not and cannot wait until research has caught up. Therefore, our energy policy must perforce be of a pragmatic nature. It more or less "muddles through". Decisions relating to energy policy are preceded by a phase of extensive negotiation between utility companies, the consumers and municipal and community governments. My staff and I are thus forced to be more versed in the skills of the animal trainer rather than those of the scientist. We frequently resort to tricks.

4. Organizational and Structural Problems of the Utility Industry

Our Energy Industry Law has resulted in the formation of monopolies and cartels. They strive for constant growth in energy consumption, tend towards excess energy generation capacity and maximum capacity utilization of the same. Decisions pertaining to energy generation investments are made far and away from the places of actual consumption.

Local feedback between production and consumption is no longer present, at least with regard to large utility companies. This problem poses a systematic obstruction to least-cost planning. A large utility company is completely indifferent to whether energy conservation investments lead to lower utility bills. Such decisions are simply not its concern. Due to the great distance between the places of production and consumption, there is hardly anybody who is capable of comparing these two types of investments. How can the social costs of inappropriate structures be computed?

5. Thinking Globally, Acting Locally

That's why we in the Saarland have long been in favor of localizing, decentralizing and regionalizing energy supply. Today we have strong utility companies which are subject to local political control, a relatively far-reaching decentralization of energy generation by virtue of the co-generation of power and heat and a surprisingly profitable "energy conservation policy" among our municipal utility companies.

This has resulted in our regional utility companies becoming subject to the political pressure of local governments. Upon the expiration of licensing agreements, many local governments have wanted to buy back the energy supply networks from the regional utility companies. The regional utility companies have thus been forced to come up with something so that they would not become unpopular and lose out on these choice tidbits. In the course of the past five years they have assumed the function of local utility companies for small communities for the most part. In licensing agreements they have assumed obligations towards local governments and energy conservation policy, things which they had heretofore consistently rejected. We now have marked competition for the most innovative energy strategy among regional utility companies as well as between regional suppliers and local utility companies.

6. Particular Energy Sources are not a Primary Issue with Us

The Saarland is a coal state, yet coal policy plays merely a secondary role. To be sure, we strive towards a globally sensible and not a solely regionalistic energy policy. Our energy policy is implemented basically independently of the question as to which primary energy source is affected. With one exception: We are adamantly against atomic energy. It appears a priori to be unacceptable to us as well as too risky on a global scale. That is why there is no nuclear power station or "nuke" to be found in the Saarland and why no one is willing to build one.

7. The Saarland Bridge Strategy

We perceive of energy policy as forming a bridge to the age of solar energy. All the instruments of energy policy are building blocks comprising this bridge.

Our energy policy instruments have been transformed into three piers or buttresses:

- energy conservation
- rational energy utilization
- introduction of renewable energy sources to the market.

We are of the opinion that energy conservation and rational utilization of energy are the necessary prerequisites for introducing renewable energy sources. The structures and level of our present energy consumption are not compatible with the massive introduction of renewable energy sources. We perceive an energy policy of continually reducing energy consumption to be of paramount importance.

8. Energy Conservation Policy

a) Scheduled Pricing Policy

Those who have discovered the external costs of the energy system perceive our energy prices to be exceedingly low. However, being the energy pricing authority we cannot raise these prices. The consequence would be that utility companies would reap larger profits. Being a state government, we have no legal authority to impose levies in the form of energy levies or taxes. Consequently, we endeavor to at least modify the pricing structures, so that there is an economic incentive to reduce and smooth out energy consumption at a lower level. We strive towards time-variable linear pricing for such forms of energy as electricity, gas and district heat. There is presently a large-scale test being conducted in the state under the auspices of the Fraunhofer Institute for Systems and Innovation Research.

Yet we don't want to wait until testing is completed in two years or so from now. That's why we have reached an agreement with our energy suppliers on introducing a temporary simple linear or quasi-linear electrical power pricing schedule as of the beginning of 1991. That means that there will no longer be any or hardly any basic rates charged. The extent of linearity of our power pricing schedule in the private household sector would amount to between 93 and 100 percent as of the beginning of next year. As consequence, we expect a reduction in power consumption in the medium term of between 5 and 10 per cent. Introducing this linear pricing schedule took five years of persuasion.

b) Production of Negawatts by the Saarland Energy Agency

Three years ago we founded Saarländische Energieagentur GmbH or the Saarland Energy Agency Co., abbreviated SEA. It is an industrial company in which the state has a majority shareholding, the minority shareholding being in the hands of a number of Saarland utility companies. The SEA is designed to demonstrate that it is possible to turn a profit in connection with low energy pricing coupled with energy conservation. It has been so successful that other utility companies in the Saarland are presently in the process of acquiring shares in this company.

The SEA makes energy conservation investments in connection with large consumers (industry and local governments) on its own account and at its own risk. The differential between the previously higher costs of energy and the new energy costs is charged to the consumer's expense to refinance investment costs. At the end of the amortization period (up to 6 years) the facility will be transferred free of charge to the consumer. In this way we have been able to profitably construct waste heat utilization, heat reclamation, biogas and small co-generation plants.

Initially our utility companies were displeased with the SEA's actions, to say the least. They lost out on megawatts without being in on negawatts. Today they have come to the realization that it is better at any rate to be involved in the least-cost planning business. We have also seen that the amortization periods of many negawatt investments are shorter than those of megawatt investments.

c) Energy Conservation Programs of Local Governments and Industry

The state government promotes the development of local and industrial energy conservation programs by way of non-repayable subsidies. 50 of 53 communities now have such programs or are in the process of developing them. As a result, a local framework of action for innovative energy policy is being created statewide. I will come back to this later.

In the industrial sector we have discovered that most companies are not familiar with their own energy structure. As a consequence, they have not developed concepts for reducing their energy costs. That just hasn't concerned them. Through the industrial energy program, however, we have alerted many companies and have brought about permanent changes in the energy structure of their production operations, the result being that completely different and in the long term state-of-the-art production techniques have been successfully introduced.

d) Earmarking of License Fees

By employing tactics verging on legal sleight of hand, we have managed to wrest away from local governments half of the energy sector license fees normally payable to them, followed by returning these funds to them, at the same stipulating that they be employed for energy conservation measures, rational energy utilization techniques and the promotion of renewable energy sources. That created a lot of trouble for us at the beginning, even with our own Minister for Internal Affairs, but no one dared to contravene the stipulations of the Minister for Economic Affairs, as the license fees as a whole would have been jeopardized.

We require that an annual report be filed concerning the use of these funds in the form of a local government energy report. In addition, the local government energy report documents the implementation of the local energy concept previously promoted by us.

As a consequence, many communities now take pleasure in dealing with the topic of energy; many local governments have discovered energy policy for the first time.

e) Formulation of License Agreements

Due to the systematic involvement of local governments in energy policy as detailed above, it was obvious that they would now want a piece of the action by completely altering license agreements with regional utility companies. For two years there was a veritable wrangle in the Saarland for more favorable license agreements. Today I believe that we have the most innovative license agreements in all of Germany. The regional utility companies were forced to commit themselves to initiating all the investments required for energy conservation, rational energy utilization, waste heat utilization and the introduction of renewable energy sources which the respective local government had shown to be relatively cost efficient in its local energy concept. In other words: the regional utility companies were forced to commit themselves to implementing the local energy concept in the license agreement. That was unheard of.

9. Rational Energy Utilization Policy

a) Waste Heat Utilization Policy

In the industrial economies of our climes we generate more utilizable waste heat, particularly in power stations, than we will ever have any practical use for. Due to the fact that waste heat utilization, i. e. district heating as opposed to oil-, gas- or electric-power-fired heating, involves practically no external costs, we in the Saarland have accorded priority to district heat in the heating market. Within the space of ten years -- of which five years have since elapsed - we will have doubled the market share of district heat in the Saarland. At ca. 25 % it will amount to approximately four times the national average.

How have we done this? First we have threatened all power station owners and all those wishing to construct new power stations with a requirement to utilize waste heat in anticipation of the legal framework still to be created. To be sure, we have issued permits for co-generation plants only, i. e. for 6 large plants during the past five years with a combined output of ca. 300 MW. We have promoted the co-generation of power and heat and utilization of waste heat given off by existing plants by way of investment subsidies of up to 30 %. In all the energy programs of local governments, the furthering of district heat has been given priority over other energy supply measures. There are district heat priority areas within communities which do not permit the encroachment of gas-fired heating.

This program has resulted in a reduction in the emission of CO_2 produced by Saarland heating systems of at least 15 %. That's quite an achievement for just one measure!

The Saar district heat system is undergoing constant extension and now covers a total length of 40 km. It collects waste heat from various sources, from industrial facilities and power plants, and conveys it to the individual communities. In this way small towns and villages can be provided with district heat as well.

Outside of the Saar district heat collection system and far and away from large power plants we have implemented a policy of small-scale local district heat systems based on gas- and oil-fired unit-type thermo-electric power plants. This is an area in which regional utility companies attempt to outbid and outdo one another. In accordance with local energy concepts, an average of two unit-type thermo-electric power stations can be realized in each Saarland community. We have implemented a small subsidy program here as well in order to help boost more plants above the economic feasibility threshold. That basically means that we, the government, assume external costs in order to ensure profitability.

b) Direct Gas Heating

Outside of the areas suitable for district heat we have been attempting to replace oil- and solid-fuel-fired furnaces with gas-fired furnaces. The stipulation of gas priority areas by villages and small towns outside of the district heat system is a topic of major significance. Three different Saarland companies are now attempting to gain access to these gas priority areas by way of gas license agreements. These companies are now involved in investment projects with an amortization period exceeding 18 years.

c) Coal Policy

Even a coal state's coal policy must be well conceived. The individual coal furnaces or coal-fired thermo-electric power plants once common in the Saarland are fossils without a future. As concerns our energy policy, coal has a future only with respect to medium and large-sized co-generation plants employing state-of-the-art flue gas scrubbing and desulfurization techniques and complete utilization or recycling of residue products. Under these circumstances, our supplies of coal, petroleum and natural gas are sure to last a long time on account of the sizable reserves present.

We no longer want any coal-fed billowing black clouds, i. e. we now only construct state-of-the-art coal-fired thermo-electric power stations. Two large coal-fired thermo-electric plants were put into operation last year. They incorporate state-of-the-art flue gas treatment for the removal of dust, SO_2 and NO_x. Their energy conversion efficiency amounts to between 70 and 90 per cent. We have achieved a residue recycling rate of over 90 per cent; 100 per cent has been achieved for the co-generation plant in Saarbrücken. The residue is used in the construction and mining industries.

In Saarbrücken, 95 % of the electricity used in the city is generated in the city's own thermo-electric plants fired almost completely by hard coal mined in the Saarland. Despite the widespread use of hard coal, CO_2 in the City of Saarbrücken has dropped by 30 % during the past ten years. This is one of the consequences of the policy I have described here.

10. Introduction of Renewable Energy Sources to the Market

The Saarland was the first German state to initiate a market introduction program for renewable energy sources. Whereas the federal government has been primarily promoting grandiose demonstration and pilot facilities not involving any follow-up projects, we have been granting subsidies to end users employing standard techniques. The user of solar collectors, photovoltaic, bioenergy or small-scale hydroelectric power facilities is awarded a non-repayable subsidy normally amounting to 50 percent of the investment and installation costs. This year the number of such facilities will amount to ca. 500. Primary importance is attached to solar collectors.

The Saarbrücken Municipal Utility Company and VSE, our largest regional electric power supplier, lend support to our market introduction program most particularly by way of favorable prices for solar energy (DM O.25/kWH) as well as by funding a team of experts providing objective advice pertaining to solar energy applications and applications for funding.

The philosophy of our market introduction program boils down to this: The mass purchase of such facilities will result in a cutback of costs. Of Course, the Saarland, small as it is, has relatively little impact on the market as a whole; moreover there have been price increases as the result of the boom demand for these systems. Despite this, we have considered the political demonstration effect to be of primary importance. Many other German states have since followed our example, and the federal government seems to be somewhat headed in our direction as well.

That wraps up the short on the Saarland's policy of "muddling through".

7.9 A Regional Strategy for Energy Efficiency - with special reference to the Co-Operation between Denmark and Schleswig-Holstein

Dr. Klaus Rave
Ministry of Health,
Social Affairs and Energy
State of Schleswig-Holstein
Kiel

Introduction

Although the project of a single European Market is strongly and controversially debated cross-border co-operation is still a fledgling. Although the European utilities all feed into the trans-european grid the EEC-member-states have not yet established regional policy-making processes which would enable them to steer the co-operation. There is a wide gap between the proposals of the Commission in Brussels and the "alternative" concept of "Think global, act local". There is another gap between the socio-technical systems - connected by one grid - concerning the way and means by which to incorporate "externalities". The Governments of Denmark and Schleswig-Holstein are set for bridging these gaps. They have started various projects under the guidance of an agreement by the respective Ministers for energy, Jens Bilgrav-Nielsen and Günther Jansen. It is supposed to be a bridge between two energy systems, which could hardly be more different: Denmark has opted against nuclear power whereas Schleswig-Holstein is heavily (up to 80 %) relying on electricity generated in three nuclear plants but is set for phasing out nuclear energy within two legislative terms.

1. The Agreement for Cross-Border Co-operation

When the two Ministers met in Kiel on the April 13, 1989, they agreed to co-operate on the following levels

- efficient use of energy
- environmentally feasible energy technology
- research and development
- training and qualification
- energy planning.

The agreement should strengthen political initiatives of the two partners within the EEC framework and thereby enhance the chances for certain projects. In addition to that both partners proclaim that it is their common task to pursue the goals as set out in the United Nations Report commonly referred to as the "Brundtland Report".

A joint steering committee is undertaking the task to initiate suitable projects and further the co-operation. Numerous working-groups have been installed. They include scientists as well as representatives from the - regional and local - utilities, engineers as well as planners. Their agenda ranges from conservation to the economic feasibility of wind energy, from district heating to the use of biomass. A network is in the making. Comparing two different systems, evaluating two different socio-technical backgrounds should lead to a new conception of regional co-operation.

A new culture in the technical as well as political meaning of the word might be established which could - or rather: should - enable us to assess what kind of context is needed to achieve the most effective ways and means to incorporate the external costs of energy production, the environmental costs of electricity into the production/generating system itself thereby making it permanently as efficient as possible.

2. The Brundtland-Town-Project

One of the numerous joint projects particularly should lead to some conclusions as to what can be achieved locally and then be transferred to the regional and international level. It is a "best case" project meaning that whatever instruments could be applied to achieve an - environmentally - supreme standard of energy efficiency should be applied. The steering committee came forward with a proposal that was later on endorsed by the Ministers to twin two towns north and south of the border which set themselves the task to reduce their energy consumption in accordance with the goals as fixed in the "Brundtland Report" to the United Nations i.e. a reduction of fifty percent of primary energy.

At first the "rules of the game" had to be fixed. A working group under the auspices of the steering committee drafted an outline of the standards possible participants would have to comply with. These "rules" then were - after having been "officially" i.e. jointly adopted - made public to all municipalities north and south of the border which might be interested in participating. Two meetings jointly chaired by a Danish and Schleswig-Holstein representative of the steering committee - were held to inform the local authorities in some detail about the project. A call for participation was issued and a deadline set. 6 Danish and 7 Schleswig-Holstein authorities applied. The selection process was up to the steering committee. Two towns, Bredstedt and Toftlund, were finally chosen on the following grounds

- comparable number of inhabitants
- manifestation of special interest in energy conversation
- manifestation of public involvement in the process of (energy) policy making
- pilot/demonstration projects successfully managed in the past
- willingness to participate in the dissemination of results and findings.

The selection was followed by the twinning of the two local authorities in a meeting again jointly chaired by the steering committee. The maximum effort of the two towns will afford maximum local support and outside help and assistance. There certainly is a competitive element to that project; how fast, by what means, to what cost can the reduction goal be achieved? There is an element of trial and error in a common learning process: a small-scale know-how transfer to be widespread. The areas of activity both local authorities have subscribed to are numerous:

- reduction of energy consumption in all public buildings
- use of whatever means of renewable energies available for electricity generation
- application of the highest possible standard of CHP-systems (combined-Heat-and-Power)
- application of maximum low-energy standards for new building sites
- experiments in the rating system to achieve consumer's energy awareness
- involvement of all local organizations and institutions in the project
- organization of special "efficiency programs" for households and small business.

A "Brundtland Newsletter" shall keep the interested public informed about the progress of the project both north and south of the border. A proposal for special EEC-funding has been drafted. Both Governments have undertaken it not only to assist that proposal but give preferential treatment to both local authorities in case of application for public grants with respect to local energy projects. These could include

- renewable energy
- district heating

- CHP systems
- energy efficient buildings
- reduction of electricity consumption in public buildings.

A local energy concept endorsed by the authorities and the utility will be the cornerstone of any future activities. The "Brundtland-Town-Project" is meant to establish a local tradition of Integrated Ressource Planning. The cost of investments and the social and environmental benefits will be evaluated. The gap between theoretical assumptions of the external cost and the local level of energy awareness could thereby be bridged. The competitive spirit should be a guarantee for good results. The supervisory function of the steering committee should ensure that results and obstacles, different standards and approaches are openly discussed, thereby implementing a continuous process of improvement. The project is a nutshell of what both governments have subscribed to: an environmentally and socially viable energy policy - a new, common socio-technical environment and culture. Whereas the respective backgrounds do differ greatly the aims correspond

- to opt out or phase out the use of nuclear power
- to reduce CO_2-emissions according to the targets set by the "Brundtland-Commission".

3. Comparative Research

Research has and is being carried out in three areas:

- the potential of CHP systems
- the potential of electricity conservation
- the comparison of CO_2-emissions in Denmark, the Federal Republic of Germany and Schleswig Holstein

3.1 The Heatatlas

The CHP potential should be included in a data basis that itself should lead to a "District Heating or CHP Atlas". That atlas would form the backbone of any future planning be it by local authorities or utilities or joint ventures of both. Danish engineering and planning standards will be adapted to the administrative framework as it exists in Schleswig-Holstein thereby making it - theoretically - transferable throughout the EEC in general.

It is a central element in the energy policy of Schleswig-Holstein to promote an energy production based on local energy sources, as well as to ensure a rational use of the various possibilities of energy supplies. As part of this objective it is the intention to carry out an energy planning project, which can form the basis for the establishment of a 'heatatlas' for the main areas of energy consumption in the region. The 'heatatlas' will serve as a basis for a political decision on the development of decentral combined heat and power production. The total potential in Schleswig-Holstein for a decentral combined heat and power production is estimated to some 400 MWe distributed over 40 - 60 urban areas.

The project will be carried out as a Danish-German cooperation project, as it is the intention to use the considerable experiences in the area of energy and heat supply planning, which have been gained in Denmark over the past 10 years. The development of the planning methods resulting from the project could very well form the basis for similar planning projects in other EG regions where the use of local energy sources is being considered.

The Ministry of Social Affairs, Health and Energy in Kiel is responsible for the project while the project management and analysis will be carried out by Danish and German consultancy companies.

3.2 The AFK-Project

Electricity saving was the second area in which research and know-how transfer took place. It was agreed to study the transferability of the Danish AKF-Project.

The background of the project is the recognition of environmental damages caused by the burning of fossil fuels. The project is furthermore part of an agreement between the Danish government and Socialdemokratiet (The Danish Labour Party) on the future development within the electricity sector and electricity savings (June 6th, 1986). The project is carried out by AKF (Local Governments' Research Institute) on behalf of the Ministry of Energy. Other institutions - universities and utilities have also been and will be involved in different phases of the project.

The research questions in the project are the following:

a. Description of the Danish electricity consumption. The distribution among sectors. Past and future trends.

b. Assessment of behavioural and technical conservation potentials.

c. Economic evaluation of the conservation potential.

d. Survey of means to obtain electricity conservation.

e. Practical experiments with means to obtain electricity conservations in selected local areas.

f. Evaluation of the national consequences of various policies directed towards reducing electricity consumption.

Table 1: Estimates of Electricity Conservation Potentials

Sector	Electricity consumption	Conservation Potential							
		Behaviour		Technology		Behav.+Techn.		Future techn.	
	in 1986 TWh	TWh	%	TWh	%	TWh	%	TWh	%
The hausing sector	9,1	0,9	10	2,6	28	3,2	35	4,5	60
Agriculture and market gardens	1,5	0,2	10	0,7	47	0,8	52	1,0	65
Industry	7,8	1,2	16	3,3	42	3,5	45	3,9	49
Trade and service	4,0	0,6	16	1,4	34	1,7	41	1,6	39
Public sector	3,3	0,8	24	1,4	43	1,7	52	1,9	57
Other + undistributed	0,9	0,1	6	0,1	13	0,2	18	0,1	13
Total	26,6	3,8	14	9,5	36	11,1	42	13,0	49

Source: Johansson, Mogens and Thomas Pedersen. Teniske elbesparelser. AKF's Forlag

The "Transferability-Study" was published in May 1989. It came to the conclusion that there exists a considerable similarity between Denmark and Schleswig-Holstein in respect of potential savings.

Table 2: Energy, 1987

	Schleswig-Holstein	Denmark	Schleswig-Holstein/Denmark
Primary Energy Consumption	458,0 PJ	807,3 PJ	0,57
Primary Energy Consumption/GDP2	6,5 TJ/DM	4,5 TJ/DM	1,44
Final Energy Consumption	302,0 PJ	594,8 PJ	0,51
Final Electricity Consumption	38,2 PJ	99,4 PJ	0,38
Final Energy Consumption Per Capita	118,3 GJ	116,1 GJ	1,02
Net Electricity Consumption Per Capita	15,0 GJ	19,4 GJ	0,77

1 DM = 3,857 Kr.

2) For Denmark: GDP

Table 3: Electricity Consumption by Sectors In Schleswig-Holstein and Denmark (1986 in %)

	Total GWh	Total %	Households %	Industry %	Agriculture %	Commerce %	Public Institutions %	Traffic %
Schleswig-Holstein	10.179	100	38	30	5	17	9	1
Denmark	26.600	99	34[1]	29	6	15	12	3

(Sources: Energy Ministry, Schleswig-Holstein und Bogens Johansson und Thomas: Tekniske elpesparelsen, AKF 1988)

1) Including consumption in the agricultural households

Figure 1: Electricity consumption per capita 1987

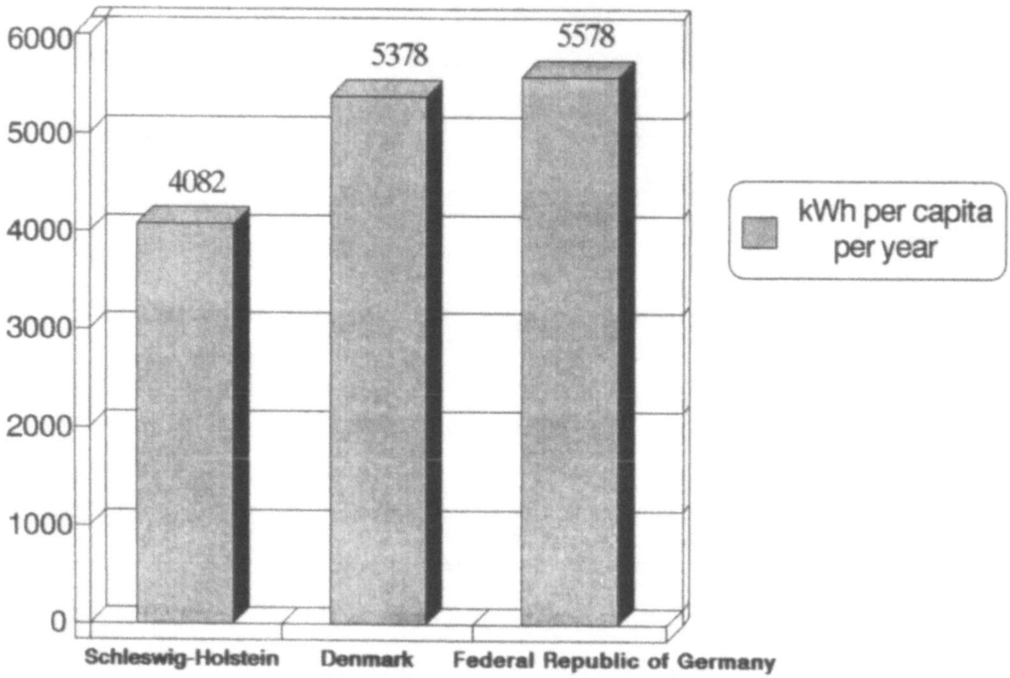

Source: Statistik Tiarsoversigt 1987, Volkszählung 1987, "Elektrizitätswirtschaft" Sonderdruck Nr. 4108, VDEW 1987. Elforsyningens Tiarsoversigt 1978-87, DEF 1988.

Table 4: Electricity Consumption by End Use in Schleswig Holstein in %. Denmark in Parenthesis, 1986

	%	(% DK)
Refrigeration/Freezers	7	(9)
Ventilation	9	(9)
Pumping	13	(11)
Lighting	11	(12)
Other Electrical Engines	37	(35)
Other Uses[2])	22	(24)

Source: "Elforbruget i industrien 1", dk-TEKNIK, August 1987.

[2]) Other uses are minor applications like pressurised air, process heat, melting and electronics

Table 5: Overview of the Saving Potentiats in Denmark by Sector

	Electricity Consumption in 1986	Saving Potential Behavior	Techno-logy	Behavior plus Technology	Future Techno-logy
Households	9,1	10	28	35	50
Agriculture	1,5	10	47	52	64
Industry and Trade	7,8	16	42	45	49
Services	4,0	16	34	41	39
Public Sector	3,3	24	43	52	57
Others	0,9	6	13	18	13
Total	26,6	14	36	42	49

Table 6: Theoretical Savings Potential for Schleswig-Holstein

	Consumption in 1986 GWh	Behavior GWh	Techno-logy GWh	Behavior plus Technology	Future Techno-logy
Households	3.899	9	25	32	45
Agriculture	499	10	47	52	64
Industry	3.016	17	43	46	49
Commerce	1.747	16	34	41	39
Public Sector	955	24	43	52	57
Others	63	6	13	18	13
Total	10.179	14	34	40	47

3.3 The CO_2-Study

Another study (IWU and Dansk Teknologisk Institut, September 1990) reveals that the Danish energy system although heavily relying on coal for its generating capacity is very efficient in its perspective of a CO_2-reduction goal compared with Germany and Schleswig-Holstein in particular.

Table 7: Comparison of the Specific Energy Demand and Emissions in the Low Temperature Heat Market in 1987

	Useful Energy kWh/(m^2 a)	Final Energy kWh/(m^2 a)	Primary Energy kWh/(m^2 a)	CO_2-Emissions kg²/(m^2 a)
FRG	161	231	281	67
Schleswig-Holstein	170	244	291	70
Denmark	131	168	188	49
Denmark/FRG	81 %	73 %	67 %	73 %

Figure 2: Share of the Different Energy Sources in the Low Temperature Heat Market for Private Households in the FRG 1987

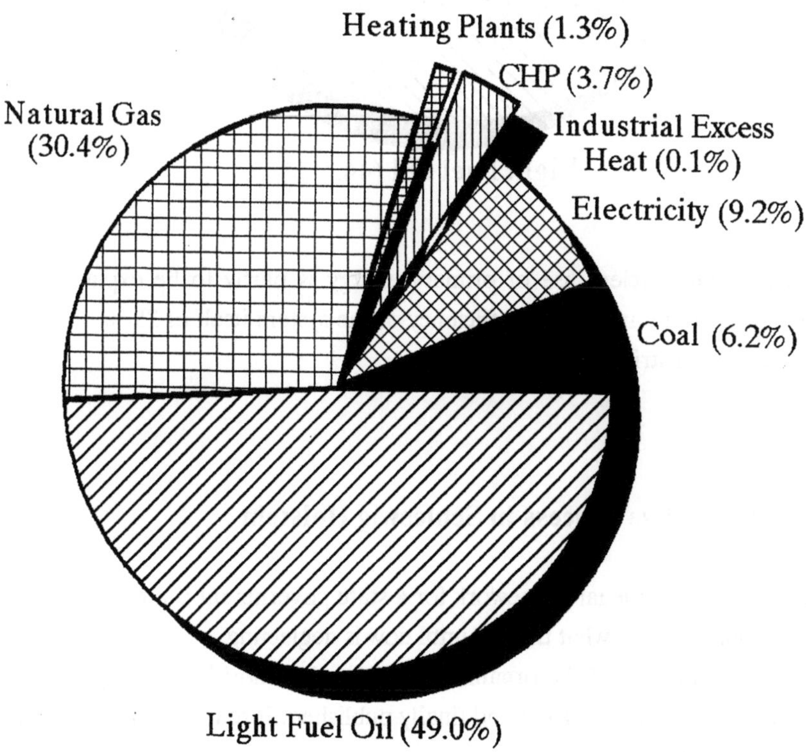

Figure 3: Share of the Different Energy Sources in the Low Temperature Heat Market for Private Households in Denmark 1987

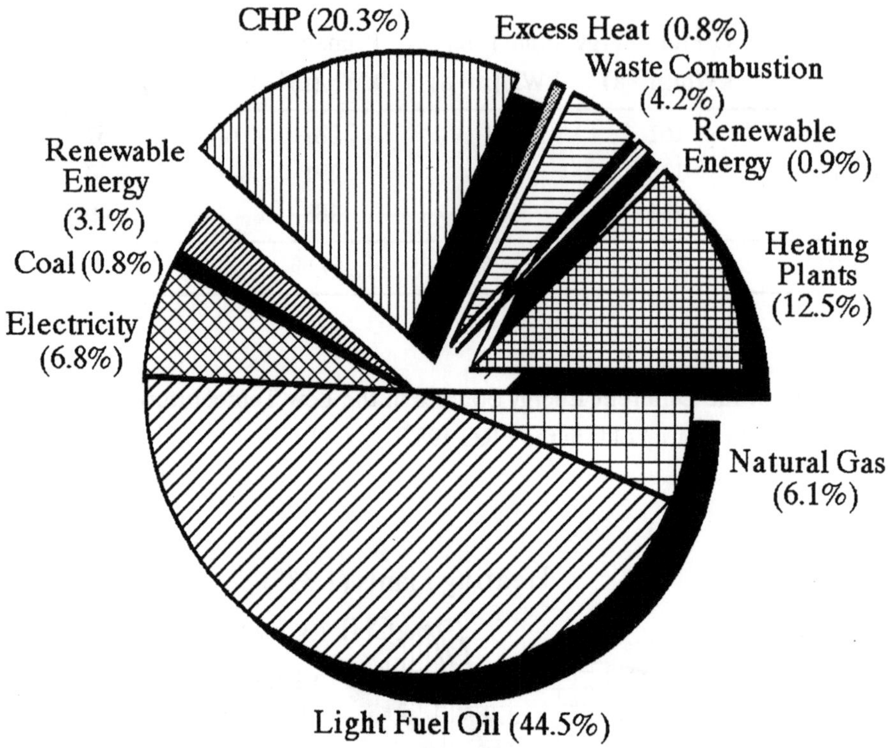

The reason for that efficient use of primary energy is partly to be explained by better building standards and partly by the efficiency of the system itself as 25 % of the energy in the heating sector is distributed in CHP-systems.

4. The Collaborative Process in Schleswig Holstein

As limited as the existing legal framework for a state government is a network of voluntary agreements should achieve what the law until now is neglecting.

At first a joint declaration of the organization of local authorities, the regional utility and the government was drafted, signed and finally published. It reaffirms the essential goals of the "efficiency strategy" as there are

- energy saving

- energy efficiency
- renewable energy.

The protection of the environment is given priority. The three institutions solemnly declare to co-operate in order to meet that common goal.

Particular reference is given to global warming or the Greenhouse effect.
This collaborative process should be followed by mainly two initiatives. The utilities are obliged to enter into special agreements with the municipalities wheres the Government is obliged to give way to the raising of an "energy levy" of 4 to 8 % of the respective electricity sales to the municipalities smaller or larger than 5.000 inhabitants. The contractual relationship between utility and municipalities follows the general principles as laid down in the joint declaration. Both partners are entitled to demand - by a joint decision-making process - whatever investment is needed to achieve the common goal. As the municipalities do not have local utilities of their own, the regional utility declares to act *as if it was such an utility*. Such a contract having been concluded the government than will give permission to the raising of the levy as mentioned above. To legalise that procedure the government has published a statutory instrument the structure of which again reflects the joint declaration as well as setting standards for the contracts that have to be concluded between utility and municipalities. This being the legal framework for the granting of the levy it has binding effect on what to spend that money for. The collaborative process is thus enriched by the "carrot-and-stick-principle".

Half of the money raised by the levy has to be invested in the local energy system, i.e. CHP, savings, renewables etc. The local contract sets the framework for that investment. No contract, no levy. (What local standards of energy efficiency can practically be achieved will be seen in the "Brundtland-Town-Project" as described above.)

In its statutory instrument the government has explicitly incorporated the principle of external costs as the guide-line for its efficiency strategy. It thereby pays special tribute to the newly enacted constitution of Schleswig-Holstein which came into effect on August 1, 1990. Article 7 thereof grants special protection to all natural resources and obliges all public institutions to act in accordance with that overriding principle.
Energy policy in Schleswig-Holstein wants to breathe life into that provision of our constitution by initiating a local as well as regional (cross-border) process of co-operation in the interest of a sustainable development.

5. Further Initiatives

Furthermore the State government has initiated a continuous process of collaboration by means of a state-wide "Energy Conference". That body consists of representatives of local industry and commerce, the utilities, trade unions, environmental groups, the society of engineers, the universities and research institutions etc. It meets and establishes various "expert conferences" dealing with matters like renewables, energy planning and forecasts, and the like. The feed-back goes into the "Energy Concept of Schleswig-Holstein" as a framework for coordinated energy planning.

Guidelines for subsidies have been published to give financial assistance to the "efficiency strategy" including
- energy efficient buildings
- renewables, particular wind, but solar and biomass as well
- Co-generation
- district heating.

A budget of roughly DM 26 Mio. annually is available to support local utilities and municipalities as well as private investors, thus creating - in parts considerable - local employment effects as well as protecting the environment.

Pilot and demonstration projects include Europe's largest wind-farm (13 MW) at the North Sea Coast, a hybrid system (wind, solar and biomass) on the island of Fehmarn, an island energy-supply-system on the island of Helgoland including a 1,2 MW-windmill, the district heating system of the town of Flensburg by which 95 % of all customers are supplied, local district heating systems based on gas engines with heat and power production. With reference to these projects local demand for the modernization of the energy system in order to comply with environmental standards is steadily increasing. The "efficiency strategy" is "on the road". There is still a long way to go.

7.10 "External Effects - A Complete Approach"

The New Prognos Study on the External Effects of Energy Procurement

Jan Keppler
Prognos AG, Basel

1. Introduction

The increasing scarcity of environmental resources in combination with growing productive capacity has sparked new and necessary interest in the nature and magnitude of so-called external effects. Of special importance in this context is the field of energy production, procurement and consumption. The energy sector is responsible for external effects of great number and importance - especially those connected with gaseous emissions. In recent years it has also become one of the major fields of debate about which general course society should take. The intensity of the discussion about atomic energy is but a sign for the fact that the choice between energy systems is considered crucial for even larger issues. This has also, but not exclusively, to do with a concern for the environment.

Both aspects in their own right contain a whole array of different external effects in the sense that they are not taken care of by existing private markets. Not only does no market exist for air (or respectively for air pollution) - much less does there exist a market for energy systems. Indeed, such a purely hypothetical market would be absurd as the discussion about energy is not so much concerned with a choice between existing forms of energy procurement but with a search for new modes of living, producing and consuming.

Especially the first aspect has led to an increased and increasing number of officially sollicited or unsollicited studies on the identification and the internalisation of external effects in the energy sector - a necessary first step on the way to action. Also the study "Identifizierung und Internalisierung externer Kosten in der Energieversorgung in Deutschland", commissioned by the German ministry of the economy with Prognos AG, Basel, is part of this concerted effort to properly define the issues at stake and to propose adequate measures to advance welfare. Yet, the Prognos study claims to have a broader approach than most predecessors. In addition to the standard problems which are dealt with in a framework interpreting pollution as the unaccounted-for use of scarce environmental resources (such as air, water, land etc.), special emphasis is given to the second group of external effects linked not so much to the misallocation of resources or pollution but with society's more complex concerns. The following pages present the theoretical framework and the specific character of the Prognos study as a coherent and comprehensive way of tackling the problem.

2. External Effects as Market Failures

The greatest challenge for the Prognos study is its claim of completeness in identifying the external effects of the German energy system. Obviously, completeness is not and cannot be understood at every level of detail, yet the study has the ambition to present a picture in which no major element of external effects is missing. Furthermore, it has to stress equally the two aspects mentioned above the allocation of scarce resources on the one hand, the search for an energy system in tune with society's fundamental values on the other. It will be shown that both aspects can be dealt with by proceeding from the same theoretical framework. This framework starts with the familiar notion that external effects constitute a failure of existing private markets.

As is well known from standard economic theory, there exists direct correspondence between a full set of perfectly functioning private markets and a state of Pareto optimality. Pareto optimality is the minimum requirement of every allocation of goods in an economy interested in serving the maximization of the well- being of its members. This connection, embodied in the two fundamental theorems of welfare economics, is the intellectual basis for the support given by libertarians for free private markets, independent of government intervention. This claim hinges on the absence of external effects.

By definition, external effects are factors influencing individuals's utility functions which are not properly allocated by existing private markets. Their existence thus negates the libertarian's claim. In other words, external effects are all the influences on the well-being of the members of a market economy not properly solved by means of the market. External effects have thus to be dealt with by different means. Be this by the formation of new markets, by direct government intervention, or by alltogether different forms of decision making. At the beginning of any analysis of these means has to stand the analysis, why external effects cannot be allocated by private markets. As will be seen, the very reasons for which an "effect" is not properly allocated by market mechanisms define in each case the most appropriate road to action.

As a first step in proceeding to identify external effects and finally propose measures for their internalisation, it is useful to repeat the necessary conditions for goods such that they can in fact be properly allocated by private markets: The first condition is the definition of property rights over the goods in question. Property rights in this context mean nothing else than the right to buy and sell definable quantities of the respective economic good. The second condition is the absence (or negligibility) of transactions costs in concluding a deal between buyers and sellers of a good. Obviously, these two conditions are linked, because we could interpret the costs (difficulties) involved in defining property rights as some sort of initial transactions cost. It is also clear, that transaction costs do not constitute a once-and-for-all barrier to trade. If due to changes in respective valuations the gains from trading become higher, it might make sense to define property rights for goods which had been previously deemed not worth the effort.

A third condition for the existence of a perfect private market is the existence of a multitude of buyers and sellers and hence the divisibility of the good in question. If prices are formed in an all-or-nothing fashion due to monopolies on the supply side, or if the goods in question are impossible to be partitioned in order to be commensurable with the incomes of the participants in the market - then a market loses its claim to be an optimal form of allocation. The last point regards in the field of energy for instance the size of recompensation payments for a low-probability-high-risk event such as an accident in a nuclear power plant.

A fourth condition is rarely mentioned but absolutely crucial for the existence of markets. It concerns the form and formation of peoples' preferences and the form they choose to articulate these preferences. The existence of markets presupposes that people have fixed preferences over a well-defined range of goods. But human well-being is influenced also by factors which cannot be grouped into such categories. For instance, markets are no good way of articulating preferences which are still in the process of formation.

People have concerns, fears, hopes, lacunae in information which all very decisively influence their well-being and therefore have to be considered in a society interested in maximizing this well-being of the greatest number of its members. Nevertheless, these influences cannot be formulated, not even on a personal level, in the form of clear choices between distinct possibilities, which is a pre-condition for the existence of markets. A major problem is information. Perfect markets imply full information of all its members. A major step towards the internalisation of external effects lies in the information about and the formulation of clear societal alternatives, which further the process of preference formation.

But this is not enough, point four reaches further. Consider the existence of intrinsic values or deficiencies in the way of preference articulation itself. Is a market really the preferred form a society wants to solve its energy problems? Do markets maybe imply ideas about forms of living which parts of a society do not share? Which markets shall there exist? Only markets of energy products, or also markets for pollution? Some markets have set-up costs, others face legal impediments, thus choices have to be made rather about than in markets. All these questions concern external effects in energy procurement - as they concern important variables in the well-being of members of a society, yet they cannot be solved by way of the market mechanism.

It probably has become clear, that the first two conditions for the existence of markets are distinctly different from the third and fourth. The first two concern the nature of the good in question, usually a scarce natural resource such as air or water or natural beauty, for which it is impossible or very difficult to define property rights due to its physical nature. On the other hand, peoples' preferences for them in comparison to other goods are rather well-defined. External effects arising out of non-fulfillment of the third and fourth condition are more difficult to grasp, yet the Prognos study is very concerned about these external effects which play a large role in the general debate about energy systems, but have not found sufficient entry into the scientific debate. It includes all those issues over which clear-cut static preferences are not yet formulated or cannot be formulated at all inside a market mechanism.

The field of energy production, procurement and consumption thus transcends the problem of "the use of scarce goods for competing ends". The energy system of a society is one of its vital elements and thus in structure and objective a reflection of the very structure of the society, and what is more, a reflection of its values and the priorities it gives to different fields of activity. Under this perspective, the concern does not lie any more with a problem of allocation inside a given system, but with the nature of the system itself.

From the above said it follows that due to the nature of the "externality" in question, two different roads have to be pursued during the study in order to do justice to the effects in question. Chapter 3 will deal with those external effects which arise out of an insufficient or non-existing pricing of scarce goods. Because these scarce goods have been usually subsumed in popular debate under the heading of "the environment", they shall be called henceforth "environmental goods". Their treatment will follow the outlines given by Pigou and Coase. The internalisation of external effects of this kind could be described as "market simulation". The outcome of the proposed internalisation measures is, ideally, again an equalisation of marginal values of buyers and sellers, users and owners. This goes along with an implicit or explicit monetization of external effects.

Chapter 4 instead, will deal with those complex phenomena where society's preferences are not yet clearly defined, and where furthermore, the very process of preference formulation and articulation is under discussion. This encompasses the problem to which extent markets are the most useful way of articulating preferences in the field of energy. In this field, there have to be discussed a whole different set of internalisation measures such as public opinion formation, court action, the political process. Any form of monetization or quantification becomes obsolete as measurable welfare evaluations do not exist. Partly, this is due to the fact that the respective preferences are not yet formed, partly it is due to the nature of these preferences which elude quantitative judgement.

Before concluding the chapter, two important assumptions that will be integrated into the Prognos study, should be mentioned. The Prognos study maintains first that there is no a priori contradiction between rigourously formulated economic theory and other ("qualitative") approaches to environmental problems. On the one hand, economic theory has to be reformulated and modified in order to adequately address the problems in their complexity, on the other hand, ecological, sociological, political or even philosophical approaches have to be subjected to tests of internal coherence and logical viability.

Second, the study maintains that the often invoked opposition between "the economy" and "the ecology" is artificially brought about by incomplete and superficial definitions of the two subjects. Both derive in the last instance from antropocentric concepts of the optimization of global welfare. Good economics will necessarily include the valuation of the environment as soon as it becomes scarce. In fact, the very moment the use of environmental goods is seen as a problem due to scarcity, the realm of economics is entered. The opposition between the production of material goods for consumption and the use of the environment for other purposes (e. g. pure pleasure) is not an opposition between economy and ecology, but the very essence of economics itself.

3. The Optimal Pricing of Environmental Goods

This chapter deals with attempts of internalisation of external effects in the case that conditions one and two are not fulfilled, i. e. we are dealing with goods (e. g. air, water, landscape) which are misallocated due to their specific physical nature. All theoretical attempts to formalise the concept of external effects in this context and to propose measures for their internalisation can be reformulated as attempts to find optimal allocations for scarce goods, which hitherto have escaped proper concern. These attempts go back at least to the British economist Arthur C. Pigou, who pointed out that in the case of a negative externality (e. g. a coal plant emitting coal dust - thus consuming clean air) the optimal point is reached when the marginal benefit for the producer of the external effect equals the marginal damage of society from the external effect.

The problem, obviously, lies with the fact that there is no inherent reason for the coal plant to stop its emissions at precisely this point, but it will go on emitting dust until its marginal benefit from doing so is zero. To remedy this sub-optimal situation Pigou proposes a "Pigou-tax" to be levied by the state on the externality. The tax should be equal to the marginal societal damage at the point of optimality (this implies that it is also equal to the marginal benefit at that point). This would force a profit-maximizing producer to restrict production at exactly this point.

So far, so good. Pigou's approach is logically coherent, but suffers from the problem that the necessary information to design an optimal tax is very hard to come by. An alternative approach has been presented by the British/American economist Ronald Coase. His thesis was, that external effects would cease to exist as long as property rights were completely defined and transactions cost were low, and their treatment was left to private individuals. This proved to be a very elegant way of looking at locally limited, precisely identifiable external effects between individuals but a rather useless approach to external effects produced by a large number of producers and influencing large numbers of people.

If external effects would be limited and property rights completely defined (imagine a clear lake, spoiled by a sewage farm, and the property right being held by the owner of a commercial lake side resort) the two parties would bargain by means of monetary payments, i. e. the sewage farm would pay the resort owner for every unit of pollution until its marginal benefit from polluting was smaller than the recompensation payment for the resort owner. In the end, the same optimal point as envisaged by Pigou would be reached. The neat thing about the Coase approach is, that exactly the same point would be reached if the sewage farm had the right to pollute and the resort owner would have to pay it to restrict output.

Another way of internalising external effects would be to put the sewage farm and the lake side resort under joint ownership, thus the operator would now restrict pollution automatically at the optimal point, in order to maximize joint welfare. Quite obviously, the

Coase approach becomes inapplicable in a case such as the previous example, where the number of people concerned would be too high and thus transactions costs would not permit effective bargaining and the coal plant would continue to emit too high a level of pollution. But the Coase approach has a very interesting side aspect to it, which connects neatly with the wider question of the organisation of energy systems of Chapter 4. It shows us that decentralisation strategies can be an effective form of internalisation for local external effects.

In this case, it is obvious that market form and external effects have a clear link. The running of energy production systems under the authority of the same local communities which would suffer from the pollution, would produce quite different effects from the running of energy systems by huge centralised entities far removed from the final effects of its externalities.

Although superficially opposed to each other, the Pigou-approach relying on the state, the Coase-approach on the profit-maximizing individual to bring about optimality, they are structurally very similar. In the end, both authors have designed ways to overcome the "tragedy of the commons", i. e. the case where a common property resource with unrestricted access is depleted due to overusage. If the state is interpreted as society's agent, the society holding the property right over a scarce resource such as air, then the Pigou-tax is nothing more than the positive price society extracts for the use of its resource. Due to transactions costs, not every single member of society can contract with the polluter, thus the state tries to infer society's preferences and to set the tax accordingly. An optimal tax, an optimal price will restrict the use of the scarce good, be it the village common, the natural water reserves or the clean air we breathe, to a point where total gain is maximized.

Thus, the Prognos study sees external effects in energy production, procurement and consumption, as long as they concern things as pollution, as the suboptimal use of scarce resources. In these cases, the failure of private markets is due to the lack of property rights and the height of transactions costs. Environmental taxes, standards, certificates can be valuable means of tackling the problem. At the same time, state interference should be limited to those cases where private transactions fail. Decentralisation, for instance, is an important subject and can be an effective way of internalising external effects. But decentralisation is effective only as long as external effects are local; state interference will be necessary for cases such as carbondioxide emissions. In those cases, information problems arise about the determination of cost and benefit functions in order to estimate the optimal tax. Here, only dilligent footwork can help to overcome the problem and most certainly the Prognos study also has the ambition to promote the state of knowledge in these fields.

At this point the problem of monetization of external effects does not pose a serious threat, as long as two conditions are fulfilled. First, monetization is a necessary evil due to lack of better information concerning society's preferences, results are proxies for better values on the way to find the right measure for policy interventions. Second, monetization is only useful as long as we are dealing with the suboptimal use of environmental goods over which there exist fixed preferences. Monetization could be interpreted as the simulation of the outcome of a market transaction, if the goods in question had possibly been given property rights and transactions had been lower. Just as in perfectly working "real" markets, the outcome would be the equalisation of marginal values. Obviously, any monetization of external effects can only mean finding a monetary value of the marginal cost at the point of optimality (equal to the height of the optimal price/-tax) - any other value is meaningless for policies interested in welfare maximisation. This seems to be a trivial point, but unfortunately it is a point not always understood.

Further work will cover the relationship between damage costs from external effects, the benefits derived from producing external effects and the costs of controlling external effects. Some interesting results are to be derived once the commonly made hidden assumption of a maintained level of energy production is abandoned and substituted with a rigourous concept of economic welfare maximisation. For instance, it can be proven that in many cases a reduction in energy output would impose smaller costs on society than an abatement technology yielding equivalent reductions in external effects. But this covers more technical aspects of the same problem. More important, at this stage, is an outline of those topics to be dealt with in entirely different frameworks, where any attempts of monetization are bound to fail - as we lack the necessary preference structure.

4. Multidimensionality in a Dynamic Context

The second group of external effects requires different approaches of internalisation. Due to the lack of clearly formulated preferences, any quantification is obsolete. It is, at this point, indifferent, whether this lack of a preference structure is due to a complete lack of clearly formulated (or formalizable) alternatives or due to difficulties in the process of defining clear-cut preferences, because of the complexity of the issues at stake. Either way, the internalisation process has to abandon the static approach of straight-forward utility maximization pursued above and has to look for alternative ways of integrating the hitherto disregarded concerns of the public.

An important example is set by the discussion about the usefulness of nuclear energy. Pure monetization assuming fixed preferences for material well-being and physical health, and even managing to attach a numerical value to human lives, multiplying these numbers with the probability of a nuclear accident, may be able to give a number by which the electricity delivered by a nuclear power plant should be taxed with. Yet, it would remain oblivious to the hidden fears of people (which influences their well-being, albeit it may be irrational), their dislike of huge uncontrollable structures, their dislike of the new security measures needed, and so on.

It should be understood, that this is not a problem of measurement, to be overcome by an elaborate preference revealing questionnaire, whose results could be appropriately quantified and converted into a financial sur-tax on electricity produced by nuclear power plants - allowing an easy comparison with electricity produced, for instance, by coal powered plants. The issue is that the alternatives are still in the process of being formulated. Do we want large nuclear power plants, small ones, how safe do they have to be, what would be an acceptable safety standard, where should they be located, who should run them, how much money should be spent into improving their safety or in the research for alternative energy? There are no clear cut choices to be made yet, and maybe there never will be. Nevertheless, decisions have to be made, problems have to be addressed, concerns have to have the possibility to be voiced and heard - and there are no markets in which each individual by voting with his money could express these concerns. The lack of a static preference structure over well-defined choices makes utility maximization through private markets obsolete.

External effects in energy consumption of this kind which hitherto have been largely disregarded by the scientific community have played an exceedingly large role in the public debate. An increasingly frustrated public found itself subjected to all sorts of external effects - besides pollution. First, it had to grapple for the words to express their uneasiness about them, and second, it had to find ways to channel their uneasiness in a way that it would become a parameter in the decision-making process. At the present stage, the Prognos study discerns three different channels through which an internalisation of the latter kind could be brought about. First, there is public awareness and opinion creation by means of public communication. These might be commercial means of communication, such as television or newspapers, but these might also be non-commercial means of public awareness creation such as demonstrations, sit-ins, public speeches and the likes. (Again, we are confronted with the fact that pure information is a first step in any internalisation of external effects.)

Second, there is the possibility to internalise through court action. Appropriately formulated motions, for instance, led to court decisions demanding stricter security measures in nuclear power plants. This way, the set of choices was thus transformed that the new choices contained more of the popular concerns. The legal approach to internalisation bears an interesting parallel to the Coase-approach mentioned above. Any internalising court action can only be undertaken as long as there is a complete legal system defining rights and responsibilities on the basis of which recommendations or court orders can be formulated. Legally not defined spheres make internalisation through the legislative system impossible. In many cases, a necessary first step of internalisation is the proper definition of the legal responsibilities according to society's fundamental values. Again, it can be seen that external effects of the non-quantifiable kind, belong to a dynamic sphere where new choices are being discovered and formulated.

The third way of internalisation is the political process. Obviously, this sphere is closely linked to the former two. In the social choice literature, voting mechanisms have been discussed at length, usually under the assumption of existing static preferences. In this context, the term "political process" is to be understood in a wider sense, encompassing those processes of opinion formation, scientific expertise, decision making, of which the actual vote is only a relatively small part. This holds all the more in cases, where all the political parties, more or less at the same time, run through similar transformations, or, in other words, deem it necessary to internalise some public concerns into their platforms.

All three internalisation mechanisms lack the theoretical rigour and the elegance of the Pigou/Coase-approach. Yet, they are the only possibilities to improve welfare in situations where factors influence individuals' utility functions and where these same factors are not taken care of by existing private markets. In conclusion, some of the aspects to be dealt with shall be presented in brief, in order to give some illustration of what kind of external effects we are actually concerned with.

International dependence is, since recently, quite obviously a very well recognized external effect of some parts of the energy system. How much international dependence is a society willing to accept in exchange for a cheap and relatively clean source of energy? Is it in last consequence willing to go to war over it, or at least internalize some of the risks in the form of higher military spending or a general draft for its young men? Is, on the other hand, a splendid isolation a renouncing of the participation in international trade a viable solution?

A society has to form some sort of opinion on this issue and decide on an acceptable degree on international dependence.

Another problem is the internal security structure of a society. Nuclear power plants, the transport and dumping of radioactive material not only pose a continuous burden on society, but also expose it to threats of terrorist attacks and accidents due to human failure inside the system. Does a society want a national security state in order to be provided with cheap energy? Is there a link between external and internal security structure? Other effects concern wealth distribution and employment effects. Can a nation which possesses large coal endowments, with a strong miner community switch suddenly to expensive imported gas because it produces cleaner heat? Probably, considerations concerning only quantifiable pollution effects will be overridden by concerns over a much less tangible form of external effects - and the country in question will most likely continue to serve its energy needs with coal. (Technically speaking, this problem could be solved as a static maximization problem involving a trade-off between society's appreciation of cleaner air and its appreciation for equity or its dislike of structural changes. Yet, in practice a solution is going to evolve along the lines outlined above.)

Again a completely different field of external effects is opened up by the fact that large numbers of people come to dislike the fact that their environment is increasingly dominated by huge impersonal structures which impede personal control and experience. This latent disdain for centralised technologies played an enormous role in the debates on nuclear energy and is independent of any objective risk analysis. The dislike of a prestructured environment with less and less possibilities of personal initiative goes along with choices concerning life-styles, habits and values. The choice between electricity produced by nuclear or fossile energy and a life style concentrating on as little consumption of energy as possible cannot be handled in a usual consumer choice framework. What is at stake is not a decision between two choices, but a decision between two entirely different choice structures. Clearly, a system solely relying on markets will only be able to accommodate the former, not the latter. As long as no alternative modes of internalisation are found, external effects will still influence peoples' welfare in a suboptimal fashion.

Closely linked to these considerations are reflections about the value of markets per se to accommodate human choices. The Prognos study at this point does not aim at a critical phenomenology of impersonal national markets as the fundamental allocation mechanism for our society, yet, it will have to take account of society's preferences regarding alternative allocation mechanisms, be those maybe only regionally limited markets for energy. Over and over again, the field of energy production, procurement and consumption shows itself to be in the midst of very complex discussions about the values of society itself.

This, at second thought, comes to no surprise as energy is literally the force which keeps the wheels of that society turning.

Last not least, there remains the issue of exhaustible resources to be discussed among the external effects of energy production. Strictly speaking, the exhaustion of a scarce resource is not an external effect concerning the present generation, but an external effect concerning future generations, and it enters present generations' welfare only in so far, as they care about the welfare of future generations. In addition, the theory of exhaustible resources has received on the basis of the work of L. C. Gray and Harold Hotelling a sound neoclassical treatment. The optimal pricing rule, that the nominal rent of the marginal unit of the scarce resource extracted (and thus, with zero extraction costs, its price) should rise with the rate of interest, seems to leave little room for discussion. Yet, the optimal pricing rule only concerns relative, not absolute, prices; as was well understood already by Gray and Hotelling, the present price is defined by the "cut-off" price, i. e. the marginal value of the very last unit of the scarce resource extracted. Once there is agreement on this cut-off price, then an optimal present price and an optimal time path of exploitation, leading to intertemporal rent maximization, can be defined.

The whole problem of scarce resources hinges on the discussion about the right cut-off price. For strict environmentalists for whom the indemnity of the environment constitutes an absolute value and who see a moral obligation to deliver an unchanged world to their children and grand-children, obviously, this cut-off price will go towards infinity, implying an infinite or at least very high value for non-renewable natural resources, already today. Most other people would disagree with the proposition that zero extraction of natural resources is an optimal path, as future generations could be more than offset for their loss in resources through the advantage of disposing over a larger capital stock. This capital stock can only be accrued over time trough the exploitation of natural resources. These issues have to be discussed, especially if large parts of the population feel that they have smaller discount rates than private markets interested exclusively in short-term profit maximization. Even if part of this opposition is mitigated by monopolistic tendencies in many markets for natural resources - the best known example being the oil market, there still remain large external effects to be taken account of. Again, market mechanisms will not be able to take care of these effects, as they concern the very structure of the agents decision parameters.

5. Conclusion

This concludes the overview of the topics to be dealt with in the Prognos study "Identification and Internalisation of External Costs in Energy Procurement in Germany". Its ambition to at least mention all relevant external effects of energy production, procurement and consumption finds its anchor in an analytical framework proceeding rigourously from the idea that external effects constitute a failure of existing private markets. This does not imply an over-critical attitude to private markets which remain the allocation method with the best efficiency record. Important at this point is an emphasis on existing private markets, because new markets, or market simulating mechanisms may well alleviate and internalize many external effects. This idea has been shown to be a very efficient solution for external effects which can be interpreted as the unaccounted-for use of scarce environmental resources. Without proper mechanisms, these resources are subjected to the so-called "tragedy of the commons", i. e. overuseage and rent-dissipation.

But it has also been shown that these market simulating mechanisms are not applicable to a large variety of external effects of an entirely different kind. This field is less concerned with the allocation of scarce resources but more with the form and function of allocation mechanisms itself. Furthermore, it is an area where both preferences and choices are not yet strictly defined, but rather in a process of formation. Here alternative mechanisms to deal with these external effects have to be found. There have been mentioned public opinion formation, the judiciary and the political process. The list is presumably far from complete, as the process of exploration has barely started and the Prognos study is indeed proceeding on relatively untouched ground.

But in this combination of reliable analytic tools and creative new thinking lies the hope for the success of this study.

7.11 Scientific Report - A Case of Emergency

Eberhard Moths
Bundesministerium für Wirtschaft
Bonn

In the analytical and political context, it is always informative to follow the way in which an "unavoidable" problem is tackled theoretically and practically. The same applies to the phases in which this problem becomes generally accepted, both semantically and by the public at large. Where does the current interest in external or "social" costs stem from? Why are the distant tropical jungle and CO_2 among all other emissions so eminently predominant nowadays? Why is the specific curiosity shown by politicians and businessmen towards the external costs of transport much more superficial, although scientists have been presenting them with "estimates on the cost of transport infrastructure" for years? Why does the sudden concern aroused for the problems of climatic change taking place so far above our heads fit much more neatly into the political landscape than the study and restoration of damage perpetrated in forest and field so close at hand and directly apparent? Why are the consequences of Chernobyl thought to be *exterritorial* as far as we are concerned, although we must expect them to migrate in our direction, if they are not here already? What lies behind such an attitude of preferences: unpredictability, a sense of over-encumbrance or helplessness, feelings of resignation, economic interest, political strategy or simply a case of misunderstanding and misconception? The latter could indeed be the case if we assume that the concept of external costs has not been clearly defined terminologically. For if we look at such costs more closely, they are not particularly *social* or *external*, but tend more to be *unsocial* and even directly *internal* for those experiencing them at first hand.

"External costs" and scientific reports have one important feature in common: their value and findings frequently "disturb" what we may call the "day-to-day operation" of industry and society. This they may already do, long before their value and findings have taken on any "binding legal force". In essence, this disturbance (conflict) lies in the fact that previously unobserved problems have to be discussed in public and on their own special terms (demands for forbearance, abolition, compensation). At the centre of these problems lie not only dangers which are studiously ignored or skilfully subliminated, but also large or small hazards of an eminently real or extremely abstract nature. These range from speed limits on German autobahns to the prevention of toxic emissions whose chemical formulae can only be understood after years of study at the laboratory bench. "Superlative forms of such problems" in this context are questions of insurability and the basic reversibility of or general responsibility for specific external costs, and by no means only for costs associated with energy or power costs.

Assuming the most favourable situation, such "breakdowns in operations" cost only time and (large sums of) money. In the worst conceivable case, the internalization of blocks of "external" costs previously not taken into account can mean business or political disaster - and this disaster usually comes overnight, unlike the more gradual process of ecological disaster. - Employees, employers and the state tend to protect themselves against encounters with "unpredictable" externalizations by entering into minor or grand coalitions with affected parties. "Shutdowns" (such as those caused by trucks at the Brenner Pass in the Fall of 1990, the closing down of pits in the Saar or of brand-new nuclear power plants in the fault trough of the Rhineland) nearly always constitute disastrous, multi-dimensional losses. The terms "emergency" or "accident", which make light of such situations, can thus be classed as a linguistic deception - distinctly separate not only in semantic terms from those small breakdowns in operations which result from human error.

External and Extreme

A feeling of sensitivity for increased costs and pending losses forms part of the businessman's psychological makeup. It includes a mistrust for expert scientific reports whose findings might exert any direct or indirect influence on his economic existence. Seen in this light, paying attention to or ignoring external costs constitutes a "battle of distribution". An additional important aspect is the strange situation of having to do battle on unfamiliar ground with people who do not constantly need to look at each penny twice or feel directly the personal effect of all market stringencies. - The distrust mechanism among political or non-political public servants works on a similar principle: everything which transcends or does not fit into the normal run of things deserves increased attention. It is often "prematurely" suspected of being destructive. Seen in terms of both politics and phonetics, therefore, the two concepts of *external* and *extreme* lie very close to each other in the course of day-to-day administration. The effects of this link on practical policy must also be included under follow-up costs in the balance-sheet common to our political culture.

The phenomena related to interests or communication outlined above occupy a key position in any closer consideration of external costs in industry and politics. "Mental" (socio-cultural) conditioning/contamination takes precedence (even if this does not find its expression in the interests currently pursued by those disciplines involved in the subject today or included in the range of topics covered by this workshop) in the case of the specific question under review here. "Ways of internalizing external costs - practical approaches". It is therefore useful and essential to focus the attention of our analysis on the practical conditions relating to implementing specialized scientific reports.

To repeat: specialized scientific reports as cases of emergency or accidents. This has nothing to do with the archaic conception that an (enlightened or aroused) scientific discipline was always welcome as a wise counsel in industry or bureaucracy. The opposite is generally the case. This usually has quite simple reasons. The publication of new scientific reports means work, a great deal of work, which is often unfamiliar. It also has to be "accomplished" in constant competition with everyday business. Not to mention the major and minor vanities of those responsible for administering (financial) funds who are generally only equipped with formalistic intelligence, and that not even to a sufficient degree. Simple dislike of any restrictions to practical everyday routine by some kind of "superior control instance" also plays a part. - Merely the careful perusal of the reports issued by a single commission of inquiry in the German Bundestag is more than enough for the attention of one man or woman.

Clausewitz Strategy

In order to avoid causing accidents or states of emergency with one's own scientific reports, it is best to remain abstemious. Only when a concrete political controversy has arisen (which is usually completely divorced from the problem in question, both thematically and in terms of time) will professionals on the political report scene advise following the Clausewitz Strategy: the active recourse to one's own scientific report as the continuation of politics with other means. For the acute state of emergency, of course, there are always emergency plans available to those who know the subject at issue (tactical counter-offensive with the aim of restricting the damage incurred). What is true for retro-fitting in nuclear power plants is also true for counter-reports in politics. It is also possible to heed the motto that attack is the most effective form of defence. The same applies to the skilful deflection of public interest to so-called "subsidiary theatres of war".

Research into cases of emergency can never function without a knowledge of generally prevailing background conditions or of the "environment" in which the accident or emergency occurs. This is also true in the case under review. Let us refer to an example for the purposes of illustration: a naive "privy-type philosophy" is accepted and practised more or less unceremoniously in all developed industrial nations. This privy-type philosophy may be understood as an attitude to production and to life in general which consistently ignores its own dirt (external costs), both in terms of preventing its occurrence and disposing of it afterwards. Or to express the same idea even more unscientifically: the main emphasis in production and consumption lies on "doing one's business" as quickly as possible before the whole thing starts "to stink". The catchword is "I must get away!" (in the sense of "I must see that I'm not caught", or made responsible, in the sense of liable) whenever the negative or cost-intensive consequences of "risky investments" become public. Otherwise, what would become of the *suitable* management bonus?

Privy-Type Philosophy

The modern crop-rotation system (gravel pit, rubbish tip, building land) proves how easy it is to live with this privy-type philosophy. Recently, a fourth form of rotation - rehabilitation/disposal measures - could also be added to the above. It all depends on your "position". And everybody knows it is impossible to see those standing on the dark side of the street. One only has to have "adopted a different position" by the time the indignation spreads through the media. As far as the normal consumer is concerned (with his high personal proportion of external costs), it is frequently sufficient in terms of changing position to express his indignation via Greenpeace and its extremely hostile campaigns against industry.

In the context of internalizing external costs, the decisive question is how far the generally prevailing privy-type philosophy is a constitutive requirement for our form of running an economy. A consistent internalization of external costs, which could scarcely be restricted to the power industry or even to separate utilities, would then be much, much more than just a computer-aided or discursively supported calculation process to restore the competitive balance between energy resources receiving different levels of subsidies from various sources. A *fairer order of the environment* would then be a *question of system*, only this time under an ecological antecedent.

Scientific reports on *systems* (*tax*, *energy*, *economic* or *social* systems) generally lead to instances of political upheaval. The strength of the tremors felt in each case depends on whether and which interests are affected (coal, nuclear power, industry or consumer). Definite indications of a political state of emergency are present when there is public outcry on the part of those affected or their representatives. There have been many "cases of emergency" or "accidents" caused by scientific reports in the sector of energy policy in recent years. They have all been associated with both erection/operation and intermediary storage or final dumping in the context of activities undertaken by the energy industry. The public associates these with such locations as Kalkar, Gorleben and Wackersdorf. Current locations also worth mentioning are Greifswald, or the brown coal deposits around Leipzig. The East German surface mining industry may be regarded as a classic example of how costs are *relocated* or *repudiated*, and then *imposed on others*.

A particularly severe state of emergency of this type occurred in the Federal Republic of Germany in the summer of 1986. Let us remind ourselves: following the disaster in Chernobyl, two scientific reports were authorized by the Federal Minister of Economics to analyze independently of each other the economic consequences of withdrawing from nuclear energy. The "emergency potential" developed by these two scientific reports after only approximately two month's preparation was enormous. This had its origin in the more or less guarded conclusion reached that neither the immediate nor the systematic, long-term renunciation of nuclear energy would cast the Federal Republic into economic chaos. This was in fact an economic platitude, as was illustrated by the statement made by the then and current Chairman of the Board of Experts, who was scientifically responsible for one of the reports. Nevertheless, the political uproar was enormous when the two reports advocating renunciation were published. It suddenly seemed as if - apparently to the complete surprise of industry and politicians - the end of the age of nuclear energy was directly at hand. And this came after years of bitter controversy concerning the correct path to be taken to guaranteeing future sources of energy supply, and above all with respect to the actual and assumed costs and losses involved (now only approximating in order of magnitude to that of a *reminder item* when these are compared with the funds required and the "payment morale" associated with reconstructing and converting the economy in the new federal states?).

Emergency Potential

The height of the emergency was reached when the Chancellor and his Minister of Economics argued for weeks in public, often on the front page of the large dailies, about how anyone could ever think of the idea of authorizing such reports. Only a few months before the last general election to the German Bundestag, the state of emergency threatened to develop into a genuine crisis for the coalition. For a short time, it was possible to have the impression that political crisis management would not succeed in mastering the state of emergency triggered by the "renunciation reports" - that withdrawal from nuclear energy might be reconstrued to signify withdrawal from the coalition. And all this was because the renunciation of nuclear energy in the Federal Republic had been found in two specialized economic reports not to be synonymous with the economic suicide of our industrial society.

The effects of this report-related state of emergency on those who had "contracted" the two reports in accordance with instructions - by no means following their own inclinations - and had "released" them for publication after checking their proper completion, are of fundamental interest here, insofar as they affect the conditions surrounding implementation of future reports having "emergency potential".

Damaged Early Warning System

It is natural for civil servants to keep a special lookout for "damage to their careers". In the specific case in question, it is fair to state that no such damage can be detected. Nevertheless - and this should not be underestimated, either in the short or long term - there is a significant drop in the willingness even to contemplate participating in such projects in future, "to subject oneself to something like that again". They are certain to cause trouble - which can ultimately be quite massive - and the results are completely open, even in the most positive of situations. Expressed pointedly, however, this means that the executive's early warning system has suffered considerable damage, insofar as it can expect willing individual support as an institutional prerequisite. Such damage has already been incurred when research projects tend increasingly to be planned, organized and focussed in terms of their "level of approvability". The general and specific susceptibility to incur risks displayed by "normal operations" in politics and industry is thus increased, rather than reduced.

However, scientific reports cannot possess the "quality of emergency" merely on the basis of their possible or actual content. It would be too simplistic and one-sided to see the cause of confirmed "individual or field damage" as lying with the person authorizing their compilation. - Neither science nor the forester escaped from the badly damaged forest by their own efforts alone. Why should this be so? But that is not our topic, today.

In the summer of 1990, the Federal Ministry of Economics authorized the completion of a scientific report on the subject of the "External Costs of the Overall Supply of Energy". Four years after Chernobyl, and certainly ten years too late. But at least it was something.

8. PARTICIPANTS OF THE WORKSHOP

<u>Bundesrepublik Deutschland/FRG</u>

Prof. Dr. Hans-Jürgen Ewers
Direktor, Institut für Verkehrswissenschaft, Westfälische Wilhelms-Universität Münster

Uwe Fritsche
Öko-Institut e.V., Büro Darmstadt

Antje Geese
Wissenschaftliche Mitarbeiterin beim IER, Stuttgart

Prof. Dr. Robert Guderian
Institut für angewandte Botanik der Universität Essen

Prof. Dr. Peter Hennicke
Fachhochschule Darmstadt

Dr. Olav Hohmeyer
Fraunhofer-Institut für Systemtechnik und Innovationsforschung (ISI), Karlsruhe

Dipl.-Ing. Ulrich Kallenbach
Universität Stuttgart, Institut für Energiewirtschaft und Rationelle Energieanwendung (IKE), Stuttgart

Prof. Dr. Reiner Kümmel
Physikalisches Institut der Universität Würzburg

Dr. E. Moths
Bundesministerium für Wirtschaft, Bonn

Dr. Klaus Rave
Der Minister für Soziales, Gesundheit und Energie des Landes Schleswig-Holstein, Abt. Energiewirtschaft, Kiel

Herrn Rennings (für Herrn Rutkowski)
Technische Universität Berlin, Institut für Volkswirtschaftslehre, Berlin

Barbara Schlomann
Fraunhofer-Institut für Systemtechnik und Innovationsforschung (ISI), Karlsruhe

Dr. Friedjof Spreer
Wirtschaftsministerium des Saarlandes, Saarbrücken

Rolf-Ulrich Sprenger
IFO-Institut für Wirtschaftsforschung, München

Dipl. Chem. Margit Weltschev
c/o Bundesanstalt für Materialforschung und Materialprüfung, Berlin

Prof. Dr. med. Herbert Wiegand
Medizinisches Institut für Umwelthygiene an der Universität Düsseldorf

Teilnehmer aus der ehemaligen DDR

Dr. Dipl.-Math. Joachim Listing
O-Berlin

Dr. oec. Gerd Schirmer
Akademie der Wissenschaften, Zentralinstitut für Wirtschaftswissenschaften, O-Berlin

Prof. Martin Weisheimer
Akademie der Wissenschaften der DDR, Zentralinstitut für Wirtschaftswissenschaften, O-Berlin

England / United Kingdom

Angelika Baumann
Newcastle Photovoltaics Applications Centre, Newcastle

Bob Hill
Newcastle Photovoltaics Applications Centre, Newcastle

Schweden / Sweden

Krister Skanberg
Member of the Swedish Parlament, Stockholm

Schweiz

Dr. Jan Keppler (kam für Herrn Masuhr)
Prognos AG, Basel

USA

Dr. John A. Beldock
Environmental Protection Specialist, Office of Planning and Assessment, Office of Conservation and Renewable Energy, Washington

Stephen Bernow
Tehus Institute, Boston

Stephen Brick
Energy Associates, Middleton

Shepard Buchanan
Bonneville Power Administration, Portland

Stephen H. Burrington
Staff Attorney, Conservation Law Foundation of New England, Boston

Ralph Cavanagh
National Resources Defense Council (NRDC), San Francisco

Emily Caverhill
Resource Insight, Boston

Daniel Dudek
Environmental Defense Fund, New York

Prof. David R. Hodas
Widener University School of Law, Wilmington

Florentin Krause
LBL Bldg. 90H, Berkeley

Mary Lou Munts
Wisconsin PSC, Madison WI

Prof. Richard L. Ottinger
Pace University School of Law, Center for Environmental Leagal Studies, New York

Donald C. Peterson
Jr., Senior Associate, Colorado

Sury Putta
NY Public Service Commission, New York

Ajay K. Sanghi
PH. D., Chief, Impact Analysis Unit & Adjunct Professor of Economics SUNY at Albany, New York

Philip Sparks
Co-Director, Communications Consortium, Media Center, Washington

Gunnar Walmet
NYSERDA, New York

Stephan Wiel
Nevada Public Service Commission,
Carson City, Nevada

John Wilson
California Energy Commission, Sacramento

David R. Wolcott
Senior Projekt Manager, New York State Research & Development, New York

Prof. David Wooley
Pace University Law School, New York

FSC
www.fsc.org
MIX
Papier aus verantwortungsvollen Quellen
Paper from responsible sources
FSC® C105338

If you have any concerns about our products,
you can contact us on
ProductSafety@springernature.com

In case Publisher is established outside the EU,
the EU authorized representative is:
Springer Nature Customer Service Center GmbH
Europaplatz 3, 69115 Heidelberg, Germany

Printed by Libri Plureos GmbH
in Hamburg, Germany